ALLE · ZEIT · WACH · 1842

Bioorganic Marine Chemistry Volume 5

Edited by Paul J. Scheuer

K.F. Albizati, V.A. Martin,
M.R. Agharahimi, D.A. Stolze

Synthesis of Marine Natural Products 1

Terpenoids

With 154 Structures and 191 Schemes

Springer-Verlag
Berlin Heidelberg New York London
Paris Tokyo Hong Kong Barcelona
Budapest

Professor Paul J. Scheuer
University of Hawaii at Manoa, Department of Chemistry
2545 The Mall, Honolulu, Hawaii 96822, USA

ISBN-13:978-3-642-76837-8 e-ISBN-13:978-3-642-76835-4
DOI: 10.1007/978-3-642-76835-4

Library of Congress Catalog Card Number 89-649318

Softcover reprint of the hardcover 1st edition 1992

Typesetting: Macmillan India Ltd., Bangalore-25;

51/3020-54210 – Printed on acid-free paper

Preface

Volumes five and six of *Bioorganic Marine Chemistry* differ from their predecessors in two respects – they deal exclusively with laboratory synthesis of marine natural products and they represent the effort of a single author and his associates.

The rationale for these departures is readily perceived. For several decades organic synthesis has without doubt been the most spectacular branch of organic chemistry. While the late R.B. Woodward's dictum – organic compounds can undergo only four basic reactions: they can gain electrons; they can lose electrons; they can be transformed with acid or with base – is still true, the wealth and variety of available reagents which will accomplish chemical transformations has reached staggering proportions. Little wonder then, that synthetic methodology has achieved a high degree of predictability and total synthesis of natural products has been successfully directed toward ever more challenging targets. As for the second point, that of single authorship, multiple authorship would invariably have led to gaps and overlaps, thus making it difficult to assemble and assess recent research in a systematic and comprehensive fashion.

These two volumes are significant not only as a testimonial to the productivity and versatility of marine biota and to the virtuosity of synthetic chemists. As the material is presented along biogenetic principles, it is ideally suited to support research into the biosynthesis of marine metabolites. The comprehensive nature of the work makes it an easy matter to compare and evaluate different synthetic approaches prior to any synthesis of labelled precursors.

The division into terpenoid (Vol. 5) and nonterpenoid (Vol. 6) compounds is a natural one not only because of bulk. Nonterpenoid, particularly amino acid-derived, metabolites have become the fastest growing group of marine natural products. As recently as a decade ago, this position was held by di-, and earlier by sesquiterpenoids. This change parallels the current trend in research emphasis. Much early work in marine natural products was the result of serendipitous collections and separations. By contrast, most of today's research is guided by biological activity, which in turn is skewed toward those activities – e.g. antitumor, antiviral, which receive funding in industrialized societies.

While reading and editing the manuscript I was struck by the large impact which marine natural product research has made on organic synthesis and indeed on contemporary chemistry. It occurred to me that these books could be valuable auxiliary texts for graduate courses in Organic Synthesis.

I am indebted to Dr. Albizati and his associates for the monumental task which this endeavor entailed. As before, I should like to express my appreciation to Springer Verlag for their prompt and expert cooperation. As always, I look forward to hearing from members of the scientific community how we can improve future volumes in the series.

August, 1991 Paul J. Scheuer

Table of Contents

Abstract

The growth and extent of chemical synthesis of marine natural products from the years 1960–1989 has been evaluated and reviewed in a near-comprehensive fashion for the first time. The rapid growth in the breadth and depth of this field in a comparatively short period of time mirrors the growth and interests of the synthesis community at large. Synthesis chemists are stimulated primarily by compounds which possess potential biomedical importance and/or provocative structures, of which there is an abundance among the metabolites from marine sources. Continued growth in this area is projected. The information in this review consists primarily of synthetic schemes and pathways which, after analysis, have been set to words. The metabolites synthesized have been organized according to broad biogenetic lines, including terpenes, alkaloids, fat-derived compounds, amino-acid-derived and miscellaneous.

1 Introduction

The isolation of natural products from marine sources [1] accelerated rapidly beginning in the early 1970s and is only now hinting that a plateau may have been reached, as measured by the number of yearly publications (Scheme 1) [2]. Chemical synthesis of marine natural products (as measured by the yearly number of publications describing *total syntheses*) has climbed steadily since the late 1970s and has reached a plateau at around 60–70 per year for the past 5 years. The annual number of publications dealing with all aspects of marine natural products synthesis is considerably higher. The rapid growth in the breadth and depth of this field in a comparatively short period of time mirrors the growth and interests of the synthesis community at large. This is not unexpected. Synthesis chemists are stimulated primarily by compounds which possess potential biomedical importance and/or provocative structures, of which there is no shortage of either in the collective metabolites from marine

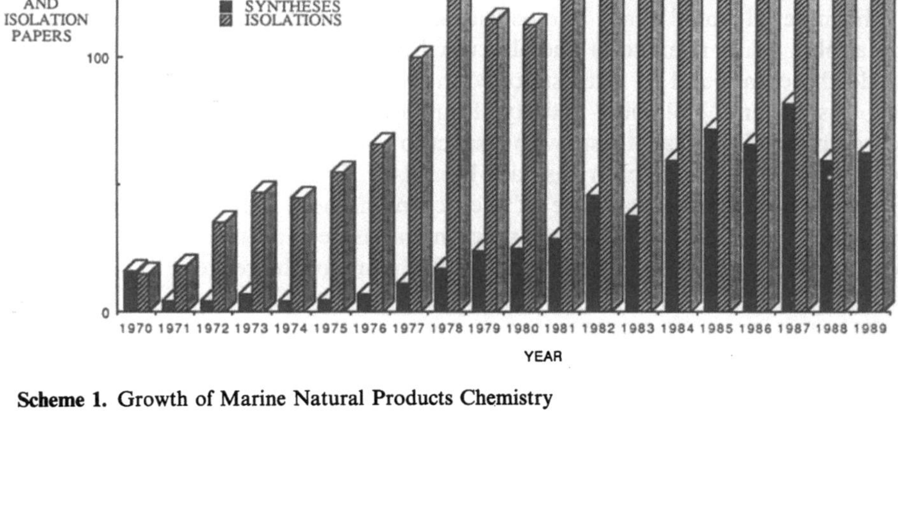

Scheme 1. Growth of Marine Natural Products Chemistry

sources. Continued growth in this area is inevitable and is a sign of the vigor and vitality of marine natural products chemistry today.

This review is devoted to a compilation of the domain of natural product synthesis that involves metabolites from marine organisms. This is meant to provide an overview of the role played by synthesis in the development of the chemistry of marine natural products and, for the first time [3], to assess the breadth and depth of activity in this field. Knowledge of the role which synthesis has played in marine natural products chemistry over the past 20–30 years provides a basis for determining future directions in this area. In initial discussions with various workers in this area, estimates of the existence of only 100–200 marine natural product syntheses were put forth. In reality, there are > 650, hence this review became a much broader undertaking than originally planned.

The intent of the review is to cover the literature from about 1960 to the end of 1989; however the coverage will not be comprehensive. We have omitted "partial" syntheses and syntheses of compounds where the relative stereochemistry at one or more centers is unknown. Also omitted is a review of cembranoid synthesis, since this has been covered previously by Tius [4]. We have not reviewed the landmark synthesis of palytoxin recently described by Kishi because this would constitute a large chapter in its own right and (as of this writing) all the necessary information for a detailed review is not available [5, 6]. We have arbitrarily left out syntheses of metabolites from fresh-water organisms, although this is a comparatively small body of literature [7].

The material has been roughly organized along structural-biogenetic lines rather than phyletic, in order to facilitate reading and searching by members of the synthesis community. There is a tendency in writing discussions of this nature to lapse into a "... compound X was converted to Y and then to Z by reaction with ..." pattern, to some extent this is unavoidable when discussing a synthesis. We have attempted to add as much critical comment and discussion as the material would allow. In most cases, we have discussed the source of each metabolite (with references) and described some of the properties that have made the compound of interest to synthesis chemists. Most of the time this revolves around biological activity. Emphasis has been placed on classical representation of the syntheses with respect to readability and as such, a number of reagent and solvent acronyms have been used. The vast majority of these are generally accepted and a table of the acronyms used will not be presented. However, in our judgment, when an acronym has been deemed uncommon, we have provided the structural formula or full name of the reagent.

2 Terpenoids

2.1 Monoterpenoids

2.1.1 From *Chondrococcus (Desmia)* and *Ochtodes* spp.

A variety of cyclic halogenated and oxygenated monoterpenoids from the red marine alga *Ochtodes crockeri* have been isolated [8]. Many of these exhibit antifeedant, sedative and antibacterial properties. The syntheses of three *Ochtodes* metabolites by Howard [9] are shown in Scheme 2. Bromination of **1**, available from dimedone, followed by addition of vinylmagnesium bromide affords *bis*-allylic alcohol **2** as a 1:1 mixture of diastereomers. Hydrolysis in aqueous acetone provides metabolites **3** and **4** in 3 steps and 15% overall yield for each. Chlorination of **7** by reaction with thionyl chloride gives the rearranged allylic chloride **5**. Treatment with one equivalent of silver acetate and separation of the isomers affords acetate **6** which is converted to metabolite **7** by saponification, thus giving **7** in 5 steps.

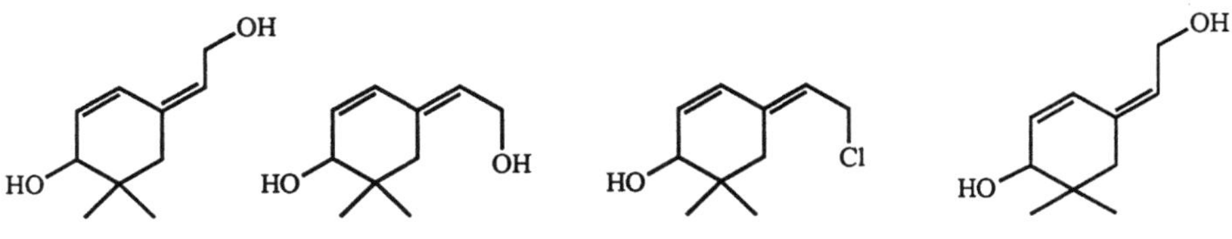

3 **4** **7** **12**

Ochtodes Metabolites

Another *Ochtodes crockeri* metabolite **12** has been synthesized by two different routes by Masaki [10] (Scheme 3). Cation-olefin cyclization of epoxide **9** [available from myrcene (**8**)] gives trifluoroacetate **10** as an 87:13 *E*/*Z* mixture of isomers. PCC oxidation of **10** and saponification followed by phenylselenylation of the ketone yields **11**. LAH reduction of the ketone **11** and selenoxide elimination affords *Ochtodes* metabolite **12** in 5 steps and approximately 13% overall yield. A second route to **12** involves addition of phenylsulfenyl chloride

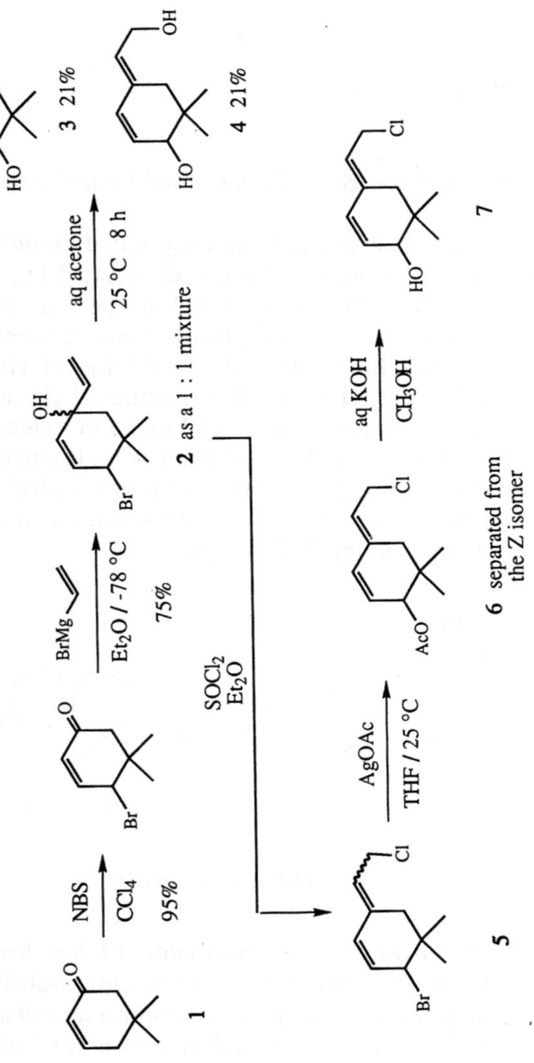

Scheme 2. Howard Synthesis of *Ochtodes* Metabolites

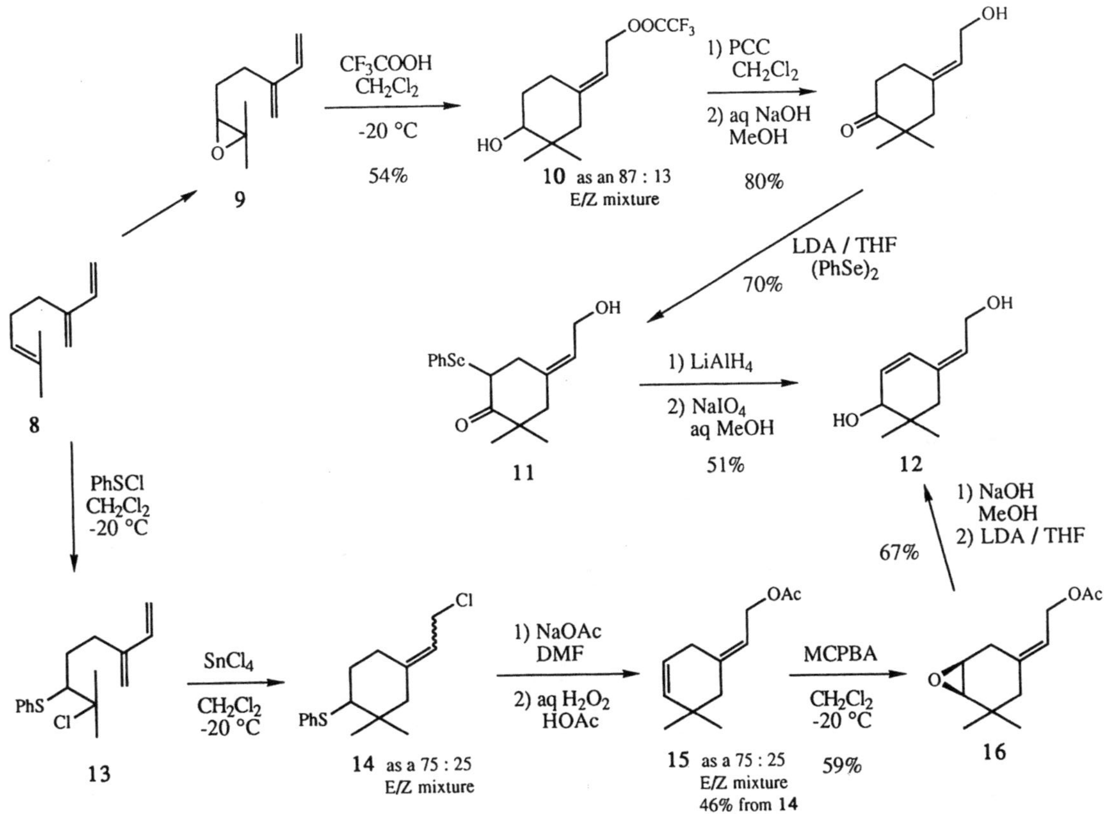

Scheme 3. Masaki Syntheses of *Ochtodes* Metabolite **12**

to myrcene **8** to obtain **13**. Treatment of **13** with tin(IV) chloride initiates cation-olefin cyclization via episulfonium ion formation to produce allylic chloride **14** as a mixture of isomers. Displacement of the chloride with acetate and sulfoxide elimination provides **15** as a 75:25 *E*/*Z* mixture of isomers. Oxidation of **15** to the epoxide **16** occurs upon treatment with MCPBA. Saponification of the acetate and eliminative opening of the epoxide with LDA proceeds with kinetic selectivity and provides the natural product **12** in 7 steps and 14% overall yield.

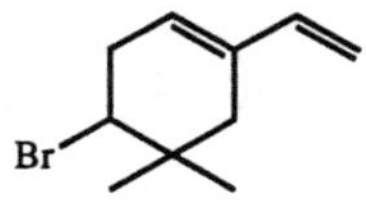

18

Chondrococcus (Desmia) Metabolite

A variety of halogenated metabolites have been isolated from the red alga *Chondrococcus* (*Desmia*) *hornemanni*. Most are acyclic myrcene derivatives. A structure has been proposed [11] for one such substance (**18**, Scheme 4) isolated from Japanese *C.* (*D.*) *hornemanni*. This structure has been confirmed through synthesis by Yoshihara and Hirose [12]. Cyclization of myrcene (**8**) by treatment with 2,4,4,6-tetrabromo-2,5-cyclohexadien-1-one produces a mixture of cyclic and acyclic brominated isomers **18–21**. Separation by GC and characterization shows **18** to be identical with the natural product.

2.1.2 From *Plocamium* sp.

2.1.2.1 Costatolide and Costatone

Costatone (**27**) and the related costatolide (**26**) are two unique halogenated heterocyclic monoterpenoids isolated from the Australian red alga *Plocamium*

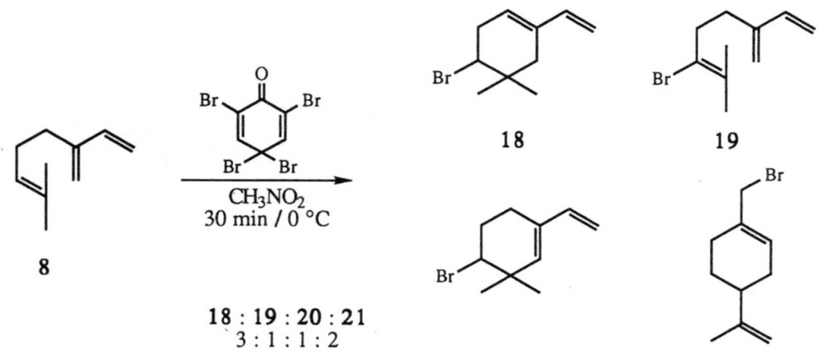

Scheme 4. Yoshihara-Hirose Synthesis of the *Chondrococcus* (*Desmia*) Metabolite **18**

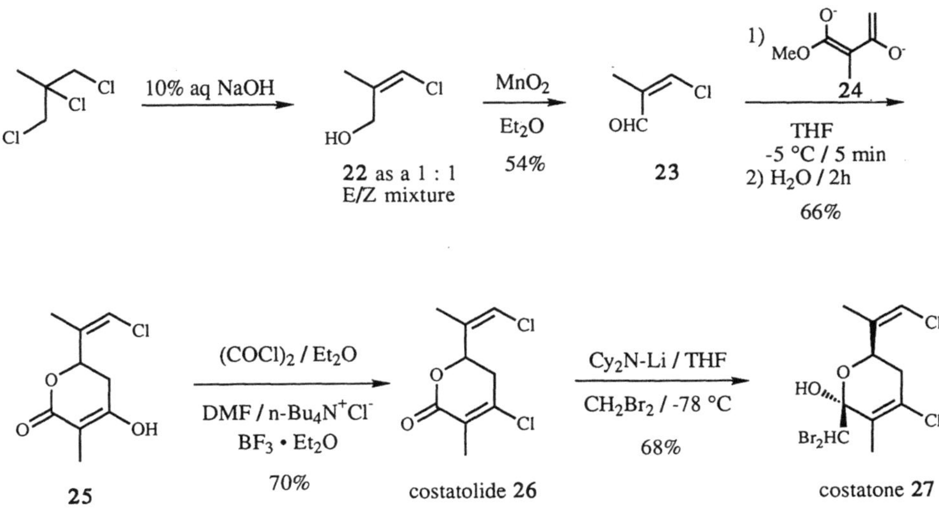

Scheme 5. Williard Synthesis of Costatolide and Costatone

Costatolide **26** Costatone **27**

costatum [13]. The synthesis of costatolide (**26**) [14] and costatone (**27**) [15] by Williard is presented in Scheme 5. Vigorous treatment of 2-methyl-1,2,3-trichloropropane with aqueous sodium hydroxide affords a separable mixture of the (*E*)- and (*Z*)-alcohols **22**. Allylic oxidation of the (*Z*)-alcohol with MnO_2 and condensation of the resulting chloroaldehyde **23** with dianion **24** yields the lactone **25** after quenching with water. Conversion of **36** into the chlorooxalate ester by addition of excess oxalyl chloride with catalytic amounts of boron trifluoride etherate produces costatolide (**26**). Addition of the anion of methylene bromide to **26** produces costatone (**27**) in 5 steps and approximately 6% overall yield.

2.1.2.2 epi-Plocamene-D and an Unnamed Metabolite

epi-Plocamene-D (**35**) and an unnamed metabolite **33** are also found [16] in *Plocamium*. They are closely related to the highly halogenated monocyclic monoterpene violacene [17]. In the course of developing an approach for the total synthesis of violacene, Williard [18] has prepared **33** and **35** (Scheme 6). As

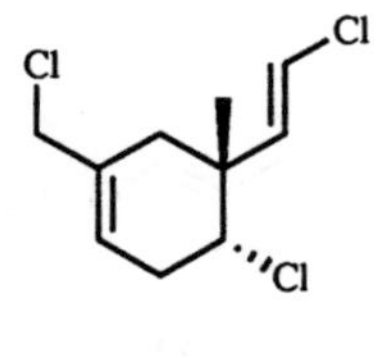

33

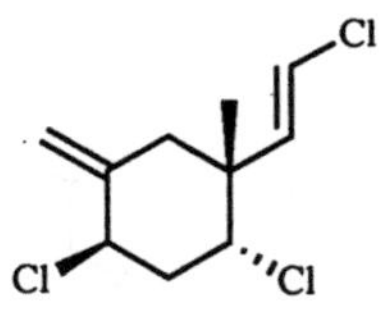

35 epi-Plocamene D

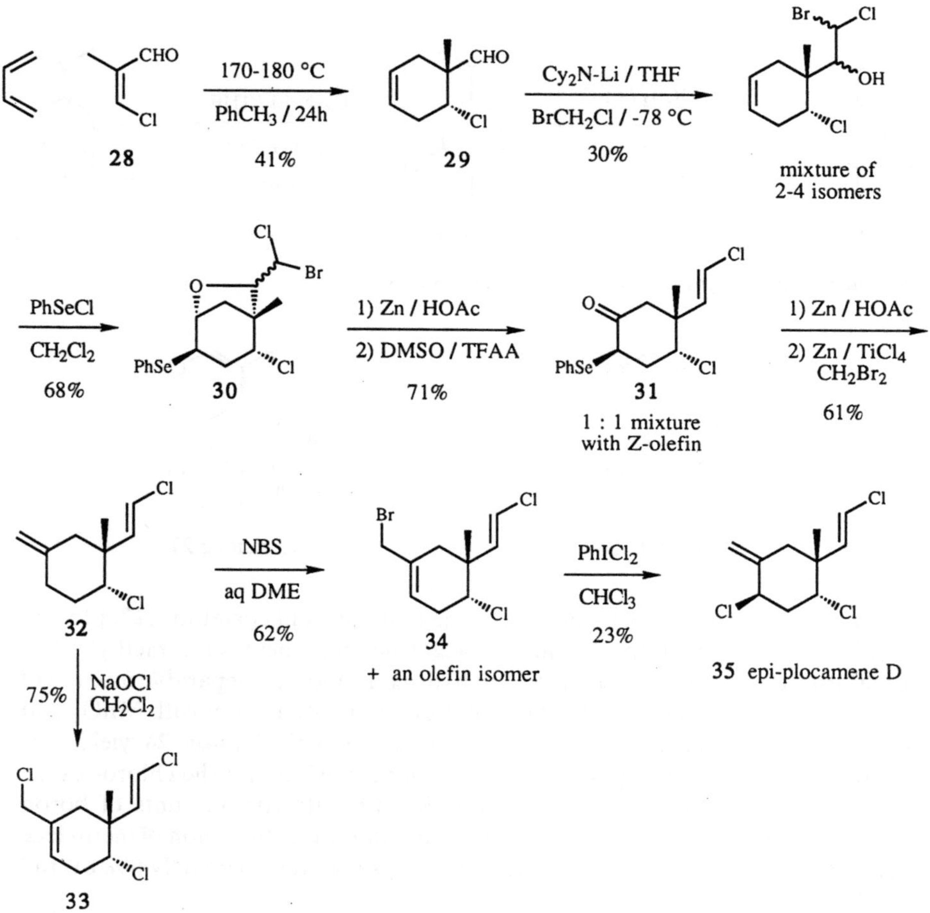

Scheme 6. Williard Synthesis of *Plocamium* Metabolites **33** and **35**

a way of circumventing many of the problems inherent in halogenation by either electrophilic addition or nucleophilic substitution, a Diels-Alder-based approach was developed for the synthesis of **33** and **35**. Diels-Alder reaction of (*Z*)-β-chloromethacrolein **28** with butadiene affords cycloadduct **29** in 41%

yield. Addition of bromochloromethylide anion to **29** gives a mixture of isomers that is cyclized to **30** by treatment with phenylselenyl chloride. Reductive elimination of **30** by treatment with Zn/HOAc and oxidation to the ketone gives **31** as a 1:1 mixture of the *E* and *Z* olefins. Attempts to convert **31** to the natural products were unsuccessful, thus necessitating removal of the phenylselenyl group and a change in strategy. Treatment with Zn/HOAc followed by titanium promoted methylenation with methylene bromide produces **32**. Allylic chlorination and rearrangement occurs upon reaction of **32** with sodium hypochlorite to give the metabolite **33** in 8 steps and 1.4% overall yield. *epi*-Plocamene D is obtained from **32** by allylic bromination with NBS to give **34** followed by treatment with $PhICl_2$. After 9 steps, **35** is obtained in 0.3% yield.

2.1.3 Prenylated Phenols

2.1.3.1 *Amaroucium Metabolite*

Amaroucium metabolite **39** is one of the active principles isolated from the colonial tunicate *Amaroucium multiplicatum* that exhibits antioxidant activities. Compound **39** exhibits properties of interest in studies toward the development of drugs to control arteriosclerosis and hypertension [19]. The synthesis of **39** by Sato [19] is illustrated in Scheme 7. Condensation of acetophenone derivative **36** with ketone **37** in the presence of pyrrolidine provides chromene **38**. LAH reduction of **38** followed by quenching of the reaction mixture with acetic acid

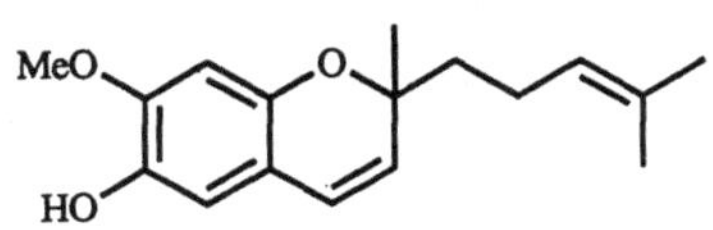

Amaroucium Metabolite **39**

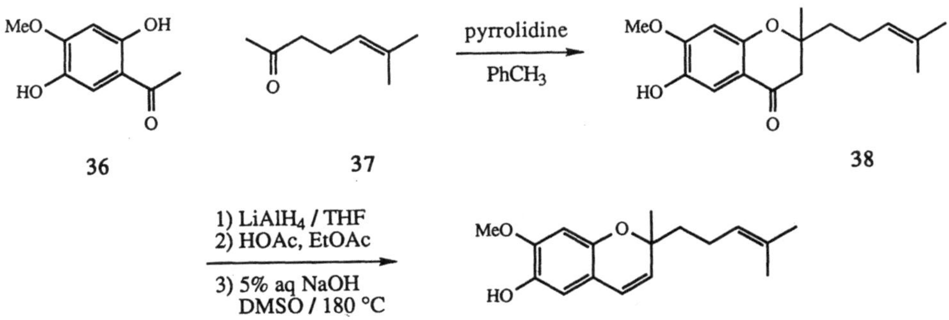

Amaroucium Metabolite **39**

Scheme 7. Sato Synthesis of *Amaroucium* Metabolite **39**

and ethyl acetate affords the acetate, which upon heating with sodium hydroxide in DMSO provides *Amaroucium* metabolite **39** in an overall 3 steps.

2.1.3.2 Hydrallmanol A

Hydrallmanol A (**45**) is a diphenyl-*p*-methane derivative isolated from the marine hydroid *Hydrallmania falcata*. It is unique in that it is the first example of a phenyl-*p*-menthane from an animal source and the first example from any source in which the menthane skeleton contains two phenyl substituents. In order to verify the proposed structure and to provide material for biological testing, compound **45** was prepared by synthesis (Scheme 8). Andersen's [20] preparation of **45** begins with copper-catalyzed conjugate addition of *p*-anisylmagnesium bromide to aldehyde **41** to afford **42**. Further addition of *p*-anisyl Grignard reagent produces alcohol **43**, as a mixture of isomers, which upon treatment with acid yields the diphenyl-*p*-menthane skeleton **44** along with other isomers. Hydrogenation followed by demethylation by treatment with TMSI yields hydrallmanol A **45** in a 5 step process.

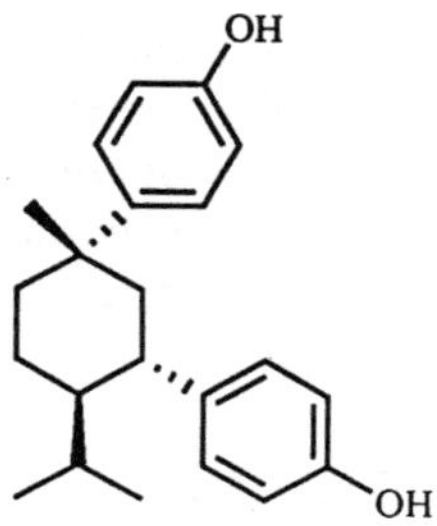

Hydrallmanol A **45**

2.2 Sesquiterpenoids

2.2.1 Lauranes

2.2.1.1 Laurene

The bicyclic sesquiterpene laurene (**50**) was first isolated from *Laurencia glandulifera*, and has been found in several *Laurencia* species, including the marine red alga *Laurencia elata* [21]. The major synthetic difficulty encountered in the preparation of laurene is the correct stereochemistry at the vicinal methyl groups. This difficulty is illustrated in the synthesis of laurene by Srikrishna [22] (Scheme 9). The key reaction in this strategy is a radical cyclization to form the cyclopentane ring with an exocyclic methylene. Allylation of ester **46** followed

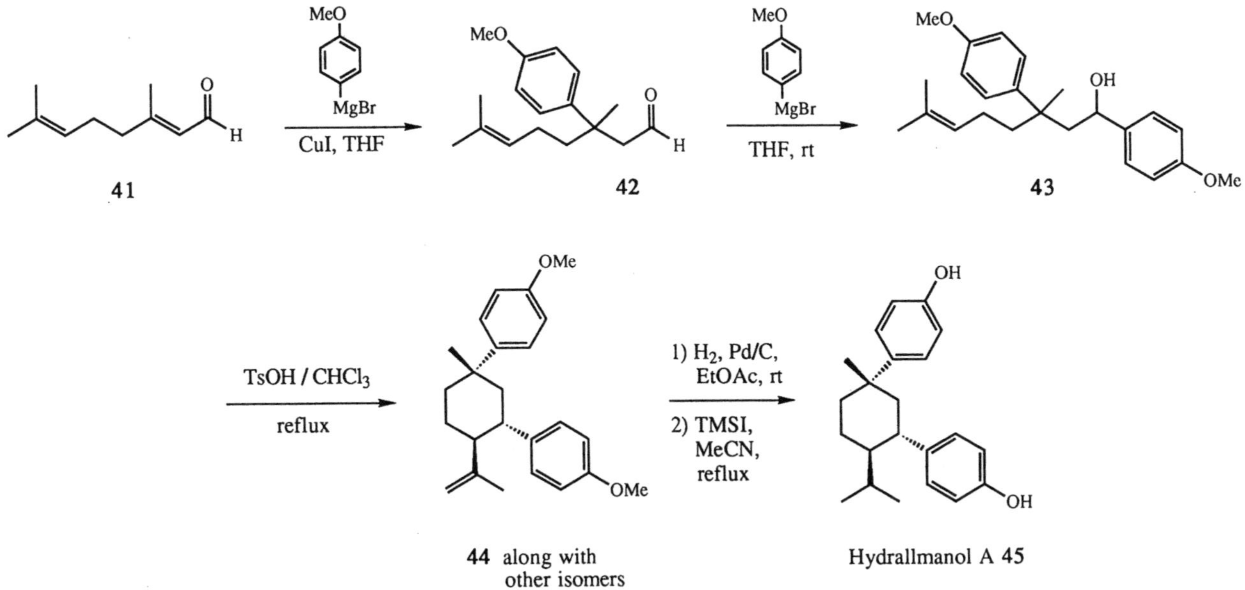

Scheme 8. Andersen Synthesis of Hydrallmanol A

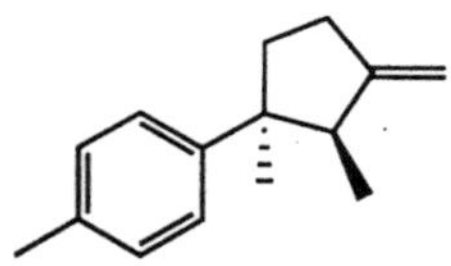

Laurene **50**

by LAH reduction, PCC oxidation and acetal formation provides the methyl-substituted quaternary center in the form of acetal **47**. Hydroboration-oxidation with subsequent PCC oxidation converts the allyl moiety to an aldehyde, which is converted to an alkyne **48** via dibromomethylene Wittig olefination and elimination. Hydrolysis of the acetal **48**, addition of methylmagnesium iodide and conversion of the resulting alcohol to the xanthate provides the needed radical precursor **49**. Treatment of **49** with tri-*n*-butyltin hydride produces a 1:1 mixture of laurene and *epi*-laurene in 12 steps and 18% overall yield.

Slightly better control of diastereoselectivity in the synthesis of laurene (**50**) is obtained by Taber [23] (Scheme 10). The key step in this strategy involves reduction of a tosylhydrazone to provide an azine. Diastereoselectivity is achieved in the ensuing azine rearrangement, in which hydride is delivered preferentially to one face of the olefin. Butylthiomethylation of ketone **52** followed by addition of methyllithium and hydrolysis gives aldehyde **53**. Formation of the tosylhydrazone and reduction with catecholborane gives a 65:35 mixture of laurene **50** and *epi*-laurene. Laurene is produced in 11% yield after the 6 steps.

The approach to laurene (**50**) that gives the greatest degree of diastereoselectivity is that of McMurry [24] (Scheme 11). This strategy for inducing diastereoselectivity depends upon the diastereoface selectivity of hydroboration of trisubstituted alkene **58** which is obtained through standard procedures. Addition of *p*-tolyl Grignard reagent **54** to cyclopentanone and elimination affords cyclopentene **55**. Epoxidation of **55** and rearrangement to the ketone **56** followed by methylation of the thermodynamic enolate produces **57**. Addition of methyllithium to **57** and subsequent elimination provides the trisubstituted alkene **58**. Hydroboration-oxidation of **58** and subsequent Collins oxidation of the resulting alcohol produces dimethyl ketone **59** as a 4:1 mixture favoring the desired diastereomer **59**. Treatment of ketone **59** with phenylthiomethyllithium, benzoylation and reductive elimination produces laurene **50** under non-epimerizing conditions. The target is obtained in 9 steps.

2.2.1.2 Aplysin, Debromoaplysin and Filiformin

Aplysin and debromoaplysin, two of the first halogenated sesquiterpenes from marine sources, were first isolated from the sea hare *Aplysia kurodai*. It was thought that these metabolites originate in the organism's diet of red algae [25]. They may act as antifeedants and antioxidants. The source of filiformin is still under discussion. It has been suggested that it occurs in extracts of *Laurencia*

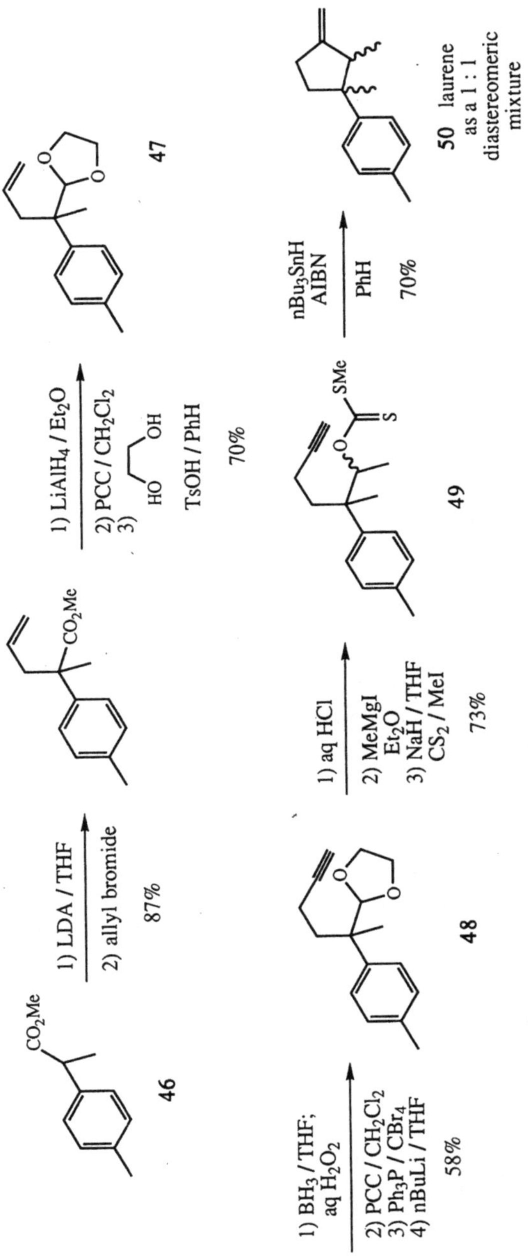

Scheme 9. Srikrishna Synthesis of (±)-Laurene

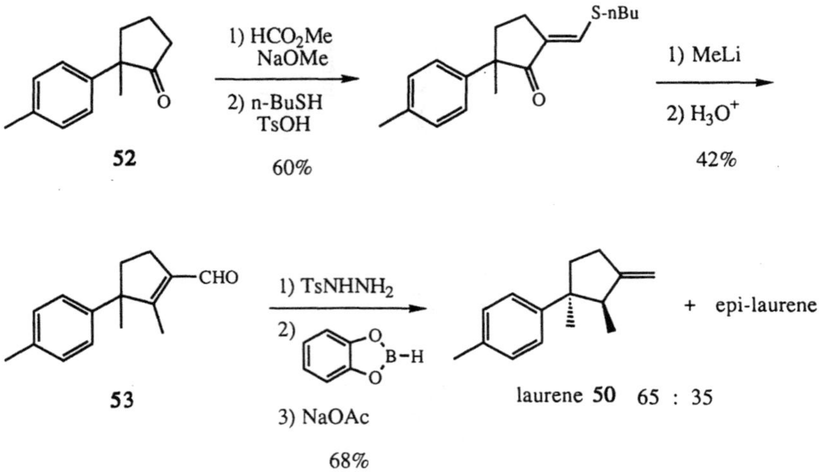

Scheme 10. Taber Synthesis of (±)-Laurene

filiformis as a consequence of in vitro cyclization of allolaurinterol during isolation [26]. Laurinterol was isolated from *Laurencia intermedia* (Yamada) [27].

The first synthesis of aplysin and debromoaplysin is that of Yamada [28] shown in Scheme 12. The strategy followed by Yamada involves coupling of an appropriately substituted aromatic fragment with cyclopentanone, elaboration of the cyclopentane moiety by sequential addition of the three methyl groups and ring closure to provide the benzo-substituted dihydrofuran system in the final step. The synthesis is initiated by *ortho*-metallation of methyl substituted *p*-bromoanisole (**61**) followed by addition to cyclopentanone with elimination to afford cyclopentene **62**. Oxidation of **62** to the ketone followed by methylation of the thermodynamic enolate produces ketone **63**, which is methylated once at the remaining open α-position in a three step process to give dimethyl ketone **64** as a mixture of isomers. Addition of methylmagnesium iodide to ketone **64** and acid-catalyzed elimination produces cyclopentene **65** in which both the aromatic and cyclopentane moieties are appropriately substituted. Demethylation of **65** by treatment with methylmagnesium iodide (with partial removal of bromine) yields a mixture of aplysin (**66**) and debromoaplysin (**67**). Aplysin is produced in 9 steps and 0.2% overall yield

An alternate procedure by Yamada [29] for completion of the aplysin synthesis is illustrated in Scheme 13. This strategy differs from that in Scheme 12 in that the order of methylation of ketone **63** is reversed. The final two methyl groups are added in an iterative process involving ketone formation, methyl addition and elimination. Addition of methylmagnesium iodide to ketone **63** and elimination provides dimethylcyclopentene **68**. Chlorination of **69** and demethylation of the anisole unit with concomitant ring closure affords the aplysin ring system **70**. Addition of methylmagnesium iodide and elimination

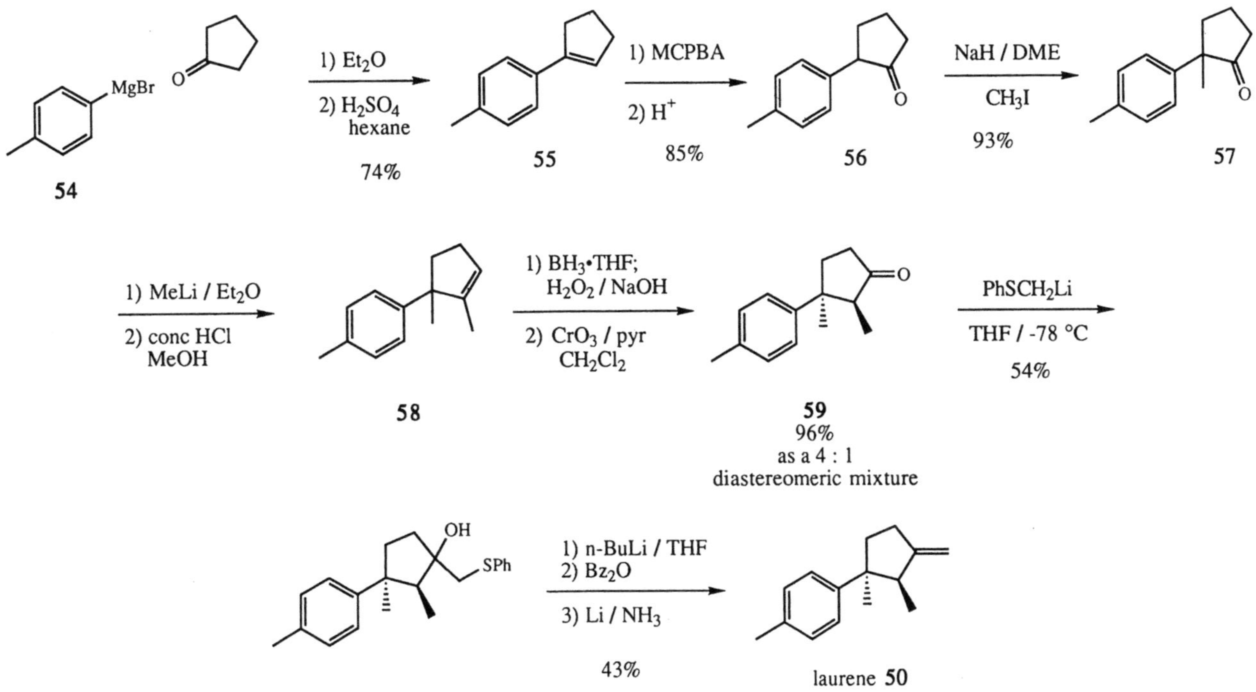

Scheme 11. McMurry (±)-Laurene Synthesis

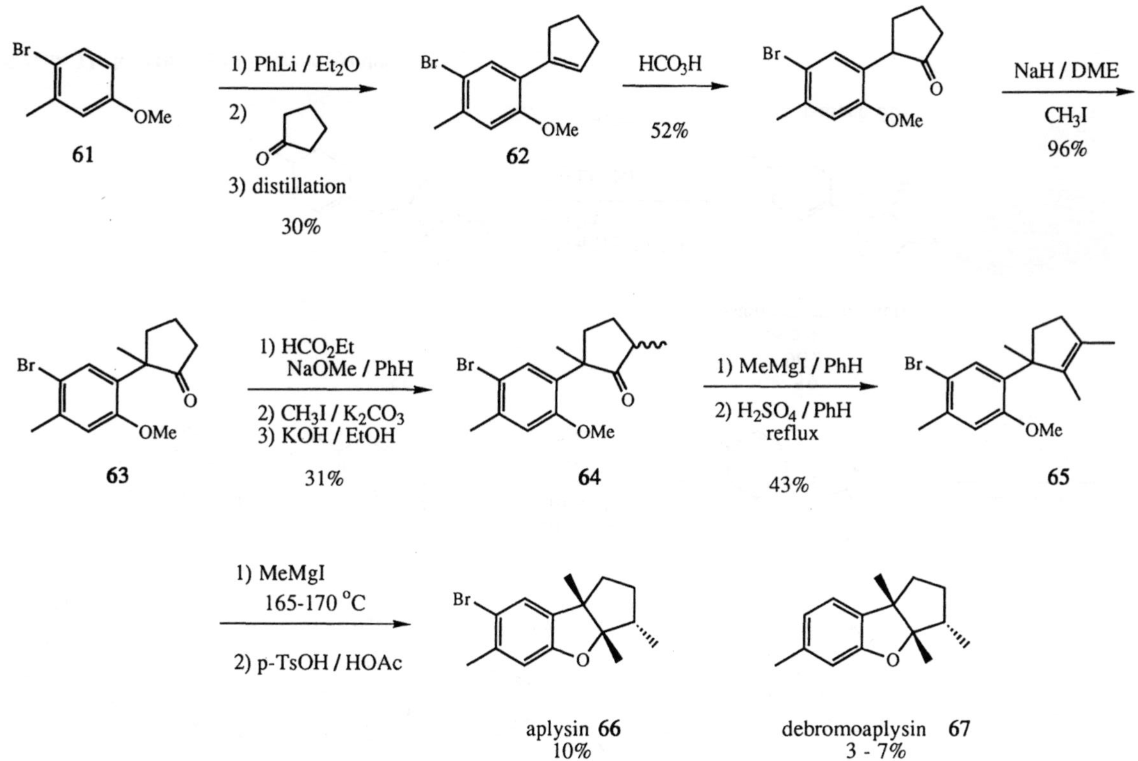

Scheme 12. Yamada Synthesis of Aplysin and Debromoaplysin

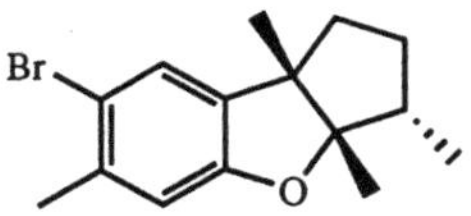

Aplysin **66**

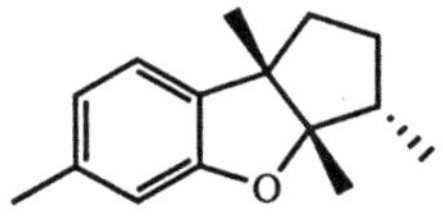

Debromoaplysin **67**

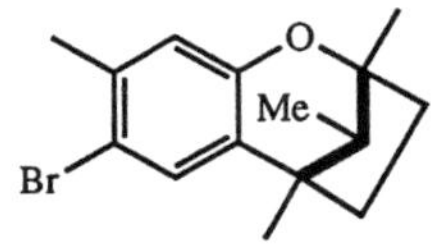

Filiformin **76**

gives alkene **71**, which is converted to a mixture of aplysin and debromoaplysin by hydrogenation. Aplysin is obtained in 10 steps and approximately 1% overall yield by this route.

Goldsmith [30] utilizes an intermediate from Yamada's first aplysin synthesis (Scheme 15) to execute a strategy that allows for the preparation of either aplysin (**66**) or filiformin (**76**) (Scheme 14). The general strategy involves preparation of a trichothecane-like intermediate that can be either further functionalized to give filiformin (**76**) or rearranged and reduced to provide aplysin (**66**). Demethylation, acylation, and chlorination of Yamada's intermediate ketone **64** provides chloroketone **73**. Treatment with wet DBN provides the tricothecane ring system **74** via an interesting ring closure process [31]. Wittig olefination provides alkene **75** which is converted to filiformin **76** by catalytic hydrogenation. Filiformin is obtained in 6 steps (11 steps from the Yamada starting material) and 36% yield (1.7% overall). Alternatively, acid-catalyzed rearrangement of alkene **75** followed by hydrogenation produces aplysin in 7 steps (12 steps form Yamada's starting material) and 12% yield (0.6% overall).

A completely different approach to debromoaplysin (**67**) and aplysin (**66**) is that developed by Ronald [32] (Scheme 15). It differs from other approaches in that the debromoaplysin ring system is attained in the initial stages of the synthesis. Further functionalization and adjustment of stereochemistry provides debromoaplysin (**67**) which can be converted to aplysin by bromination. This allows preparation of **66** and **67** separately, rather than as mixtures. Addition of aryllithium **77** to α-chloroketone **78** provides alcohol **79**. Treatment of **79** with refluxing potassium methoxide in methanol gives the cyclized product, via a process that is formally a base-catalyzed methanolysis of a mixed formaldehyde acetal. Conversion of alcohol **80** to the bromide and simple Grignard coupling (without rearrangement) allows insertion of the third methyl group in low yield. Isomerization of alkene **81** to the trisubstituted olefin followed by catalytic hydrogenation provides the proper relative stereochemistry giving debromoaplysin as a 95:5 mixture favoring the desired isomer. Bromination of **67** provides aplysin in 7 steps and 4% overall yield. Use of the aryllithium derived from the optically pure isopinocampheol mixed acetal **82** gives **83** as a mixture of diastereomers that can be separated and carried on to provide optically pure (−)-debromoaplysin and (−)-aplysin.

Another approach that involves early formation of the aplysin ring system and bromination as the final step is that of Venkateswaran [33] (Scheme 16). Coupling of α-bromopropionic acid with phenol **85** affords phenoxyacetic acid

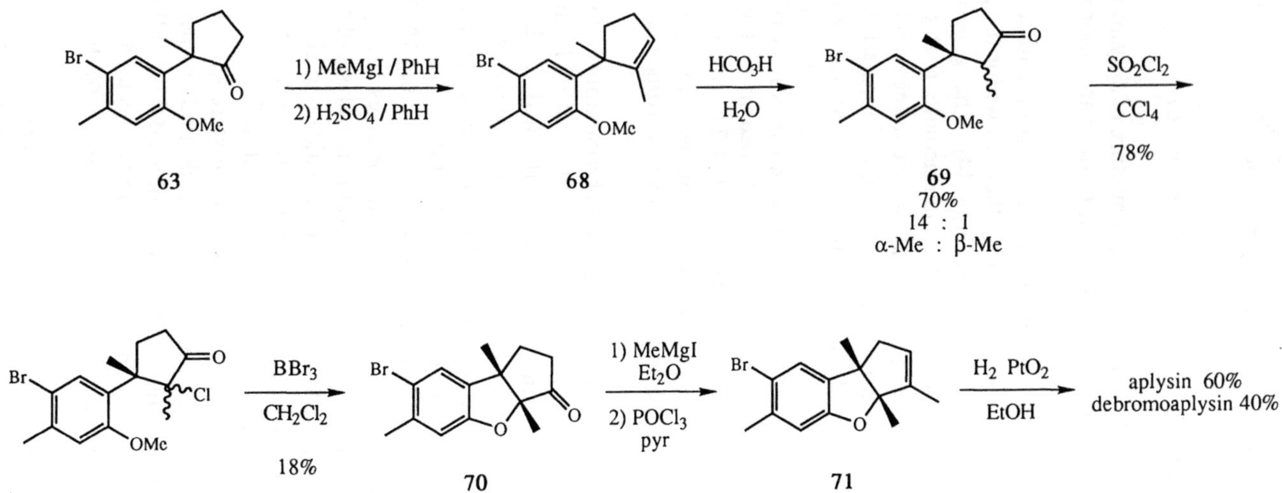

a single unassigned isomer

Scheme 13. Yamada Synthesis of Racemic Aplysin and Debromoaplysin

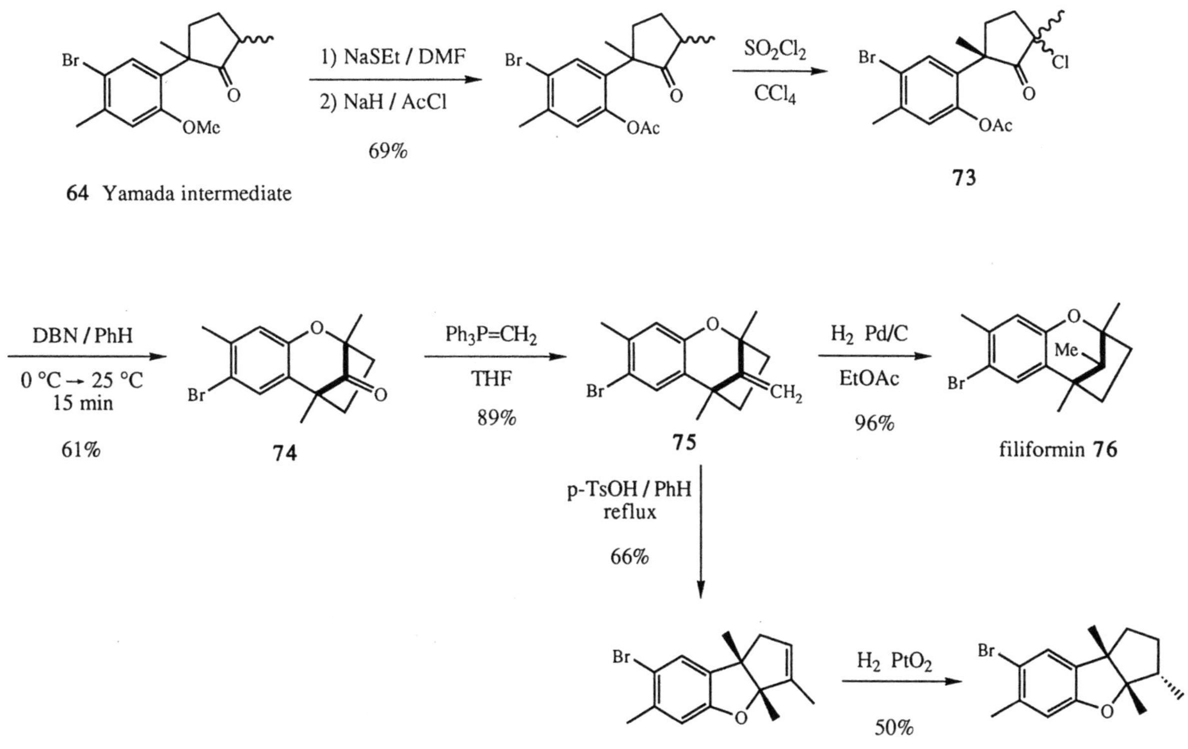

Scheme 14. Goldsmith Synthesis (±)-Aplysin and Filiformin

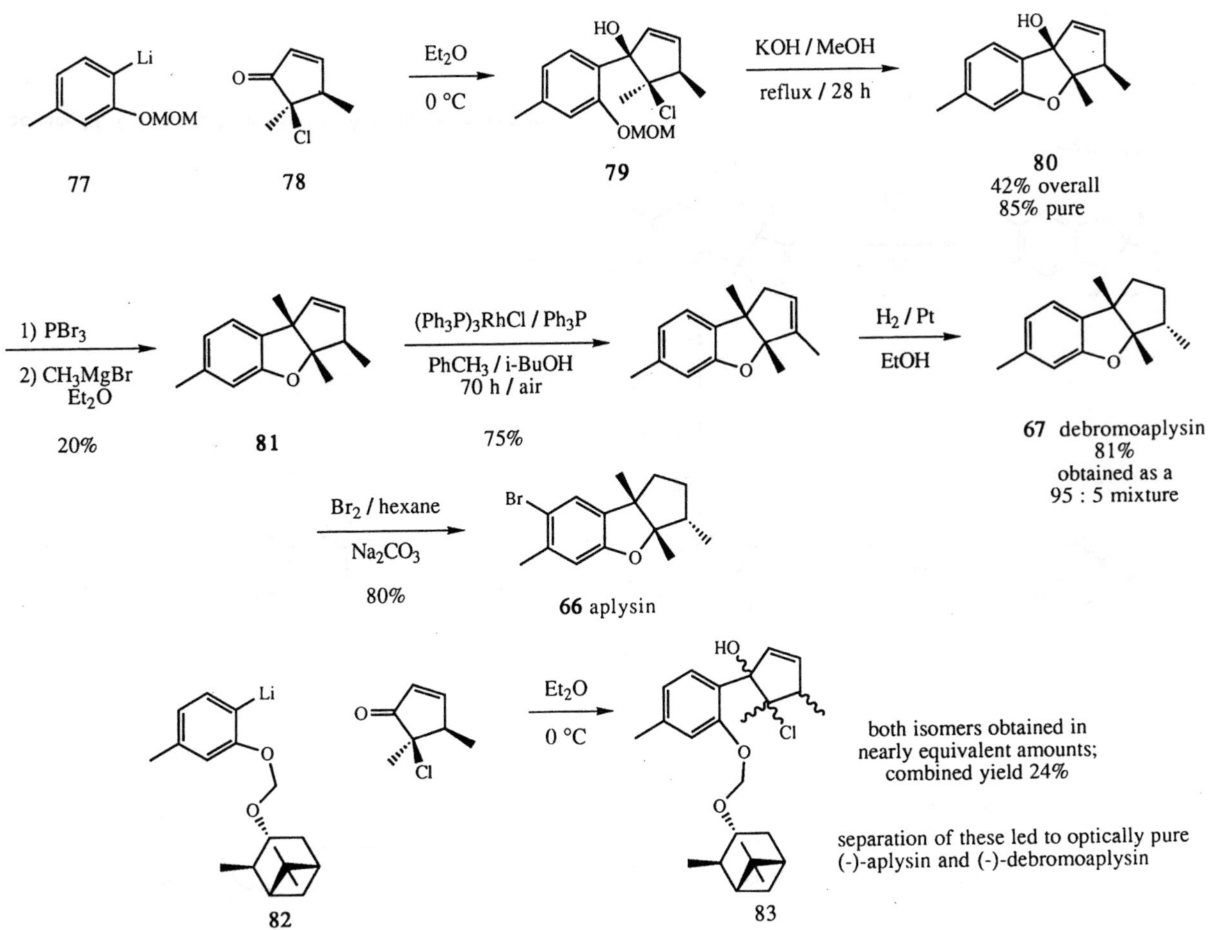

Scheme 15. Ronald Synthesis of Racemic Aplysin and Debromoaplysin

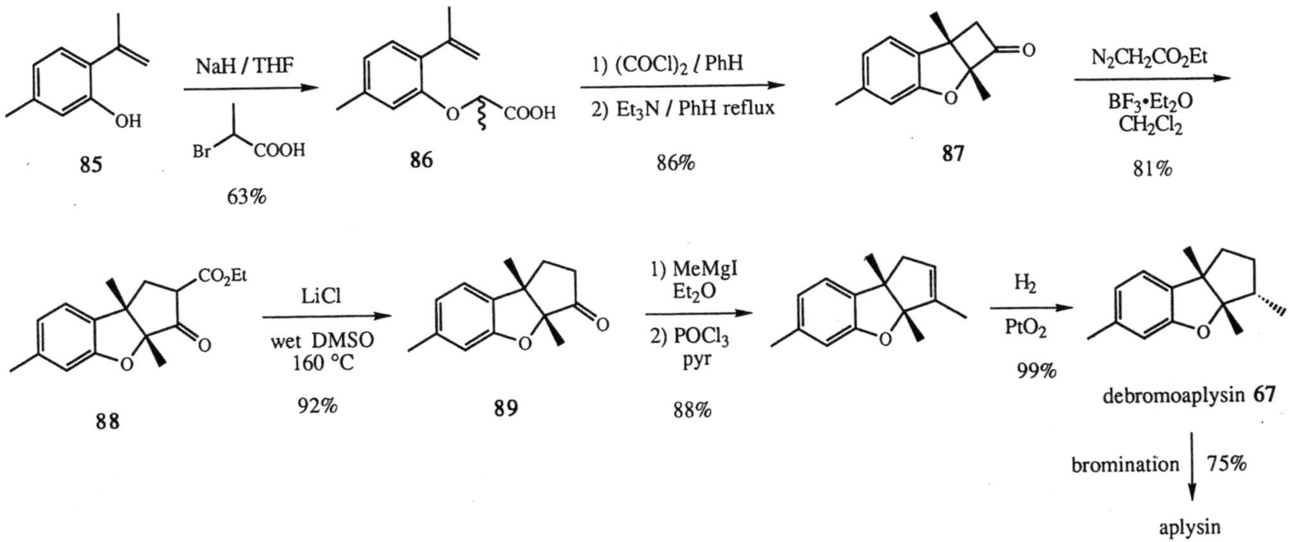

Scheme 16. Venkateswaran Synthesis of (±)-Aplysin and Debromoaplysin

86. Conversion of **86** to the acid chloride and treatment with triethylamine generates a ketene which undergoes intramolecular ketene-olefin cycloaddition to provide the linear 6-5-4 ring system **87**. Ring expansion of **87** by treatment with ethyl diazoacetate in the presence of $BF_3 \cdot Et_2O$ produces ester **88** which is decarboxylated to afford the nor-methyl aplysin skeleton as in **89**. Methylation with methylmagnesium iodide, elimination and catalytic hydrogenation affords debromoaplysin (**67**). Bromination following the procedure of Ronald (Scheme 15) produces aplysin (**66**) in 9 steps and 26% overall yield.

The procedure developed by Laronze [34] produces both aplysin (**66**) and filiformin (**76**) (Scheme 17). Reaction of *O*-substituted hydroxylamine **90**, prepared by a known but lengthy procedure [35], with 2,5-dimethylcyclo-pentanone affords oxime **91**. Treatment of **91** with acid in refluxing methanol provides aminal **92** along with its regioisomer **93**, via the Sheradsky rearrangement of α,α′-disubstituted cyclopentanone aryloximes [36]. Hydrolysis of **92** produces the hemiacetal **94** which is converted to the tertiary alcohol by treatment with

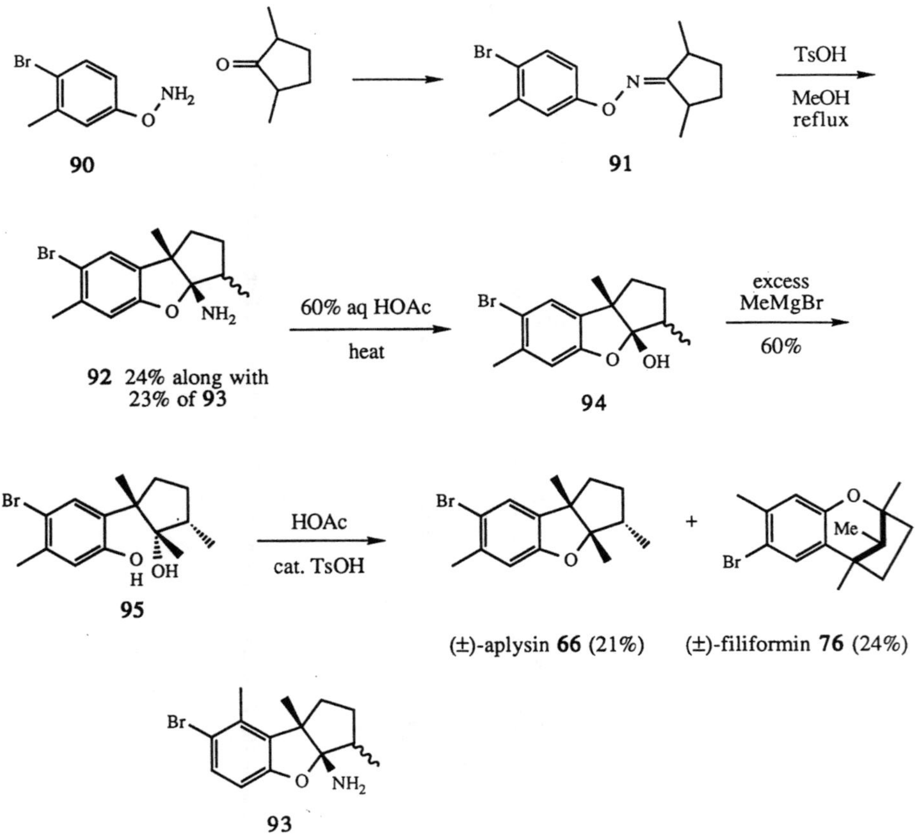

Scheme 17. Laronze Synthesis of Racemic Aplysin and Filiformin

excess methylmagnesium bromide. Acid catalyzed cyclization of **95** produces both aplysin (**66**) and filiformin (**76**) as a 1:1 mixture in 5 steps from **90**.

2.2.1.3 Laurinterol and Allolaurinterol

Allolaurinterol (**100**) has been synthesized by Ronald [37] (Scheme 18) utilizing a strategy developed for the synthesis of aplysin (Scheme 15). The initial steps in the synthesis are nearly identical with the aplysin synthesis, producing tricyclic ether **96** in five steps. Eliminative ring opening of **96** by treatment with LiTMP yields phenol **97**. Silylation of the phenol and hydroboration-oxidation yields alcohol **98**. Oxidation of **98** with PDC and treatment with bromine forms bromide **99**. Treatment of **99** with methylene bromide and $TiCl_4$ in the presence of zinc metal produces the exocyclic olefin without epimerizing the adjacent stereocenter. Desilylation with TBAF completes the synthesis to provide allolaurinterol in 12 steps and 3.6% overall yield.

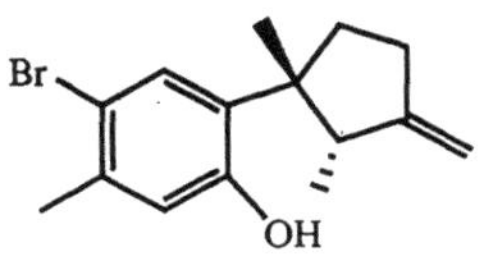

(±)-Allolaurinterol 100

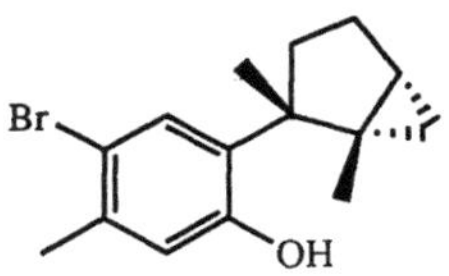

(±)-Laurinterol 105

Laurinterol (**105**) has been prepared by Mirrington [38] as shown in Scheme 19. Five years after the Yamada synthesis of aplysin (Scheme 12) and four years after his second synthesis of aplysin (Scheme 13), Mirrington uses a nearly identical strategy to prepare the common intermediate ether **63** for this synthesis of laurinterol. Cyclopropanation of **101** with methylene iodide and diethylzinc affords cyclopropane **102** as a 3:1 mixture of diastereomers. The demethylation of **102** does not proceed smoothly. Therefore, the bromide was removed by treatment with LAH yielding methyl ether **103**. Reaction of **103** with sodium thioethoxide and acylation produces the acetate **104** in low yield. As **104** has been converted to laurinterol previously [39], this completes the formal synthesis of **105**.

2.2.2 Chamigranes and Related Spirocyclics

The chamigranes are a series of metabolites isolated from various red algae of the genus *Laurencia* and from herbivorous molluscs such as *Aplysia* that consume *Laurencia*. The basic carbon skeleton is that of the terrestrial natural product β-chamigrene, which was isolated from the leaf oil of *Chamaecyparis taiwanensis* [40]. The mono-brominated chamigrane 10-bromo-α-chamigrene (**109**) has been isolated from an unidentified *Laurencia* species [41]. The epimeric α-chamigrene (**123**) has been found in *Laurencia pacifica* [42]. The

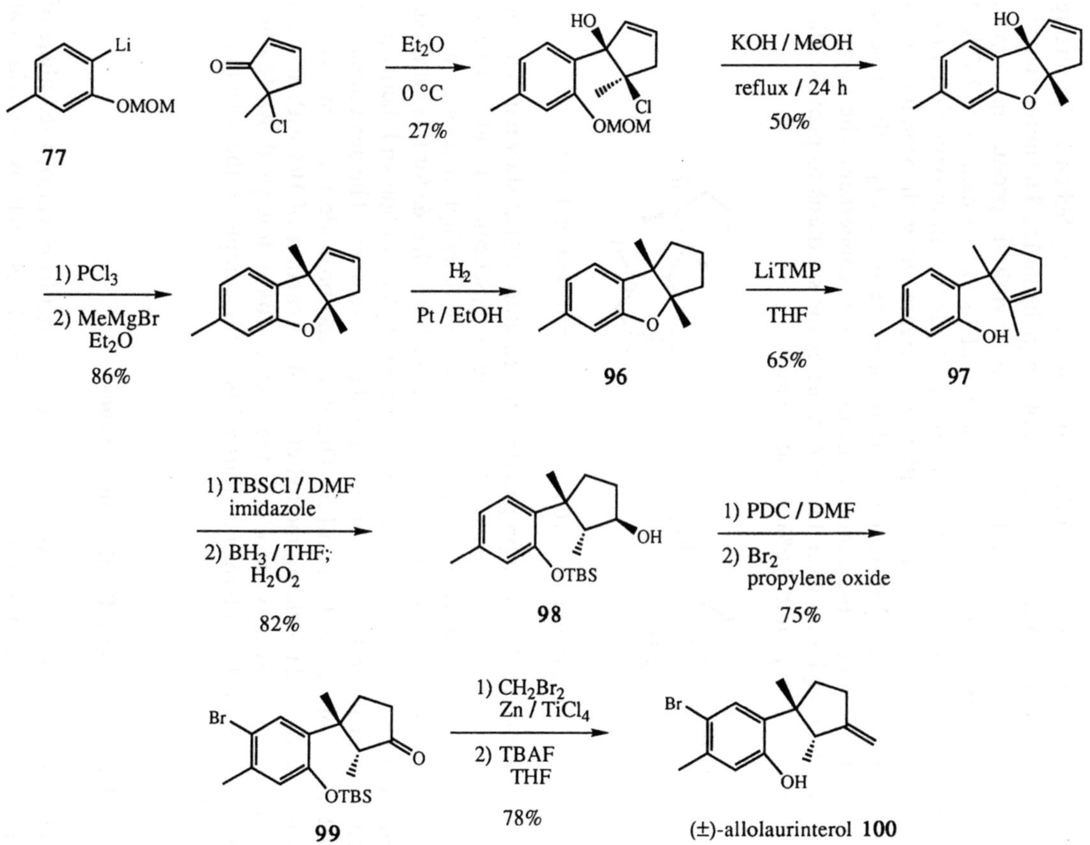

Scheme 18. Ronald Synthesis of (±)-Allolaurinterol

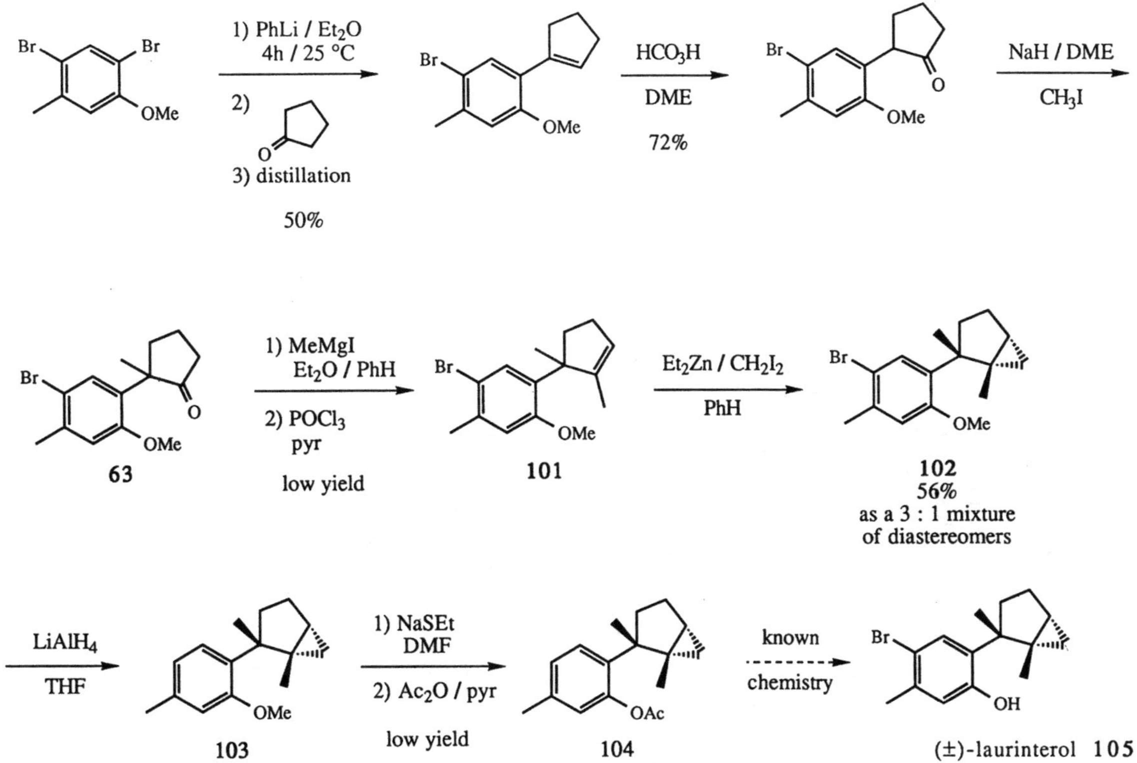

Scheme 19. Mirrington Synthesis of (±)Laurinterol

unnamed brominated chamigrane **130** was isolated from *L. majuscula* [43]. The structurally related compound spirolaurenone **155** from *L. glandulifera* [44] possesses the spirolaurane skeleton and exhibits antifungal properties. An example of a rearranged chamigrane is sesquiterpene **145** isolated from *Aplysia dactylomela* [45].

2.2.2.1 10-Bromo-α-Chamigrene

The first synthesis of 10-bromo-α-chamigrene (**109**) is that of Faulkner [46] (Scheme 20). The strategy utilized is biomimetic and involves a bromonium ion initiated cyclization reaction. This approach is widely utilized in the synthesis of bromine-containing marine terpenoids. Treatment of geranylacetone with bromine and silver tetrafluoroborate in nitromethane provides the brominated bicyclic ether **106**. Rearrangement of **106** with *p*-toluenesulfonic acid affords ketone **107** which is converted to the allylic alcohol **108** upon addition of vinylmagnesium bromide. Acid catalyzed cyclization of the allylic alcohol **108** provides 10-bromo-α-chamigrene (**109**) in 4 steps and 4% overall yield.

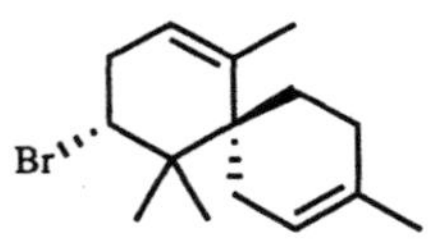

10-Bromo-α-chamigrene **109**

A second synthesis of **109** that follows a biogenetic strategy is that of Kato [47] shown in Scheme 21. Although very short, this approach is not overly efficient. Bromonium ion-induced cyclization of methyl 2,3-*cis*-farnesoate by treatment with 2,4,4,6-tetrabromocyclohexadienone produces the monocyclic bromoester **110**. Reduction of **110** to the alcohol **111** with aluminum hydride

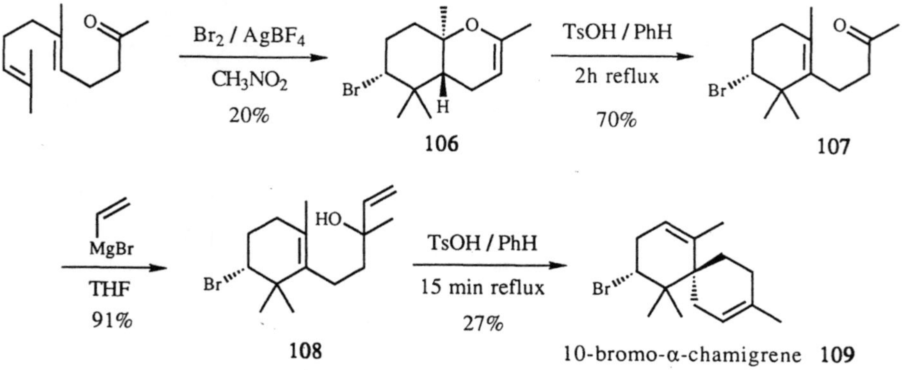

Scheme 20. Faulkner Synthesis of 10-Bromo-α-Chamigrene

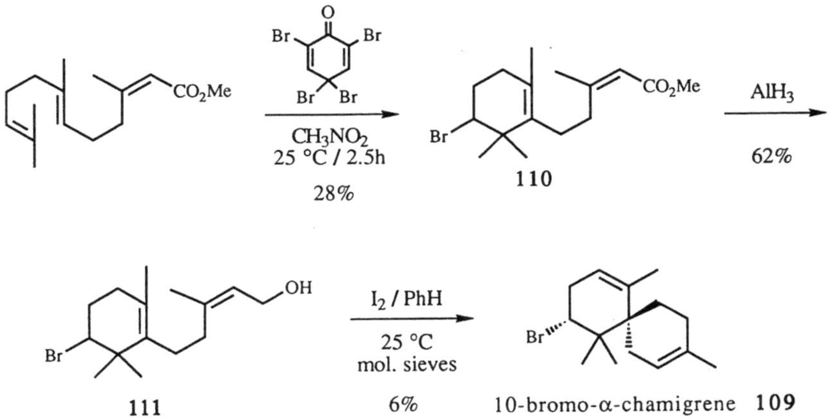

Scheme 21. Kato Synthesis of 10-Bromo-α-Chamigrene

and treatment of the alcohol with iodine in the presence of molecular sieves gives 10-bromo-α-chamigrene **109** in only 3 steps in 1% overall yield.

2.2.2.2 *10-Bromo-β-Chamigrene*

Martin [48] has developed a biogenetically inspired synthesis of 10-bromo-β-chamigrene (**123**) and related metabolites which is highlighted in Scheme 22. Resolution of a key intermediate allows the enantioselective preparation of **123**. The synthesis is initiated by iodolactonization of diastereomerically pure β-hydroxyacid **113** to give iodolactone **114**. Saponification of the iodolactone produces epoxide **115** which is then opened in an intramolecular fashion by treatment with *p*-toluenesulfonic acid to give bicyclic acid **116** after acetylation. Acid **116** is resolved by recrystallization of its quinine salt from acetone. Protection of the alkene by bromination and decarboxylation with lead tetra-acetate and NCS affords the trihalogenated ether **117**. The alkene is regenerated by reaction with zinc in acetic acid and the resulting chloroether **118** is purified by reacetylation. Reductive cleavage of the chloroether **118** with sodium produces diol **119**, which is converted to epoxide **120** by mono-tosylation and treatment with base. The epoxide is opened with dilithium tetrabromonickelate and cyclization is initiated by reaction with a source of positive bromine to give naturally-occurring bromohydrins **121** and **122** in poor yield [49]. Reductive

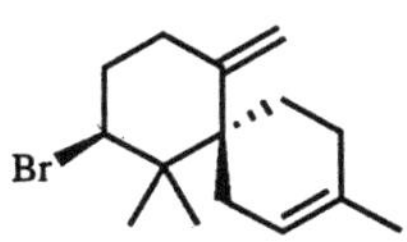

(+)-10-Bromo-β-Chamigrene **123**

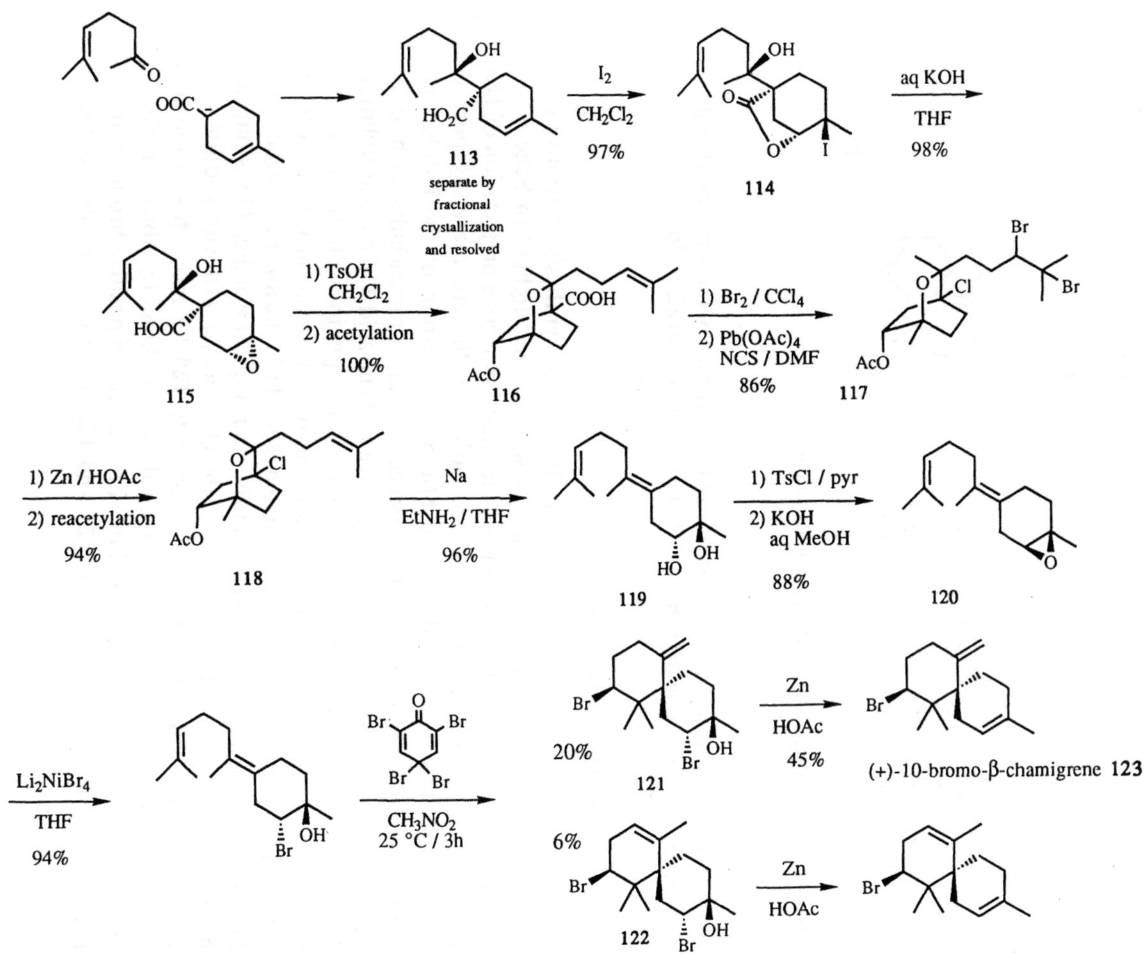

Scheme 22. Mariń Synthesis of (+)-10-Bromo-β-Chamigrene **123** and Related Metabolites

elimination of bromohydrin **121** by treatment with zinc yields (+)-10-bromo-β-chamigrene in 15 steps and approximately 3% overall yield. The enantiomeric series of compounds was also produced starting from the enantiomer of hydroxy acid **113**.

2.2.2.3 *Various Laurencia Metabolites*

The first total synthesis of the *Laurencia* metabolite 9-(*Z*)-bromomethylidene-1,5,5-trimethylspiro[5.5]undeca-1,7-dien-3-one (**130**) is that of Yamada [50] in 1985 (Scheme 23). The key step involves spiroannulation of aldehyde **125**. Friedel-Crafts alkylation of tertiary alcohol **124** with anisole followed by Birch reduction and hydrolysis provides the aldehyde **125**. Acid catalyzed cyclization of **125** with 6N HCl affords spiro alcohol **126** as a 5:1 epimeric mixture, which is converted to diketone **127** by treatment with PCC. Site specific bromomethylenation of diketone **127** is achieved by addition of the lithium anion of methylene bromide to **127** followed by reduction of the adduct with zinc in acetic acid to afford a 1:1 mixture of the *E*- and *Z*-isomers. Addition of methyllithium to the *Z*-isomer and elimination produces triene **129** as a 3:1 mixture with the exocyclic olefin. Allylic oxidation of the mixture of olefins with chromium trioxide-3,5-dimethylpyrazole complex yields **130** in 10 steps and 6.6% overall yield.

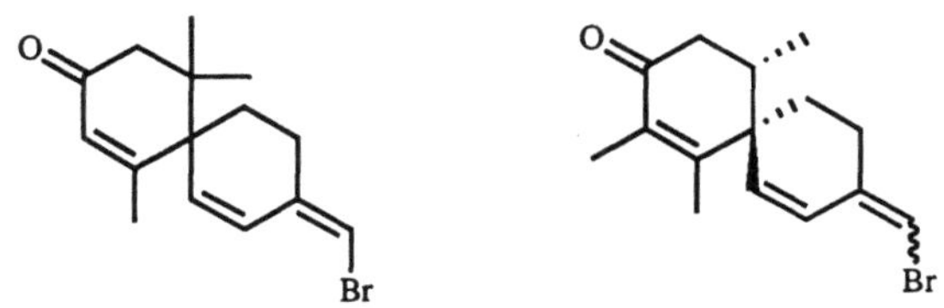

130

144 *E*
145 *Z*

Laurencia Metabolites

This metabolite has also been synthesized by Iwata [51] as shown in Scheme 24. The key step in the synthesis is the copper-catalyzed decomposition of a phenolic α-diazoketone to provide a spirocyclic ring system with a variety of manipulatable functionality. Chlorination of phenolic acid **132** to produce the acid chloride, followed by reaction with diazoethane affords the diazoketone **133**. Copper-catalyzed decomposition of the diazoketone **133** in refluxing chloroform produces the spirocyclic ring system in 22% yield. Partial reduction of **137** by reaction with lithium in ammonia followed by tosylation and elimination of the resulting alcohol **135** produces dienone **136**. Olefination with bromomethylidene triphenylphosphorane gives the brominated triene **137** as a 3:2 mixture of *E*/*Z*-isomers. Allylic oxidation of the mixture of olefins, with chromium trioxide-3,5-dimethylpyrazole complex, affords the natural (*Z*)-isomer of 9-bromomethylidene-1,5,5-trimethylspiro[5.5]undeca-1,7-dien-3-one (**130**) in 8 steps and approximately 2% overall yield.

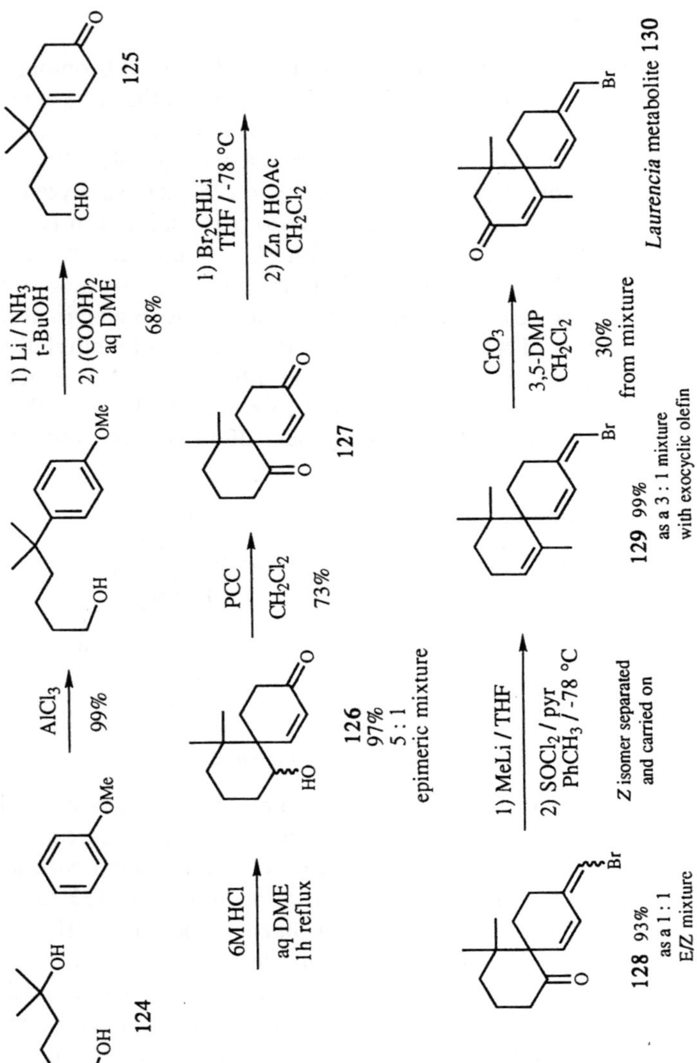

Scheme 23. Yamada Synthesis of *Laurencia* metabolite **130**

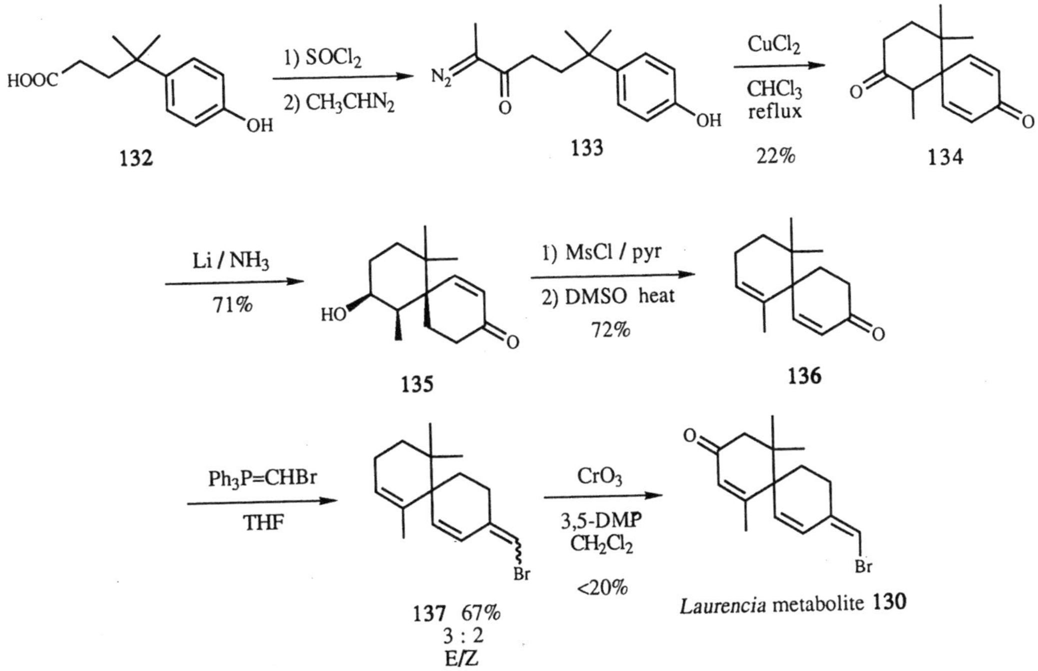

Scheme 24. Iwata Synthesis of *Laurencia* Metabolite **130**

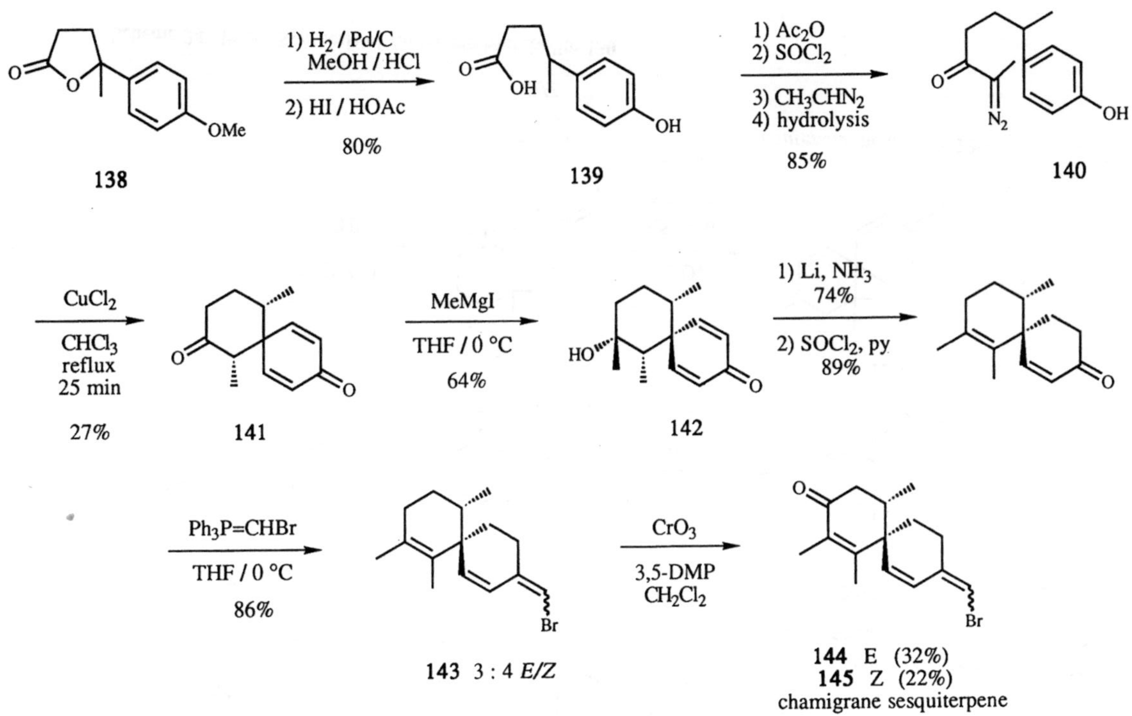

Scheme 25. Iwata Synthesis of Chamigrane Sesquiterpene **145**

The same general strategy is followed by Iwata [52] in the synthesis of the *E*- and *Z*-isomers of the rearranged chamigrane type metabolite 9-(bromomethylene)-1,2,5-trimethylspiro[5.5]undeca-1,7-dien-3-one **145** (Scheme 25). The α-diazoketone **140** required for spirocyclization is obtained from the known lactone **138** in six steps. Hydrogenolysis of the benzylic C–O bond followed by demethylation with HI affords the phenolic carboxylic acid **139**. Acylation of the phenol, conversion of the carboxyl functionality to an acyl chloride and reaction with diazoethane provides the needed α-diazoketone. Hydrolysis of the phenolic acetate provides the phenol **140**. Copper-catalyzed decomposition of the phenolic diazoketone in refluxing chloroform produces the spirocyclic dienone **141**. Addition of methylmagnesium iodide to the saturated carbonyl of **141** followed by controlled Birch reduction affords enone **142**. Dehydration with thionyl chloride and pyridine and subsequent olefination with bromomethylidene triphenylphosphorane secures triene **143** as a 3:4 mixture of *E*/*Z* isomers. Allylic oxidation of triene **143** as before produces the natural *Z*-isomer **145** in 22% yield along with 32% of the *E*-isomer. (*Z*)-9-(bromomethylene)-1,2,5-trimethylspiro[5.5]undeca-1,7-dien-3-one (**145**) is produced in 1.5% overall yield over 12 steps. It has been suggested that the *Z*-isomer is not a natural product but rather an artifact of the isolation process.

2.2.2.4 Spirolaurenone

Masamune's synthesis [53] of spirolaurenone (**155**) is presented in Scheme 26. The synthesis begins with the Lewis acid catalyzed cyclization of the bromohydrin of homogeranonitrile to afford bromonitrile **147** in 51% yield along with 13% of the Δ^2 isomer. Reduction and hydrolysis of **147** to the aldehyde and oxidation with PDC produces the unsaturated acid **148**. Conversion of **148** to the α-diazoketone and treatment with copper yields the angularly fused tricyclic 6-3-5 system **149** via intramolecular carbenoid cycloaddition onto the alkene. The diastereomer **150** is also obtained in 11% yield. Addition of 1,3-dithienium tetrafluoroborate to the kinetically formed TMS enol ether of ketone **149** from the less hindered β-side stereospecifically provides the α-alkylated ketone **151**. Addition of acid cleaves the C-5–C-7 bond and thus secures the unsaturated spiro-system **153** together with 11% of the isomeric exocyclic alkene. Removal of the carbonyl proceeds in a standard manner to give dithiane **153**. The protected formyl moiety is revealed by reaction with methyl iodide in aqueous acetone to afford aldehyde **154**. Addition of methyllithium and Jones oxidation completes the synthesis, yielding spirolaurenone **155** in 15 steps and approximately 2% overall yield.

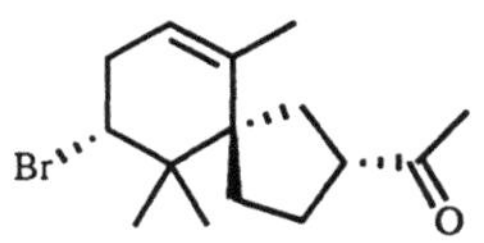

Spirolaurenone **155**

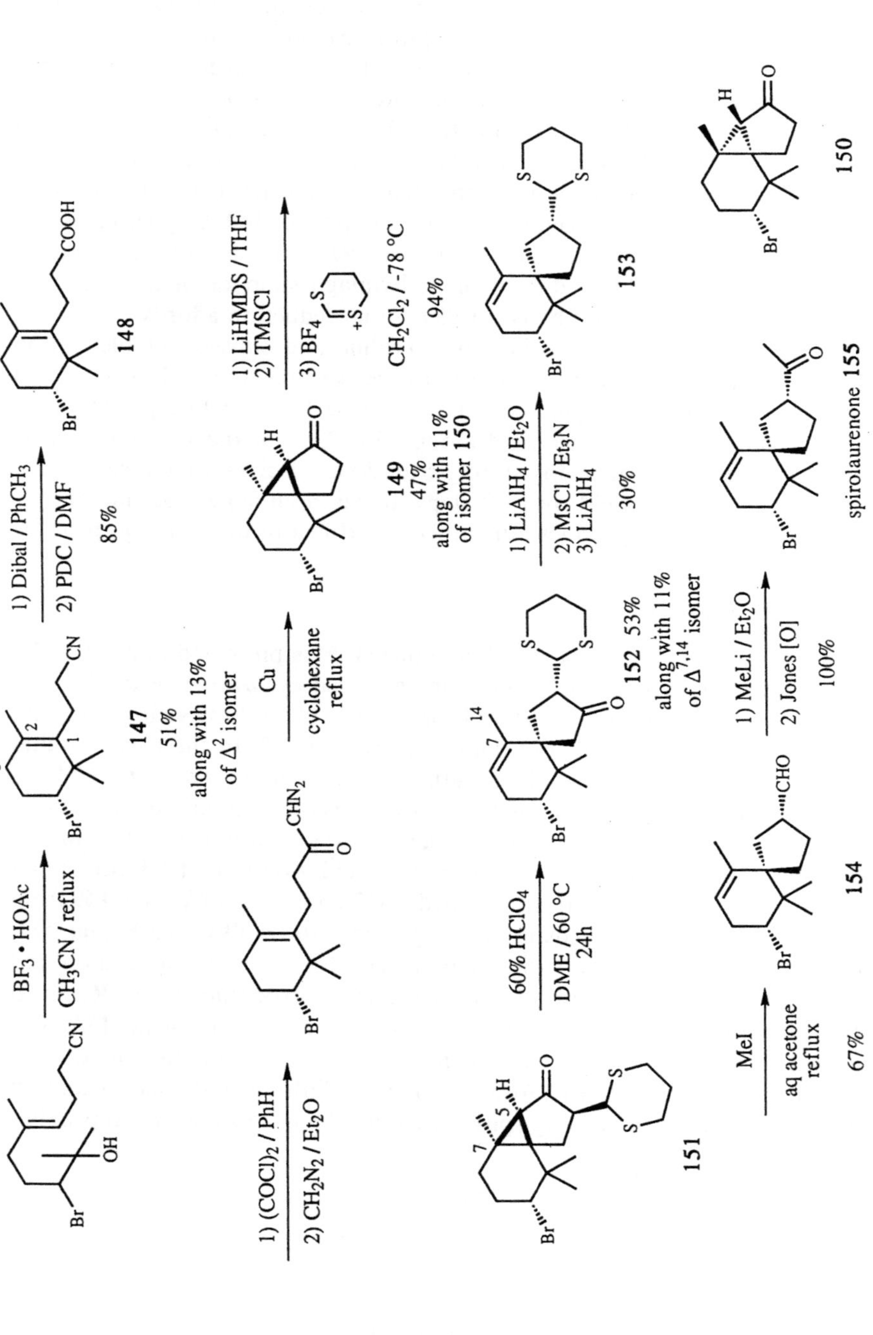

Scheme 26. Masamune Spirolaurenone Synthesis

2.2.3 Furanosesquiterpenoids

2.2.3.1 Pleraplysillin-1

Pleraplysillin-1 is a furylsesquiterpene isolated [54] from the marine sponge *Pleraplysilla spinifera* in 1972. It possesses a carbon skeleton, which is structurally similar to the ochtodane monoterpenes. The first synthesis of **161** was developed by Masaki [55] (Scheme 27). The strategy followed involves the coupling of ochtodane and 3-furylmethyl derivatives to obtain the pleraplysillin-1 framework. Diol **158** is prepared by epoxidation of the known *E*-alcohol **156** followed by Lewis acid-catalyzed eliminative ring opening of the epoxide **157**. Selective tosylation of the diol **158**, allylic oxidation, and displacement of the tosyl group provides the α-ketosulfone **159**. Coupling of **159** with 3-furylmethyl bromide provides ketone **160**, which is reduced with sodium borohydride to give the alcohol. Acetylation of the alcohol and reductive elimination of the β-acetoxysulfone provides pleraplysillin-1 (**161**) with the correct *E*,*E*-diene geometry. The metabolite is obtained in 9 steps and 24 % overall yield.

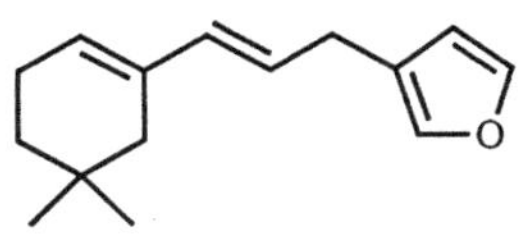

Pleraplysillin-1 161

Stille [56] developed a very short and efficient synthesis of pleraplysillin-1 based on the palladium-catalyzed coupling of vinyl triflates with organostannanes (Scheme 28). Addition of the lithium (*E*)-vinyltin cuprate **163** to 3-furylmethyl bromide provides the furfuryl vinyl tin reagent **164** in 73% yield. Vinyl triflate **166** is prepared as one regioisomer in 93% yield by conjugate reduction of ketone **165** with lithium tri-*sec*-butylborohydride and trapping with *N*-phenyltriflamide. Palladium catalyzed coupling of stannane **164** and triflate **166** affords pleraplysillin-1 (**161**) in 3 steps and 70% yield from 5,5-dimethyl-2-cyclohexenone.

2.2.3.2 Euryfuran

Euryfuran (**172**) has been isolated in (−)-form from the nudibranchs *Hypselodoris californiensis* and *H. porterae* [57] and in (+)-form from the sponges *Dysidea herbacea* and *Euryspongia* species [58]. Interest in euryfuran stems primarily from its structural similarity to the drimane sesquiterpenoids, such as warburganal and polygodial. The first syntheses of **172** appeared before the natural product had been isolated. The first of these was that of Oishi [59] in 1979 (Scheme 29), in which **172** was prepared as an intermediate in a synthesis of confertifolin. The synthesis begins with the partial hydrogenation of β-ionone to afford the unconjugated enone **167**. Reaction of **167** with trimethyl orthoformate and 70% perchloric acid affords a mixture of products which is converted to

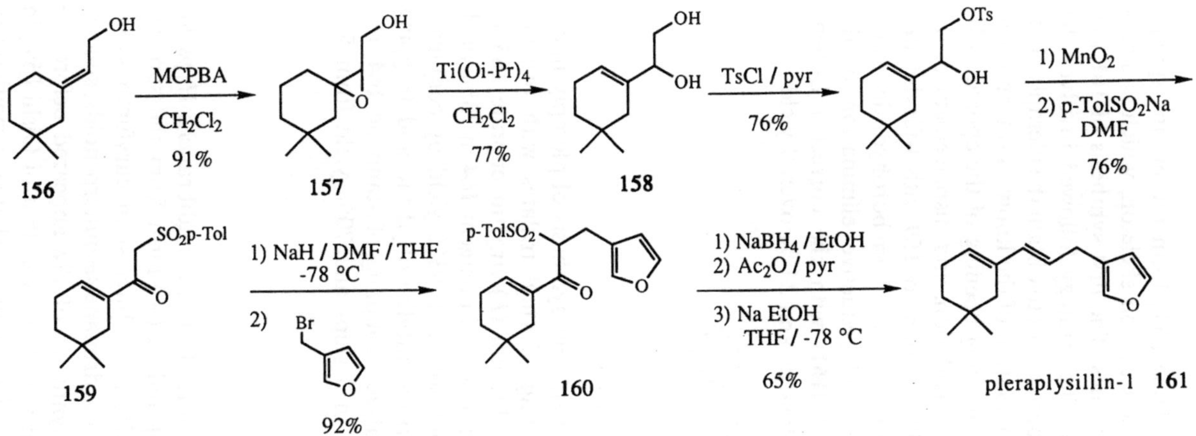

Scheme 27. Masaki Synthesis of Pleraplysillin-1

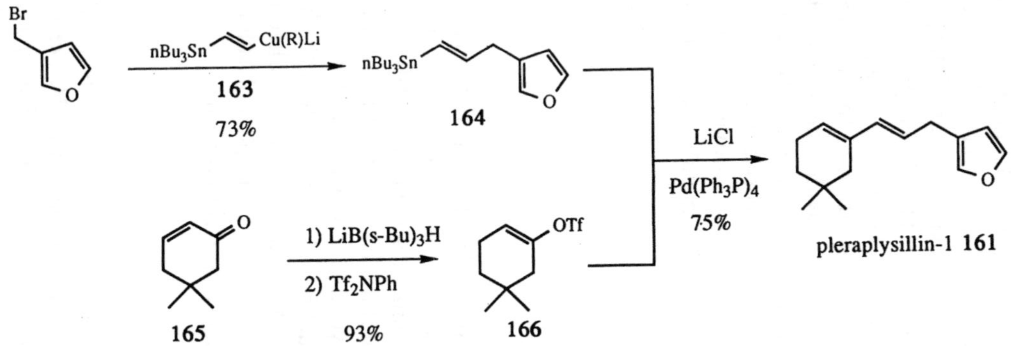

Scheme 28. Stille Pd-based Synthesis of Pleraplysillin-1

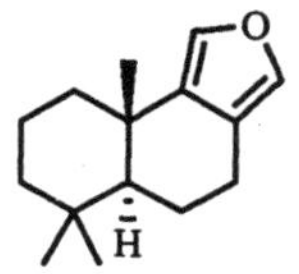

(±)-Euryfuran **172**

acetal **168** upon treatment with pyridinium hydrobromide in aqueous methanol. Darzens condensation of **168** with methyl chloroacetate yields the epoxide **169** as a mixture of isomers. Rearrangement of epoxy acetal **169** with catalytic amounts of TsOH produces carbomethoxy substituted furan **170**. Lewis acid-catalyzed cyclization produces the euryfuran skeleton in the form of ester **171**. Saponification of the ester followed by decarboxylation gives euryfuran **172** in 8 steps and 19% overall yield.

An enantiospecific synthesis of euryfuran was also developed before the natural product was isolated. In the process of developing a synthesis of confertifolin, Nakano [60] prepared peroxide **177**, which was converted to euryfuran (Scheme 30). Permanganate oxidation of manool yields ketone **175** in 60% yield. Upon irradiation in pentane **175** undergoes a Norrish type II cleavage to give diene **176** in 78% yield based on recovered ketone. Photooxidation in the presence of *meso*-tetraphenylporphine provides peroxide **177**. Upon treatment with basic alumina or ferrous sulfate **177** is converted to (+)-euryfuran, thus providing a three-step enantiospecific synthesis of **172**.

Nakano has developed an alternative procedure [61] for the preparation of **172** (Scheme 31). Oxidation of diene **178** (a common intermediate in Nakano's first synthesis of **172**; Scheme 30) with thallium(III) acetate affords diol **179** in 35% yield. Oxidation-dehydration of diol **179** by treatment with PCC produces (+)-euryfuran **172** in 4 steps and 12 % overall yield.

Ley's synthesis [62] of euryfuran is outlined in Scheme 32. Decalone **181** is formylated and converted to the *n*-butylthiomethylene derivative **182** using standard methods. Reaction of **182** with trimethylsulfonium methylide at −78°C provides dihydrofuran **183** (via rearrangement of the initially formed epoxide). Brief treatment of **183** with mercuric sulfate provides euryfuran in 60% overall yield from the *trans*-decalone **181**.

2.2.3.3 *Furoventalene*

The benzofuran sesquiterpene furoventalene has been isolated from the sea fan *Gorgonia ventalina* by Weinheimer [63]. It possesses an isoprenoid but non-farnesyl skeleton. It was first synthesized by Weinheimer in a non-regioselective fashion in order to confirm the proposed structure (Scheme 33). The strategy for the preparation of the benzofuran system involves acid-catalyzed intramolecular cyclization of a ketone onto an aromatic ring, followed by elimination. Treatment of bromophenolic ether **184** with polyphosphoric acid produces benzofuran **185** as a minor product. Grignard condensation of **185** with the

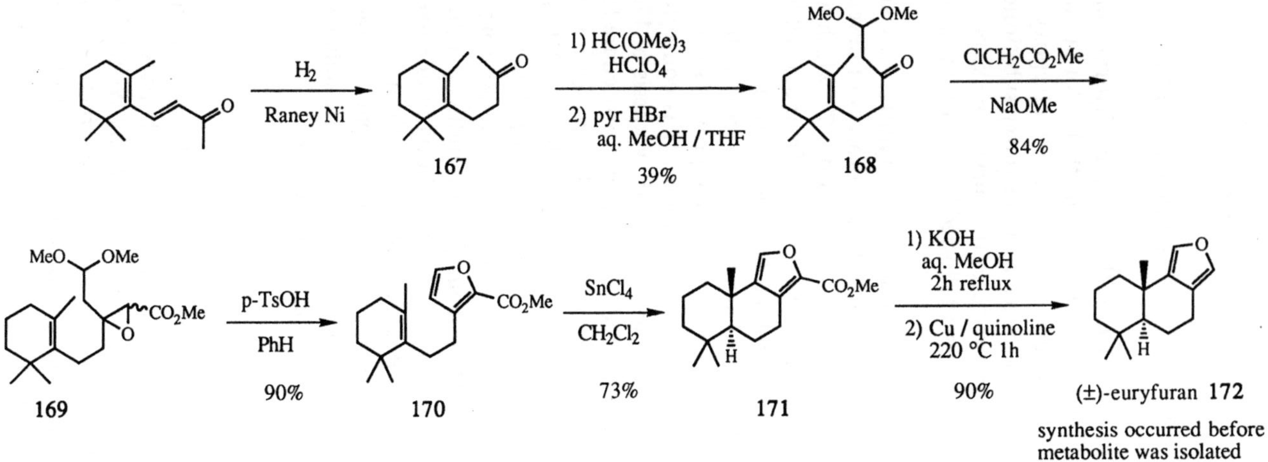

Scheme 29. Oishi Synthesis of Racemic Euryfuran

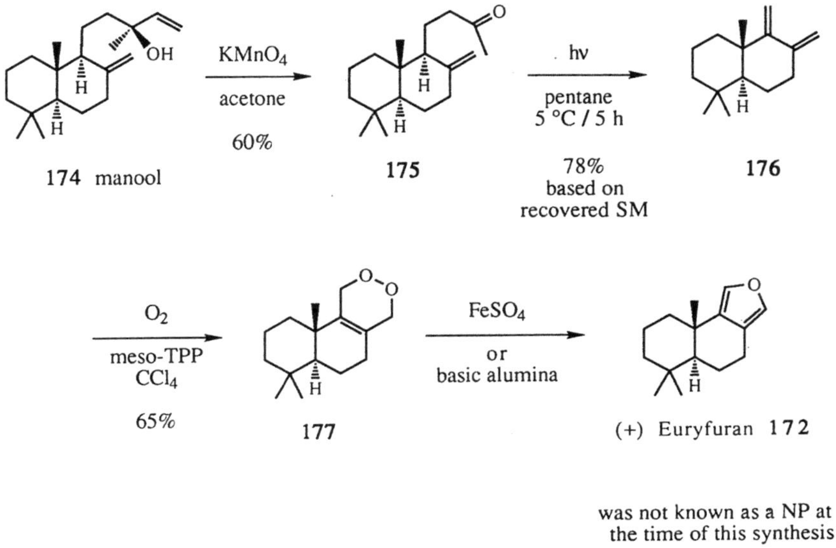

Scheme 30. First Nakano Synthesis of (+)-Euryfuran

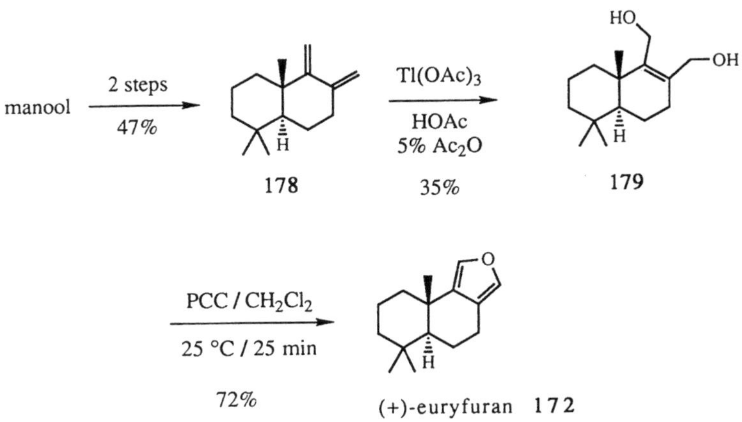

Scheme 31. Second Nakano Synthesis of (+)-Euryfuran

ethylene ketal of levulinaldehyde followed by hydrogenolysis of the benzylic alcohol **186** yields ketone **187** as a mixture with the reduced furan. Dehydrogenation of the mixture affords only ketone **187**. Addition of methylmagnesium iodide to **187** and elimination provides furoventalene (**188**) as a mixture of isomers.

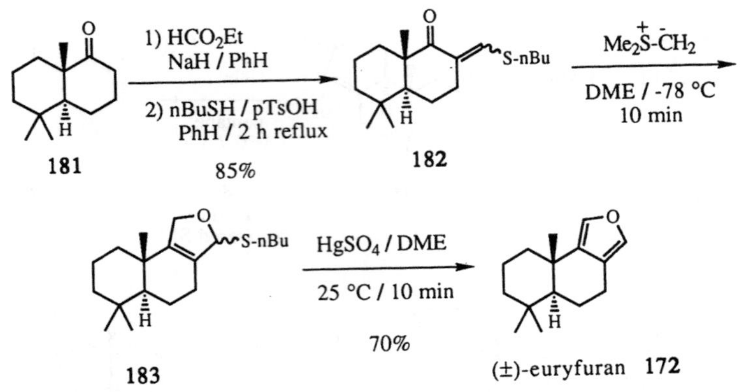

Scheme 32. Ley Synthesis of Racemic Euryfuran

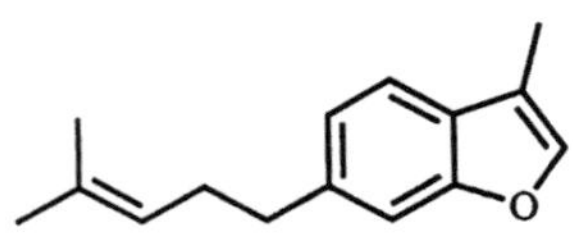

Furoventalene **188**

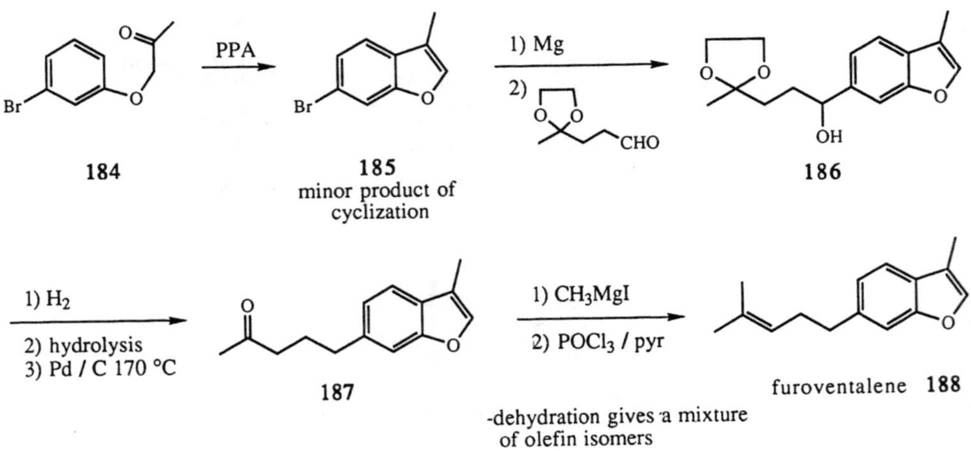

Scheme 33. Weinheimer Synthesis of Furoventalene

Yoshikoshi [64] has developed a second synthesis of furoventalene (**188**), as shown in Scheme 34. The general strategy utilized involves the annulation reaction of a 1,3-dicarbonyl compound **191** with a β-vinylbutenolide **192** to provide the furoventalene skeleton. The furan ring is obtained by reduction of the butenolide and elimination. The synthesis begins with the conversion of ester

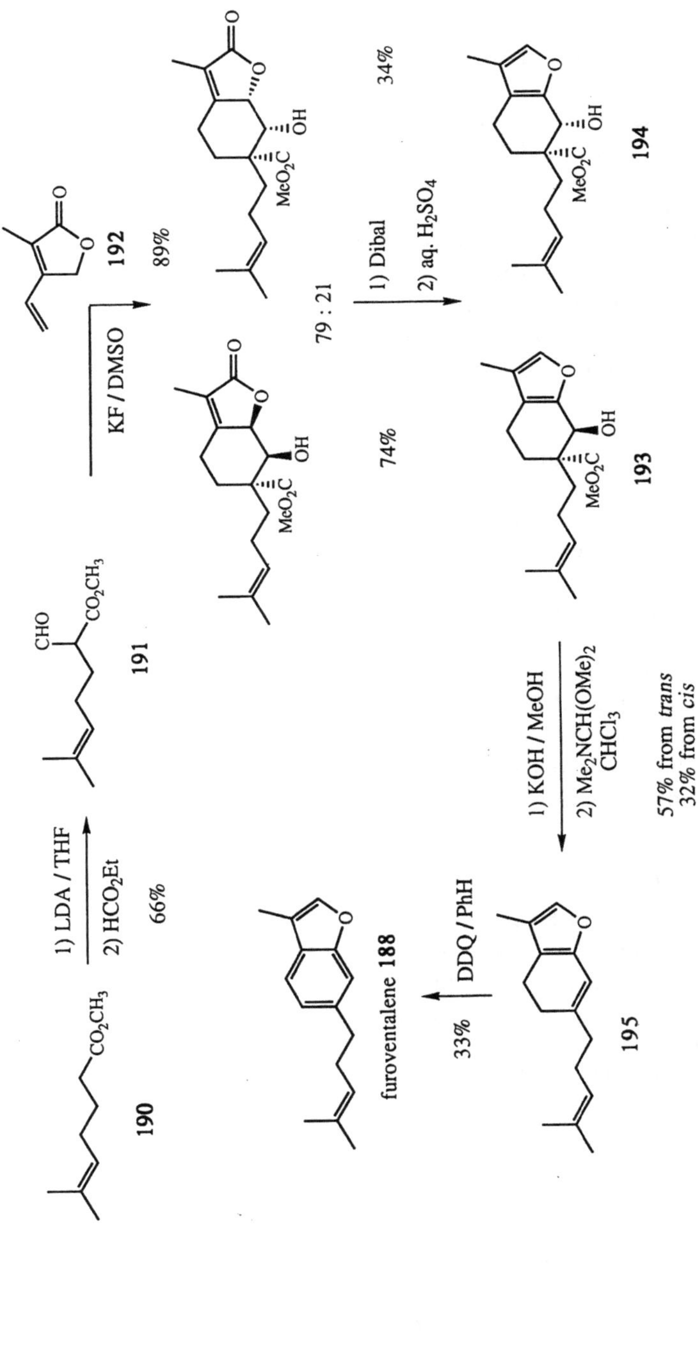

Scheme 34. Yoshikoshi Synthesis of Furoventalene

190 to aldehyde **191** by treatment with LDA and ethyl formate. In the key transformation, aldehyde **191**, which exists in the enol form, is condensed with butenolide **192** by treatment with KF in DMSO to give the annulation product as a mixture of two diastereomers. Separation of the diastereomers followed by DIBAL reduction and acid catalyzed elimination provides furans **193** and **194**. Saponification of either **193** or **194** and elimination by treatment with *N,N*-dimethylformamide dimethyl acetal produces dihydrobenzofuran **195**, which is aromatized to give furoventalene (**188**) by reaction with DDQ. An overall yield of 8% over 7 steps is obtained.

Most recently, Bergstrom [65] has prepared furoventalene (**188**) as shown in Scheme 35. The benzofuran system is prepared via an intramolecular anionic aldol-type condensation process between an acid and a ketone, followed by decarboxylation to give the furan ring. The second key step is the nickel-catalyzed coupling of a Grignard reagent with a aryl chloride to append the unsaturated side chain. The synthesis is initiated by acetylation and Fries rearrangement of *m*-chlorophenol to produce the acetophenone derivative **196**. Ketoester **197** is obtained via Williamson ether synthesis between phenol **196** and ethyl chloroacetate. Saponification of **197** affords an acid, which is cyclized and decarboxylated by heating with fused sodium acetate and acetic anhydride to produce benzofuran **198**. Nickel catalyzed coupling of the chlorobenzofuran **198** with Grignard reagent **199** affords furoventalene (**188**) in 6 steps.

The pallescensins are a series of related furanosesquiterpenoids isolated [66] mostly from the sponge *Dysidea pallescens*. Pallescensin A (**204**) has been postulated to act as part of the defense mechanism of the opisthobranch molluscs. The molluscs concentrate sponge metabolites in skin glands and release them in their defense secretions when in danger [67]. Dihydropallescensin-2 was first isolated [68] from the nudibranch *Cadlina luteomarginata* by Faulkner in 1982. Three years later the same substance was isolated [69] from the marine sponge *Dysidea fragilis* by Pietra and named penlanpallescensin.

2.2.3.4 Pallescensin A

The first syntheses of both pallescensin A (**204**) and pallescensin-1 (**203**) is that of Matsumoto [70] shown in Scheme 36. The route is enantiospecific but not stereospecific, as the final step in the synthesis produces both **204** and **205** as a 2:1 mixture. The synthesis begins with the Wittig reaction of (*R*)-(−)α-cyclocitral (**200**) with 3-furylmethylidenetriphenylphosphorane (**201**) to give the *trans*-alkene **202** in 70% yield. Selective reduction of the less substituted double bond gives pallescensin-1 (**203**). Lewis acid initiated cationic cyclization of **203** produces a mixture of pallescensin A (**204**) and the diastereomer **205** in 11% and 5.5% yields over 3 steps.

A more stereospecific synthesis of racemic pallescensin A is shown in Scheme 37. The strategy followed by Liotta [71] involves as a key step an electrocyclic ring closure of a furyldiene to obtain the pallescensin A skeleton. The synthesis is initiated by the addition of furylacetylide to 2,2,6-trimethylcyclohexanone to

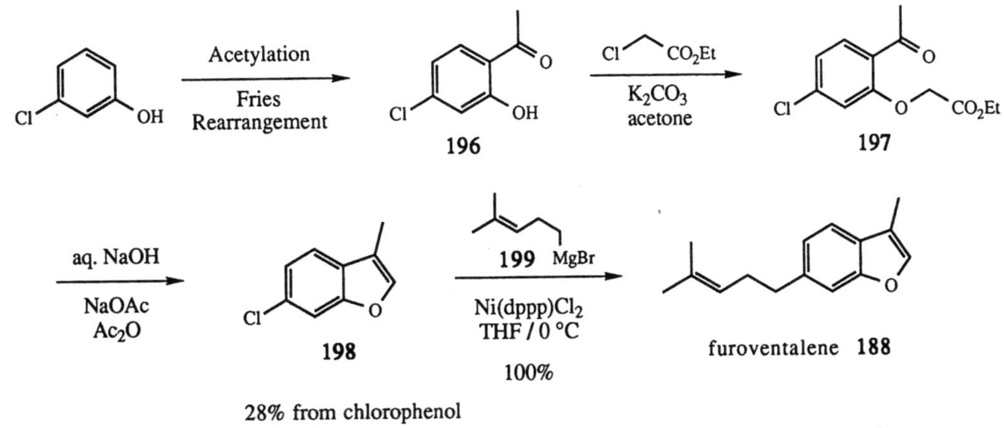

Scheme 35. Bergstrom Synthesis of Furoventalene

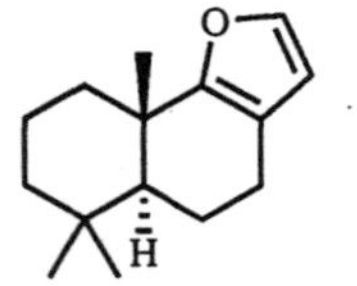

Pallescensin A 204

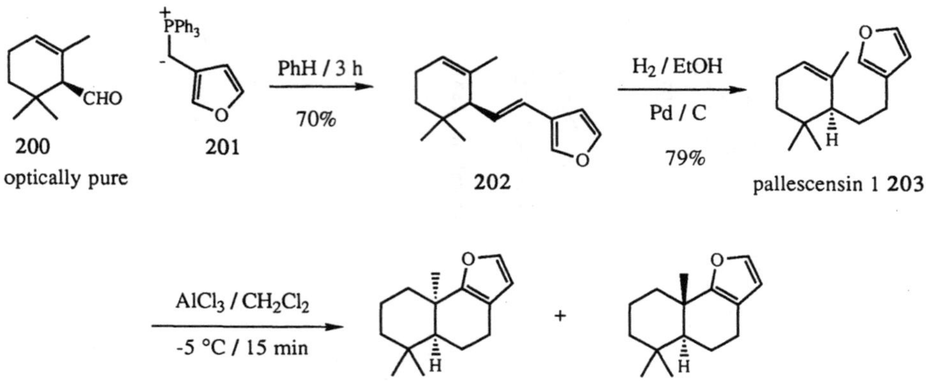

205 pallescensin A 204

20% 10%

Scheme 36. Matsumoto Syntheses of Pallescensins 1 and A

give propargylic alcohol **206**. Acid-catalyzed elimination of water with $BF_3 \cdot Et_2O$ produces enyne **207** in 77% overall yield. Reduction of the alkyne with Lindlar catalyst affords (*Z*,*Z*)-diene **208**. Over-reduction of the furan ring is minimized by the presence of the TMS group. Desilylation of **208** with TBAF produces the key intermediate **209**. Thermal ring closure of **209** produces dihydrofuran with an exocyclic diene system. A thermally allowed suprafacial [1,5]-hydrogen shift produces a *cis*-decalin which isomerizes under the reaction conditions to give the more stable *trans*-decalin system **210**. Selective catalytic hydrogenation of **210**, using less than one equivalent of hydrogen, affords pallescensin A (**204**) in quantitative yield based on recovered alkene. An overall yield of approximately 23% is obtained in 6 steps.

A biomimetic approach to pallescensin A, beginning from dendrolasin (**213**), has been developed by Naispuri [72] (Scheme 38). Coupling of 3-furylmethylmagnesium chloride **212** with geranyl chloride in the presence of cuprous iodide affords dendrolasin (**213**) in 32% yield. Treatment of **213** with $BF_3 \cdot Et_2O$ provides pallescensin A (**204**) and an olefinic impurity in an 84:16 ratio. Removal of the impurity by chromatography on silver-impregnated silica gives pallescensin A in 17% overall yield in 2 steps. The synthesis can be completed in

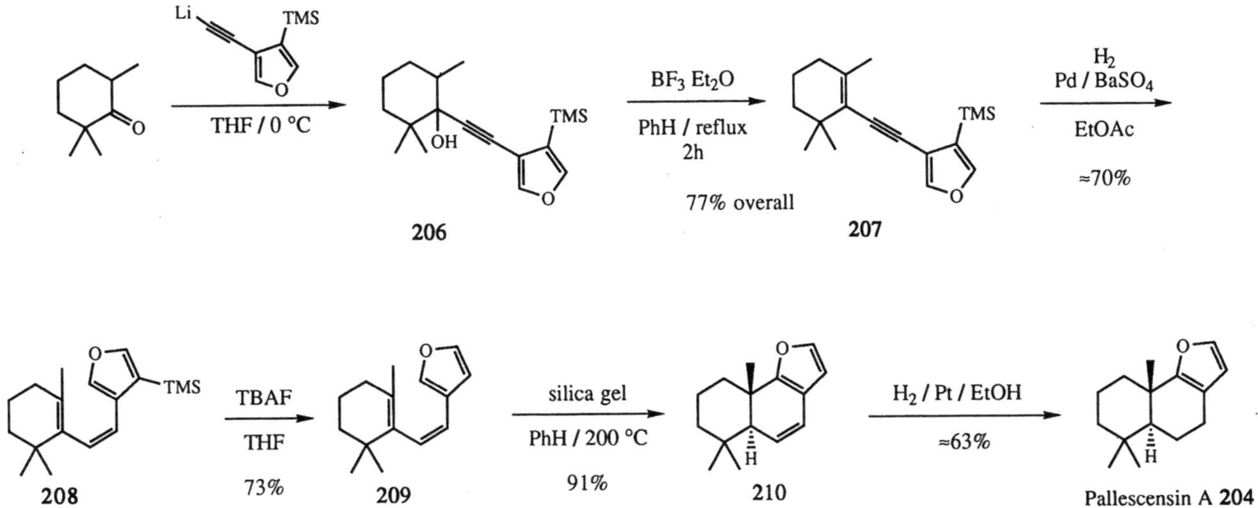

Scheme 37. Liotta Synthesis of Pallescensin A

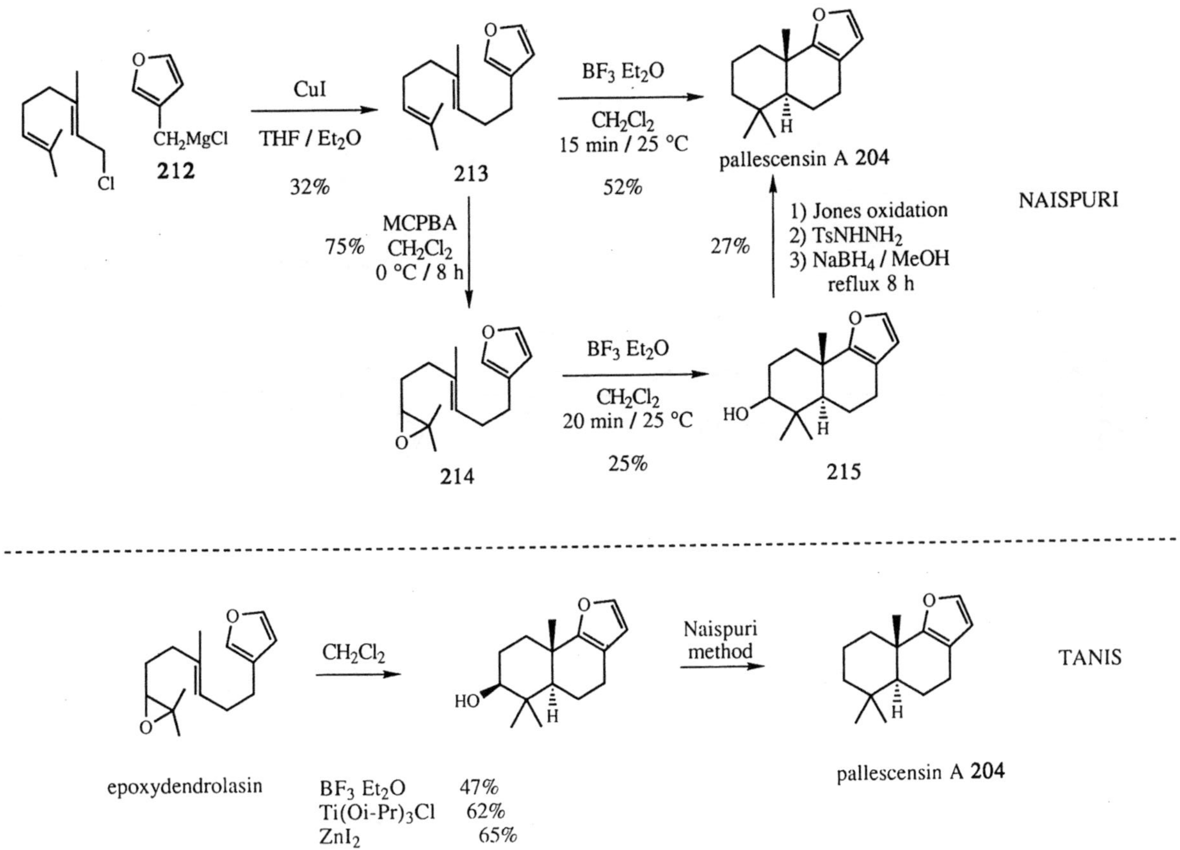

Scheme 38. Tanis and Naispuri Approaches to Pallescensin A via Dendrolaisin

a more stereospecific manner by cyclization of epoxydendrolasin (**214**). Treatment of **214** with $BF_3 \cdot Et_2O$ yields alcohol **215**. Deoxygenation of the alcohol is achieved by Jones oxidation, tosylhydrazone formation and sodium borohydride reduction to provide pallescensin A in 6 steps and 1.6% yield. An improvement on the cyclization step has been made by Tanis [73]. Higher yields in the cyclization of epoxydendrolasin can be obtained through the use of zinc iodide or $Ti(Oi\text{-}Pr)_3Cl$.

A short enantiospecific synthesis of (+)-pallescensin A has been developed by Oishi [74] (Scheme 39). Ozonolysis of a readily available dehydroabietane derivative **216** provides the hydroperoxybutenolide **217**. Sodium borohydride reduction of **217** affords butenolide **218** which upon further reduction with DIBAL and treatment with aqueous sulfuric acid stereospecifically provides (+)-pallescensin A (**204**). An overall yield of 11% is obtained over 3 steps.

The most recent synthesis of pallescensin A, by Smith [75], is shown in Scheme 40. The general strategy followed by Smith involves preparation of an optically pure *trans*-decalone that can be converted to pallescensin A by appending the furan ring via an annulation process. Synthesis of **204** begins with Wolff-Kishner reduction of ketone **220**, available from the Wieland-Miescher ketone, and hydrolysis to afford ketone **221**. Attempted alkylation of **221** using standard methodology results in the formation of a mixture of monoalkylated and dialkylated products, along with recovered starting material. Monoalkylated **222** is obtained via the TMS enol ether by treatment with a fluoride source and allyl bromide. A 1:1 mixture of diastereomers is obtained which is equilibrated to **222** by treatment with base. Ozonolysis of **222** to give the aldehyde **223** and treatment with $BF_3 \cdot Et_2O$ affords partially racemized pallescensin A in 9 steps from the Wieland-Miescher ketone and 16% overall yield from ketal **220**.

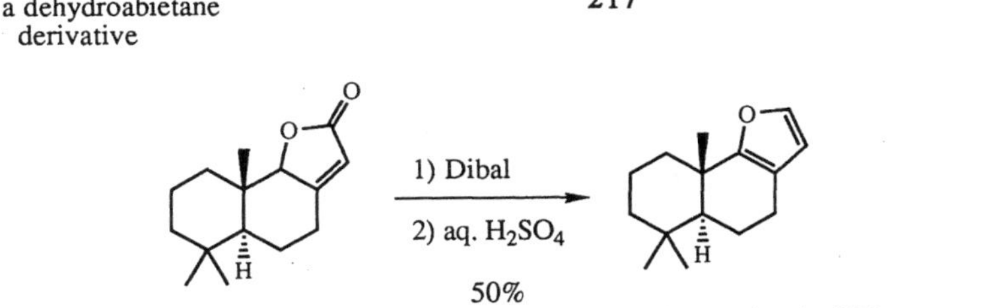

216 a dehydroabietane derivative

217

218

(+)-pallescensin A **204**

Scheme 39. Oishi Synthesis of (+)-Pallescensin A

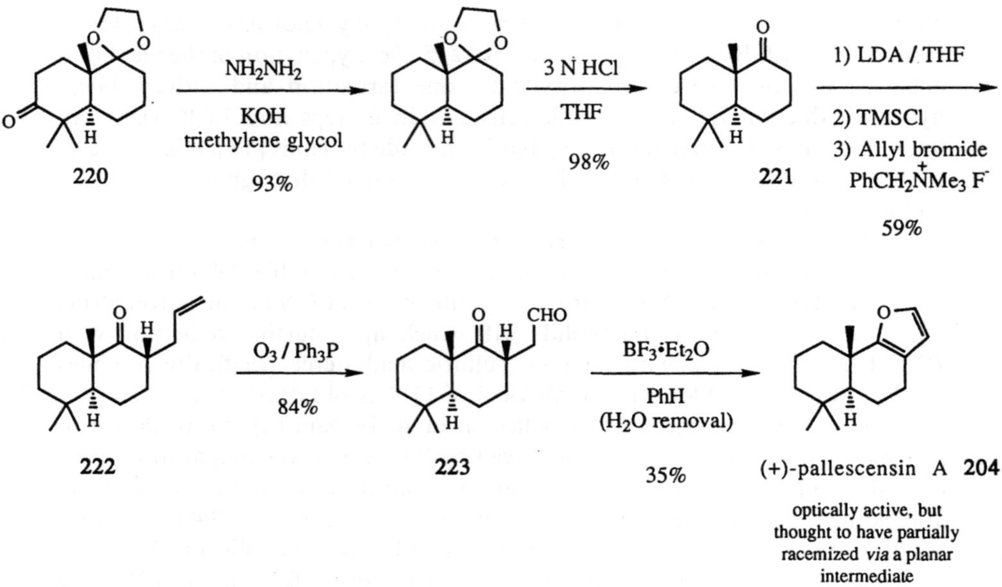

Scheme 40. Smith Synthesis of (+)-Pallescensin A

2.2.3.5 Pallescensin-1

Pallescensin-1 was first prepared as a mixture with pallescensin A in an enantiospecific, but nonstereospecific, synthesis by Matsumoto and has already been detailed (Scheme 36).

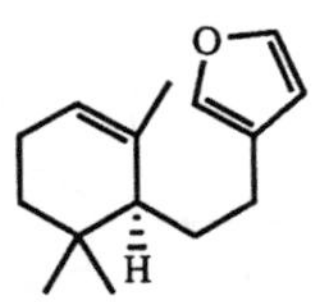

Pallescensin 1 **203**

The second synthesis of pallescensin-1 is that of Tius [76], outlined in Scheme 41. The strategy followed is biomimetic in nature, involving a cation-olefin cyclization to give the six-membered ring in a key step. The synthesis begins with formation of the *O*-trimethylsilyl cyanohydrin **225** of aldehyde **224** (available from farnesol). PDC oxidation of **225** in dry DMF affords diene butenolide **226**. Treatment of **226** with tin(IV) chloride produces the cyclized product **227** in 60% yield, along with 10% of the $\Delta^{6,13}$-isomer and a trace of the $\Delta^{5,6}$-isomer. The synthesis is completed by reduction of the butenolide with DIBAL and elimination in acetic acid to afford pallescensin-1 in 6 steps.

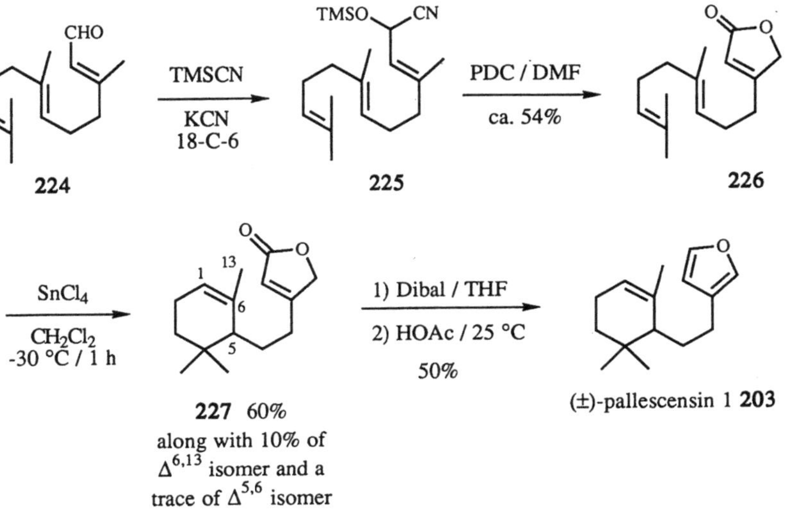

Scheme 41. Tius Synthesis of (±)-Pallescensin 1

2.2.3.6 *Pallescensins-2, F and G*

Matsumoto has extended his enantiospecific synthesis of pallescensin-1 (Scheme 36) to provide pallescensins-2, F and G, as shown in Scheme 42. Epoxidation of pallescensin-1 (**203**) and eliminative epoxide opening by treatment with lithium diethylamide produces alcohol **229** as a mixture of diastereomers. Addition of **229** to refluxing HMPA results in elimination to give pallescensin-2 in 44% yield from pallescensin-1 (6 steps and 2.4% overall yield from optically pure cyclocitral). Pallescensins G and F are also prepared from alcohol **229**. Oxidation of **229** with PCC followed by acid catalyzed cyclization with phosphoric acid gives tricyclic ketone **231**. Unsaturated ketone **232** is obtained in 50% yield by selenoxide elimination. Reduction of **232** to the alcohol and elimination in hot HMPA affords Pallescensin G in 11 steps and 0.5% overall yield from cyclocitral. Further heating of **233** in HMPA equilibrates the diene system to give pallescensin F in 12 steps and 0.36% overall yield.

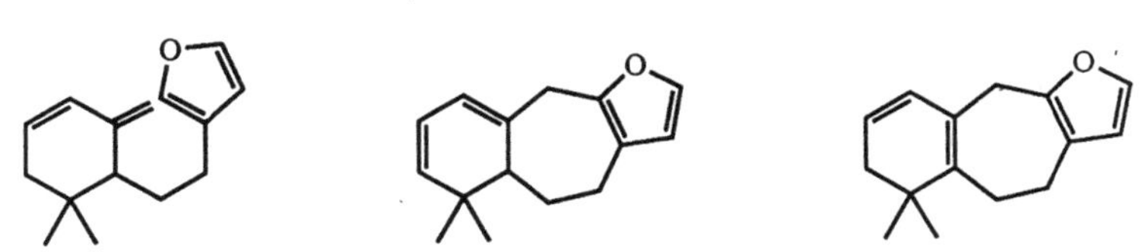

Pallescensin 2 **230** Pallescensin G **233** Pallescensin F **234**

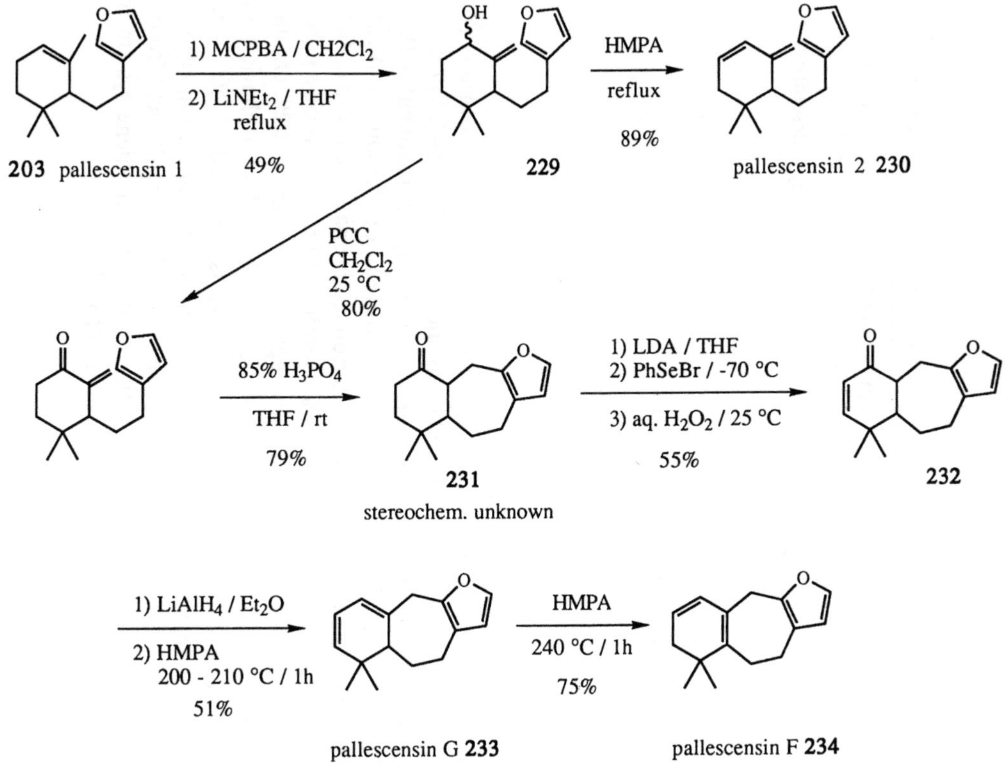

Scheme 42. Matsumoto Syntheses of Pallescensins 2, F and G

2.2.3.7 *Dihydropallescensin-2 (Penlanpallescensin)*

Kurth [77] has prepared (+)-dihydropallescensin-2 (penlanpallescensin) in an enantioselective fashion (Scheme 43). The key step around which the synthesis was built involves an asymmetric aza-Claisen rearrangement using an oxazoline as a chiral auxiliary. Allylic alcohol **235** (prepared from 2-methylcyclohexanone) is mesylated in 99% yield. Conversion of the mesylate to **238** in one flask is accomplished by conversion to oxazolinium salt **236**, titration with *n*-butyllithium to give neutral **237** and heating in decalin to give **238** in 80% ee. Removal of the chiral auxiliary without lactonization of the free acid is achieved by methylation of the oxazoline and basic hydrolysis to afford acid **239**. Conversion of the acid to the primary iodide **240** occurs via reduction, mesylation and treatment with sodium iodide. Coupling of the iodide with 3-lithiofuran provides dihydropallescensin-2 in 8 steps and 23% overall yield.

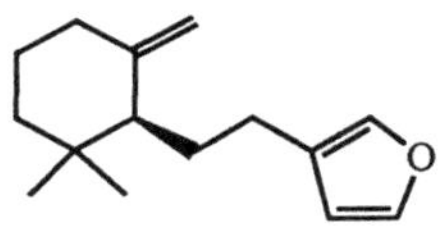

(+)-Dihydropallescensin 2 **241**

2.2.3.8 *Pallescensin E*

The only synthesis of pallescensin E is that of Baker and Sims [78], illustrated in Scheme 44. The key steps in their strategy are a crown ether-accelerated Wadsworth olefination, to couple the aryl and furan moieties, and an intramolecular Friedel-Crafts acylation to afford the tricyclic ring system. Basic hydrolysis of 2,3-dimethylbenzonitrile and a subsequent reduction-oxidation sequence provides aldehyde **243**. The Wadsworth olefination reaction of phosphonate **244** with aldehyde **243** was found to occur only in the presence of catalytic amounts of 15-crown-5. Hydrogenation of the stilbene **245** that was formed, followed by saponification of the ester provides acid **246**. Formation of the acid chloride and reaction with aluminum chloride in nitrobenzene gives ketone **247** via an intramolecular Friedel-Crafts acylation. Removal of the carbonyl by formation of the tosylhydrazone and reduction with sodium cyanoborohydride yields pallescensin E in 10 steps.

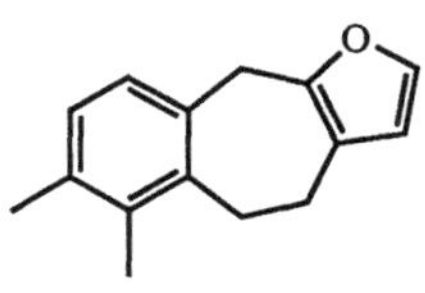

Pallescensin E **248**

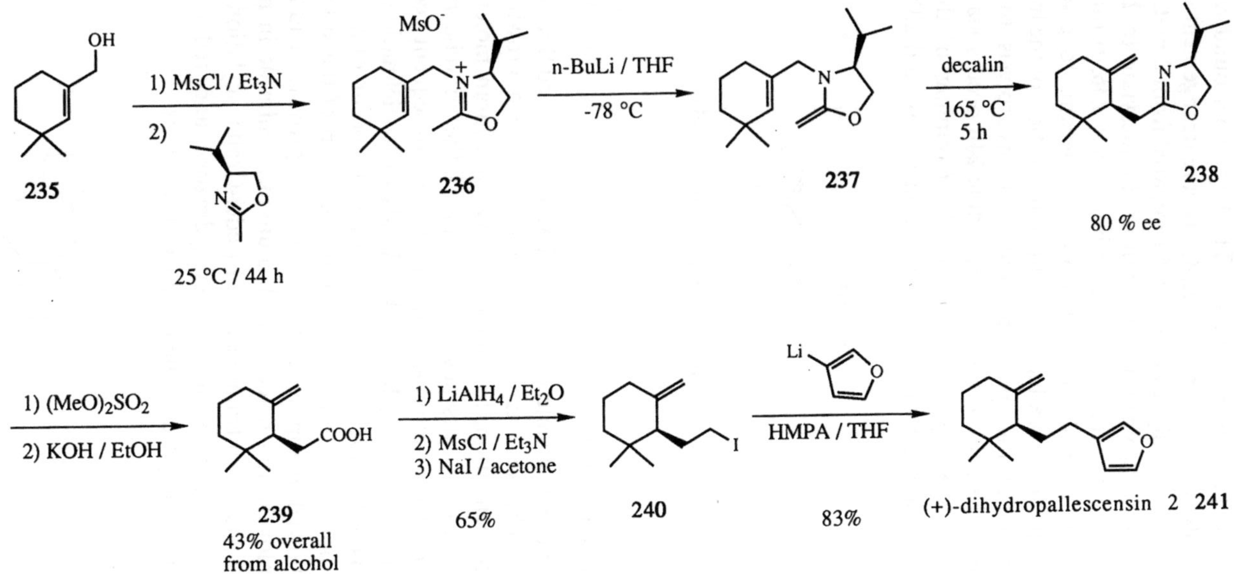

Scheme 43. Kurth Synthesis of (+)-dihydropallescensin 2 [(+)-Penlanpallescensin]

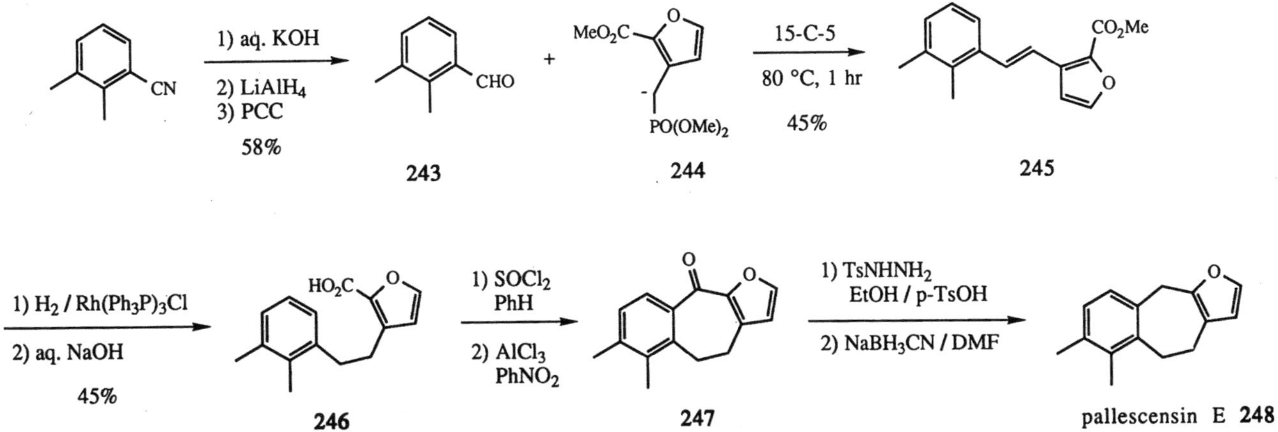

Scheme 44. Baker Synthesis of Pallescensin E

2.2.3.9 *Furodysin and Furodysinin*

Furodysin and Furodysinin and their thioacetyl analogs have been isolated from various species of *Dysidea* from several geographic regions [79]. Interestingly, metabolites have been isolated in both enantiomeric series from the same sponge genus. Compounds in this series exhibit significant antiparasitic activity. Only the parent compounds (**249** and **251**) have been synthesized so far, in both racemic and optically pure form.

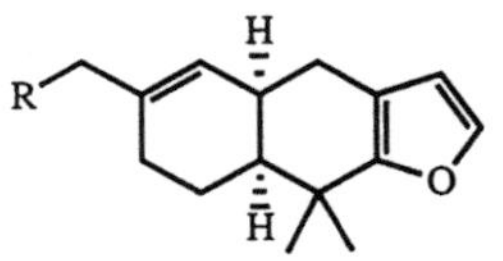

R = H Furodysin **249**
= SAc Thiofurodysin acetate **250**

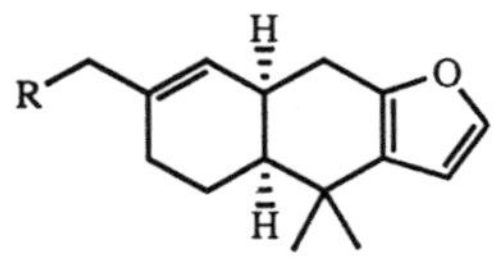

R = H Furodysinin **251**
= SAc Thiofurodysinin acetate **252**

The first syntheses were reported by Hirota [80] (Scheme 45) using the common intermediate **255**. Anionic oxy-Cope rearrangement of **253** led to the *cis*-fused bicyclic ketone **254**. Gem-dimethyl alkylation led to the branchpoint compound **255**. Alkylation with methyl bromoacetate appended the final two carbons. Conversion of the ketoester **256** to racemic furodysin was straightforward. Synthesis of furodysinin required α-oxidation of **255** → **257**. The two carbon appendage was added as lithio *t*-butylacetate to the carbonyl giving **258** as an inconsequential mixture of diastereomers. Conversion of the **258** to racemic furodysinin was again straightforward.

A second approach to (−)-**249** and **251** by Albizati [81], began from the readily available (+)-9-bromocamphor and led to the two metabolites in 4 steps (Scheme 46). Aldol reaction of the enolate of **260** with 2-furaldehyde and trapping of the enolate with acetyl chloride led to primarily the exo-threo adduct **261** which could be isolated by recrystallization. Treatment of **261** with the electron transfer agent sodium naphthalenide in THF at low temperature resulted in fragmentation of the system to the enolate **263**, presumably by way of the anion **262**. The enolate **263** could be trapped *in situ* with diethyl chlorophosphate to give the vinyl phosphate **264**. Reductive cleavage of both the phosphate and acetate groups was accomplished by Li/NH_3 to give the limonene derivative **265**. Cation-olefin cyclization initiated by mercuric nitrate followed by a reductive workup led to optically pure (−)-furodysinin in 14%

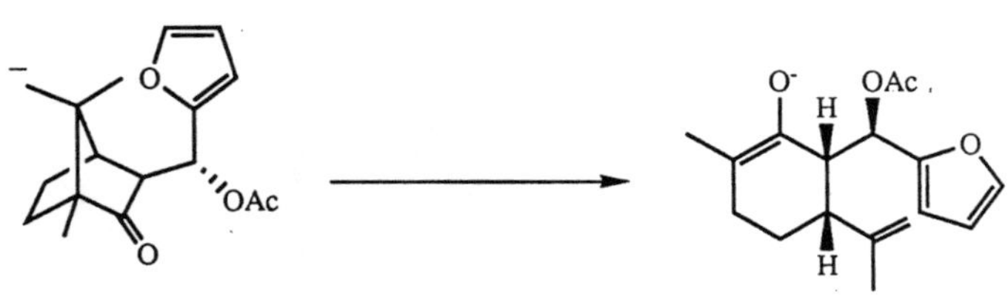

262 **263**

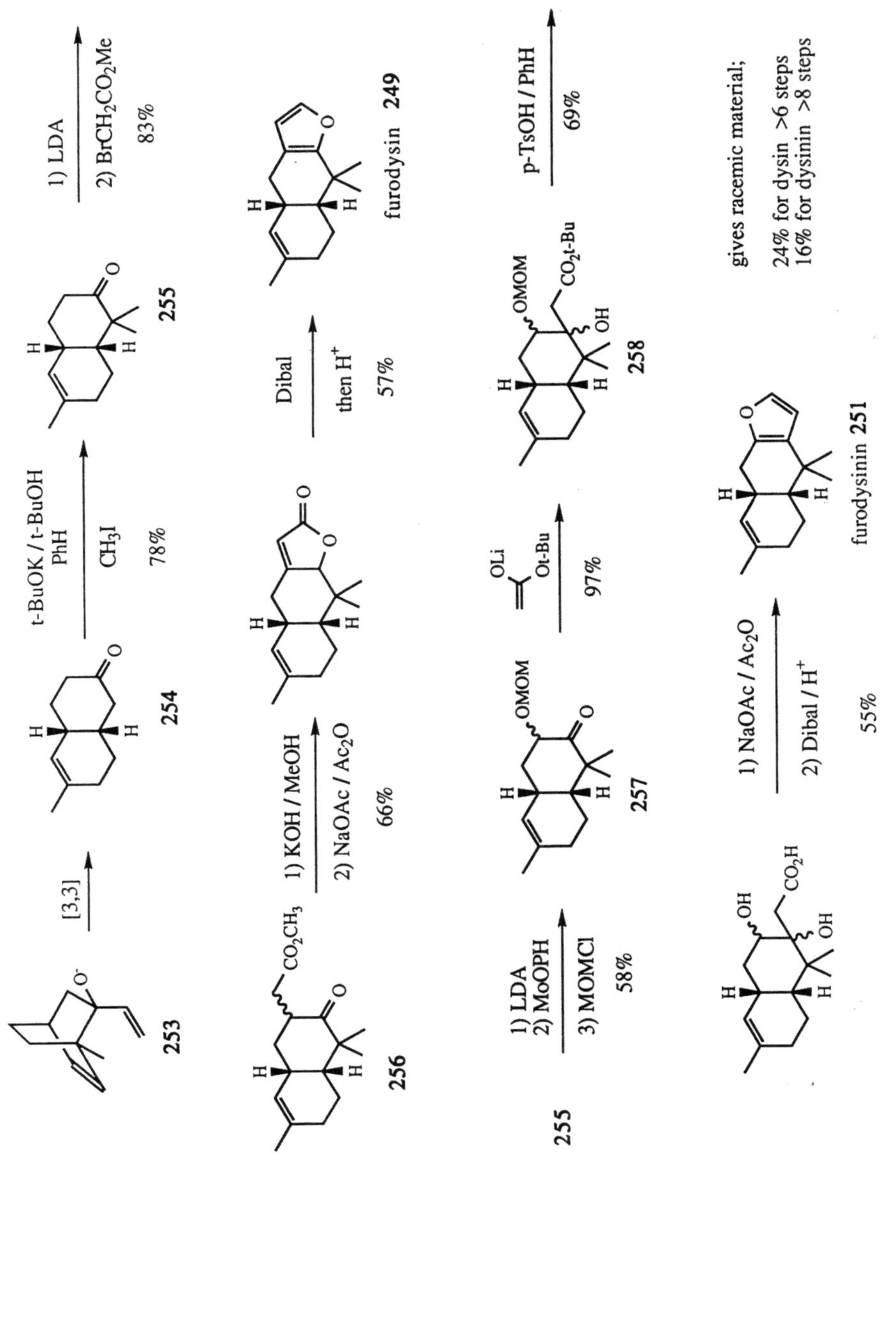

Scheme 45. Hirota Syntheses of (±)-Furodysin and Furodysinin

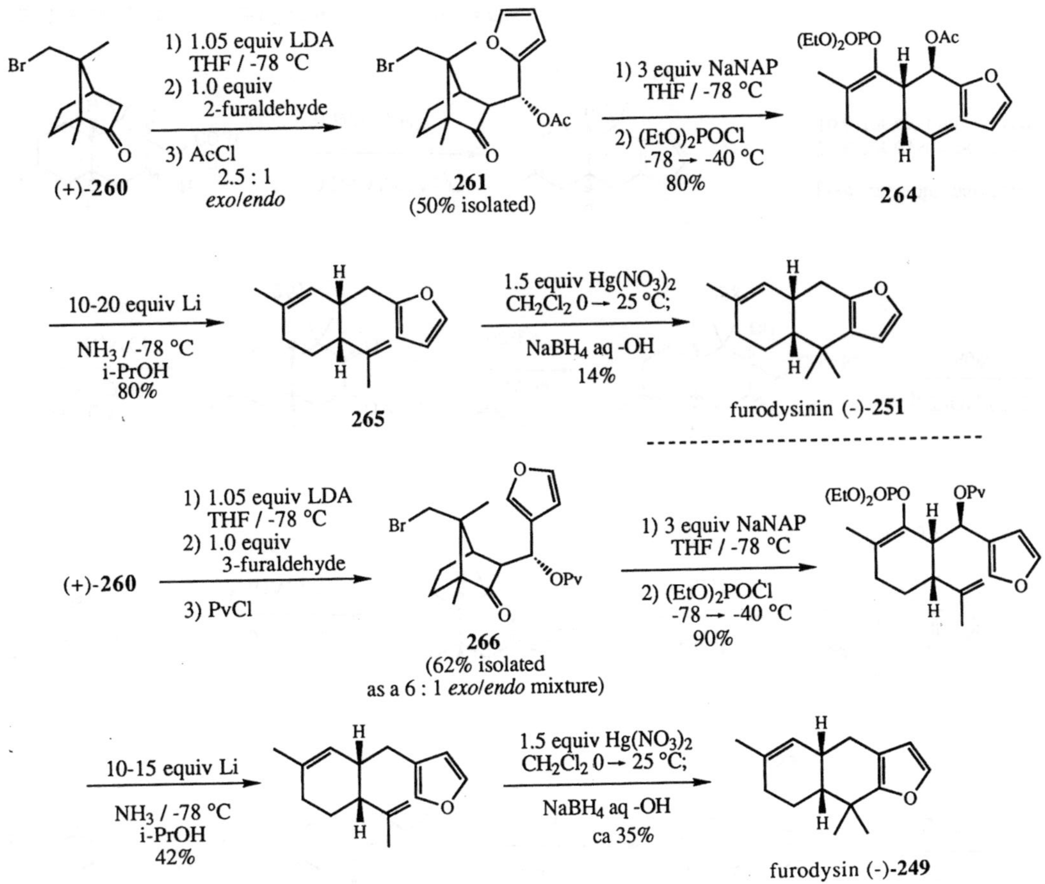

Scheme 46. Albizati Synthesis of (−)-Furodysinin **399** and (−)-Furodysin **397**

yield. In an analogous fashion, aldol reaction of **260** with 3-furaldehyde led to adduct **266**. This adduct was transformed in a near identical sequence to (−)furodysin. These syntheses established the absolute configuration of furodysin, furodysinin and presumably of the related metabolites **250** and **252**.

2.2.3.10 Nakafurans-8 and -9

Nakafuran-8 and -9 have been isolated [82] from the marine sponges *Dysidea fragilis* and *D. etheria* and from the nudibranchs *Hypselodoris godeffroyana* and *Chromodoris maridadilus*, which graze upon *D. fragilis*. Both compounds possess fish antifeedant properties. The total synthesis of nakafuran-8 by Yamamoto [83] is shown in Scheme 47. The key points of Yamamoto's strategy involve formal replacement of the bridgehead methoxy substituent on Diels-Alder adduct **267**, sequential ring enlargements and construction of the furan moiety via butenolide reduction. The synthesis begins with Lewis acid-catalyzed rearrangement of readily available ketone **267** to produce **268**. DIBAL reduction of the ketone and further rearrangement by treatment with *p*-toluenesulfonic acid gives ketone **269** in 74% yield from **267**. Tiffeneau-Demjanov ring expansion of **269** by sequential formation of the *O*-trimethylsilyl cyanohydrin, LAH reduction and treatment with nitrous acid gives **270** in 56% yield. A second ring enlargement is carried out by reaction of ketone **270** with trimethylsilyl diazomethane and $BF_3 \cdot Et_2O$ to produce a mixture of two isomeric ketones in a 2:1 ratio. Alkylation of **271** with LDA and ethyl iodoacetate affords a mixture of diastereomeric ketoesters **272** which are converted to butenolide **273** by saponification of the ester and treatment of the resulting ketoacid with TsOH. Reduction of the butenolide **273** and dehydration produces nakafuran-8 in 12 steps and 7% overall yield.

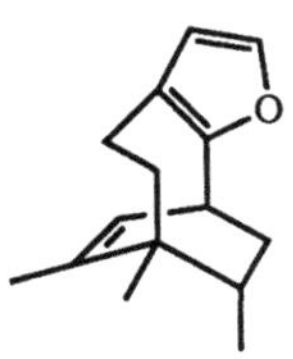

Nakafuran-8 **274**·

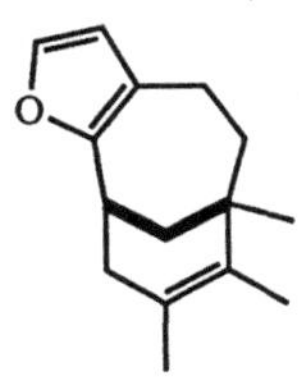

Nakafuran-9 **279**

The strategy utilized by Tanis [84] in the synthesis of nakafuran-9 involves an acid catalyzed cyclization onto a pre-existing furan to provide the nakafuran-9 skeleton. The synthesis begins (Scheme 48) with the copper-catalyzed conjugate epoxide opening of **276** with 3-furylmethylmagnesium chloride. PCC oxidation and boron trifluoride-assisted addition of methylcopper to the derived enone gives ketone **277** as a mixture of diastereomers. Selenylation of the kinetic enolate and selenoxide elimination, in the presence of triethylamine to suppress premature cyclization, affords furyl substituted enone **278**. Treatment of **278** with formic acid in cyclohexane produces the crucial bicyclo[4.3.1]decanone in

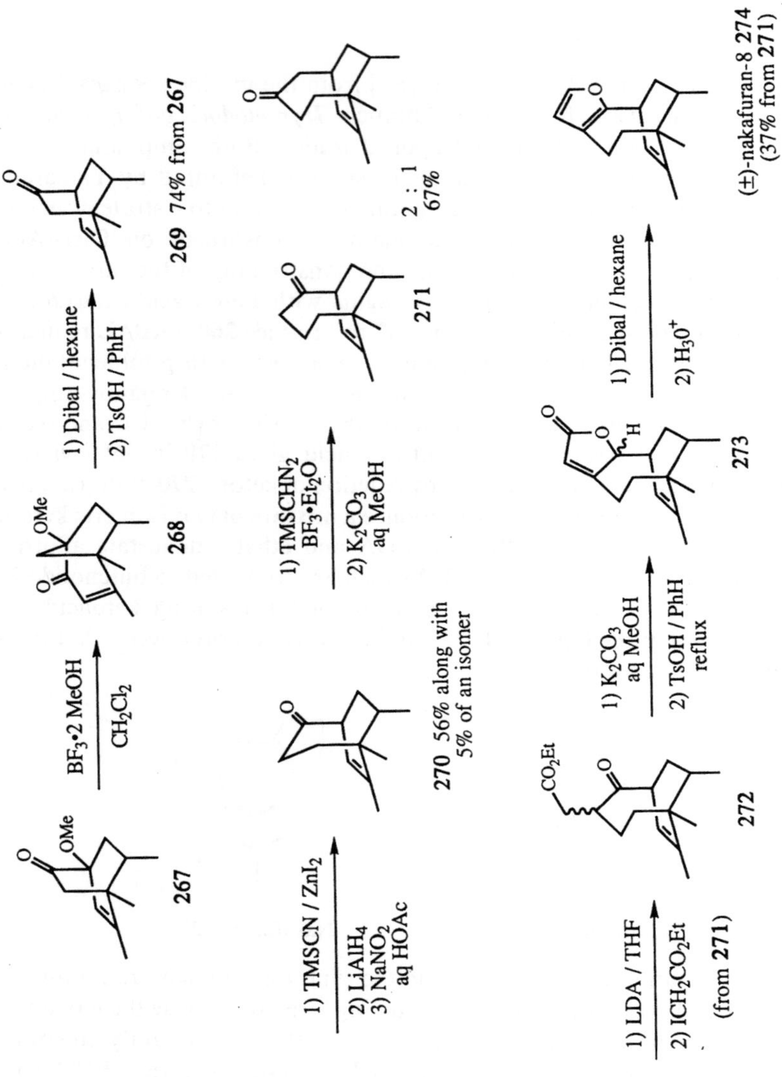

Scheme 47. Yamamoto Synthesis of (±)-Nakafuran-8

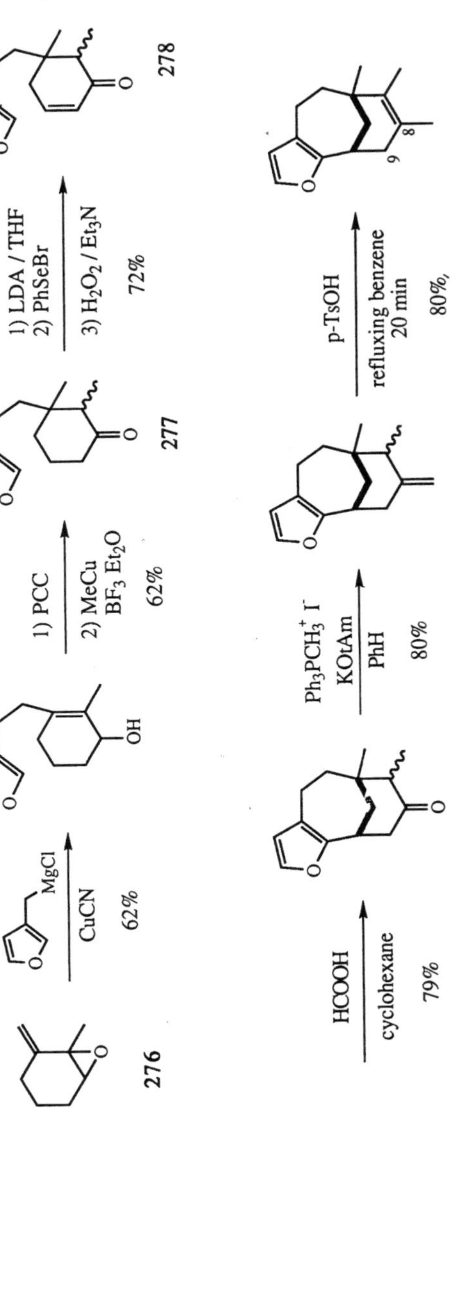

(±)-nakafuran-9 **279**

also obtained 5% of 8,9-olefin isomer

Scheme 48. Tanis Synthesis of (±)-Nakafuran-9

79% yield. Nakafuran-9 is obtained, via Wittig olefination and acid-catalyzed olefin isomerization, in 8 steps and 14% overall yield.

2.2.3.11 Spiniferin-1

Spiniferin-1 is an unstable furanosesquiterpene isolated [85] from the Mediterranean sponge *Pleraplysilla spinifera.* It appears to be the first known natural product containing the novel 1,6-methanol[10]annulene carbon framework. The only synthesis of spiniferin-1 is that of Marshall [86] shown in Scheme 49. The key transformation in the synthesis is the introduction of the methanoannulene structural unit via a norcaradiene-cycloheptatriene-type electrocyclic rearrangement (**287** → **288**). The synthesis begins with stepwise Robinson annulation of β-ketoester **280** with MVK to afford decalone **281**. Protection of the ketone, with concomitant olefin migration, and ester reduction gives **282**. Epoxidation of **282** and mesylation gives epoxide **283**. Hydrolysis of the ketal **283** with eliminative ring opening of the epoxide with aqueous HCl, followed by mesylation of the alcohol provides **284**. Elimination of the mesylate **284** to afford dienone **285** and Birch reduction with intramolecular trapping yields the cyclopropyl decalone **286** in 19% overall yield. Formylation of **286** by Claisen condensation with ethyl formate followed by careful dehydrogenation with DDQ yields the key intermediate **287**. Base-catalyzed rearrangement of **287** gives rise to an enolate that is immediately trapped with ethyl iodoacetate to give ester **289**, thus avoiding isolation of the very unstable ketoaldehyde. In a fortuitous development, it was found that addition of base to aldehyde-ester **289**, followed by acidification of the aqueous phase during extractive workup produces a mixture of ester **290**, spiniferin-1 carboxylic acid (**291**) and trace amounts of spiniferin-1 (**292**). The yield of spiniferin-1 was increased by heating the unstable carboxylic acid **291** in quinoline with copper oxide to effect decarboxylation. Although the yields are low due to the instability of the product, spiniferin-1 is obtained in 0.75% overall yield in 15 steps.

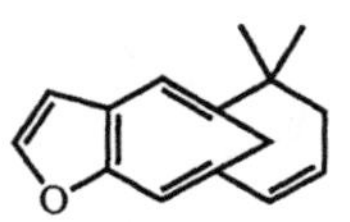

Spiniferin-1 292

2.2.4 Bicyclo[4.3.0] Ring Systems

2.2.4.1 Brasilenol

Brasilenol (**299**) has been isolated [87] from the sea hare *Aplysia brasiliana,* an opisthobranch mollusc that is known to feed on red algae of the genus *Laurencia,* and from the red alga *Laurencia obtusa,* thus suggesting a dietary source of the metabolite. It possesses a novel nonisoprenoid [4.3.0] skeleton.

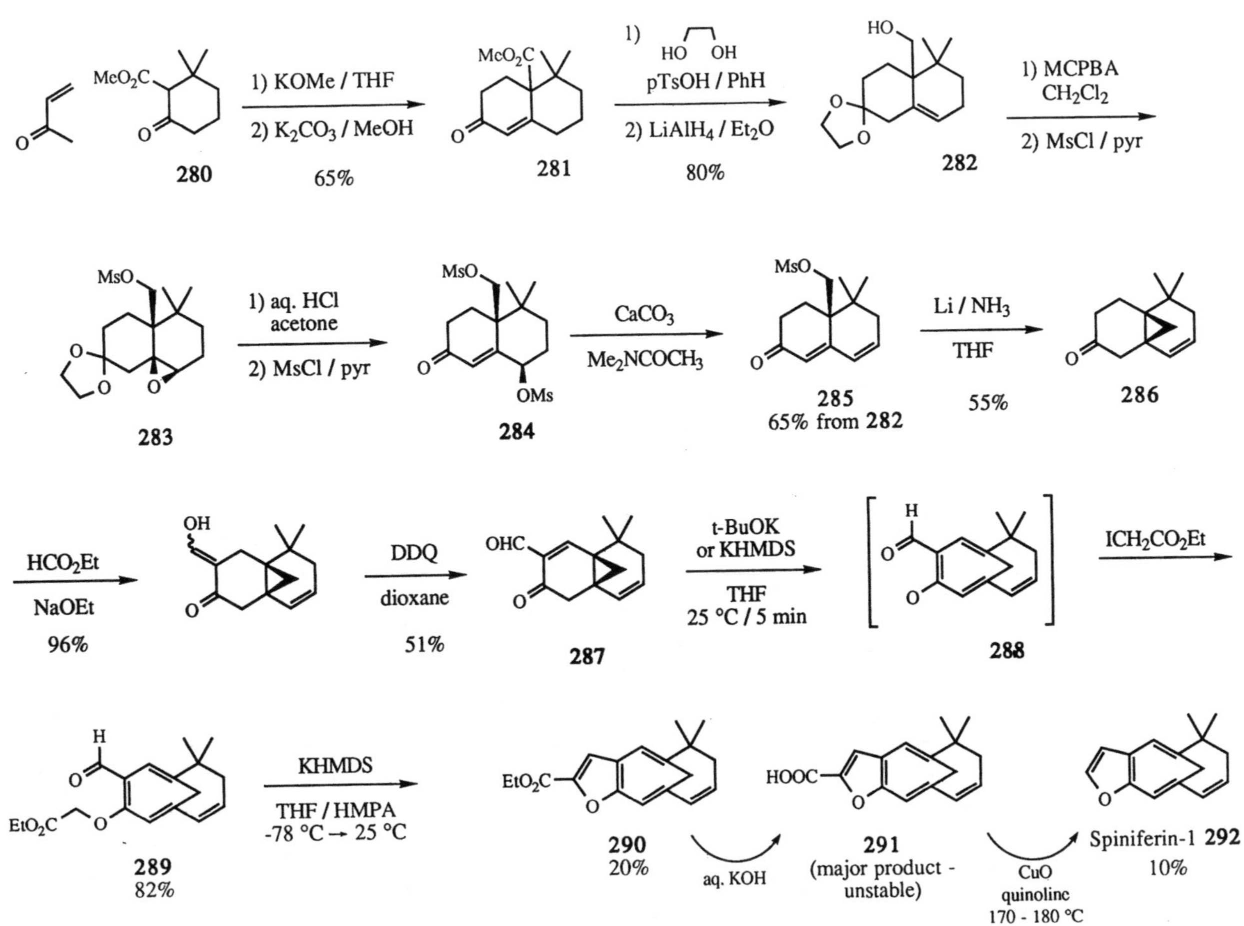

Scheme 49. Marshall Synthesis of Spiniferin-1

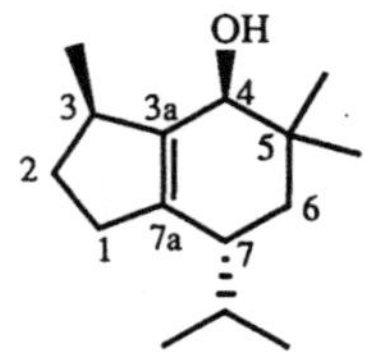

Brasilenol **299**

From a synthetic point of view the major difficulty encountered in the preparation of brasilenol is the correct relative stereochemistry between the C-3 methyl and C-7 isopropyl groups. The correct relative stereochemistry at C-4 is readily obtained by stereospecific carbonyl reduction.

The first synthesis of brasilenol, by Greene [88] is shown in Scheme 50. Substituted anisole **294** is obtained via Claisen rearrangement of the crotyl ether and methylation of the resulting phenol. Indane **295** is obtained by a sequence involving hydroboration-oxidation to obtain a carboxylic acid, cyclization in neat polyphosphoric acid and removal of the excess oxygen functionality by reduction and hydrogenolysis of the resulting benzyl alcohol. Despite problems with the Birch reduction, enones **296, 297** are obtained using a modification of the procedure of Hendrickson and DeCapite [89]. Treatment of the indane **295** with a large excess of Li/$MeNH_2$/tBuOH/THF at $-40\,°C$ gives an easily separable mixture of enones **296** and **297** in a 1:2 ratio. The Δ^{3a} isomer **296** is converted to the target enone 84% yield by treatment with LDA and methyl iodide. Likewise, the Δ^6-isomer can be converted to enone **298** by treatment with LDA and methyl iodide followed by isomerization over rhodium trichloride. Mixtures of the *cis-trans* isomers are obtained in both cases. The pure *cis*-isomer **298** is obtained by chromatography. The *trans*-isomer can be recycled by isomerization over rhodium trichloride. Stereospecific reduction of enone **298** with lithium triethylborohydride produces brasilenol **299** in 12 steps and 16% yield.

An enantiospecific synthesis of (+)-brasilenol involving an interesting transfer of chirality has been developed by Greene [90] (Scheme 51). Copper-catalyzed stereospecific conjugate addition of 3-butenylmagnesium bromide to optically pure (−)-cryptone followed by Wacker oxidation affords the diketone **300**. Aldol condensation of **300** produces only the conjugated product **301** in 84% yield. By taking advantage of the fact that the entering and leaving hydrogens are generally cofacial in palladium hydrogen-induced migration of double bonds, it was possible to transfer chirality from the 7a position to C-3. Treatment of **301** with hydrogen and palladium on carbon produces the necessary *trans* relationship between the secondary methyl and isopropyl functionalities. Methylation with LDA and excess methyl iodide yields the trimethylated ketone **302**. Optically pure (+)-brasilenol is obtained from **302** by reduction with lithium triethylborohydride. An overall yield of 31% is obtained in 6 steps.

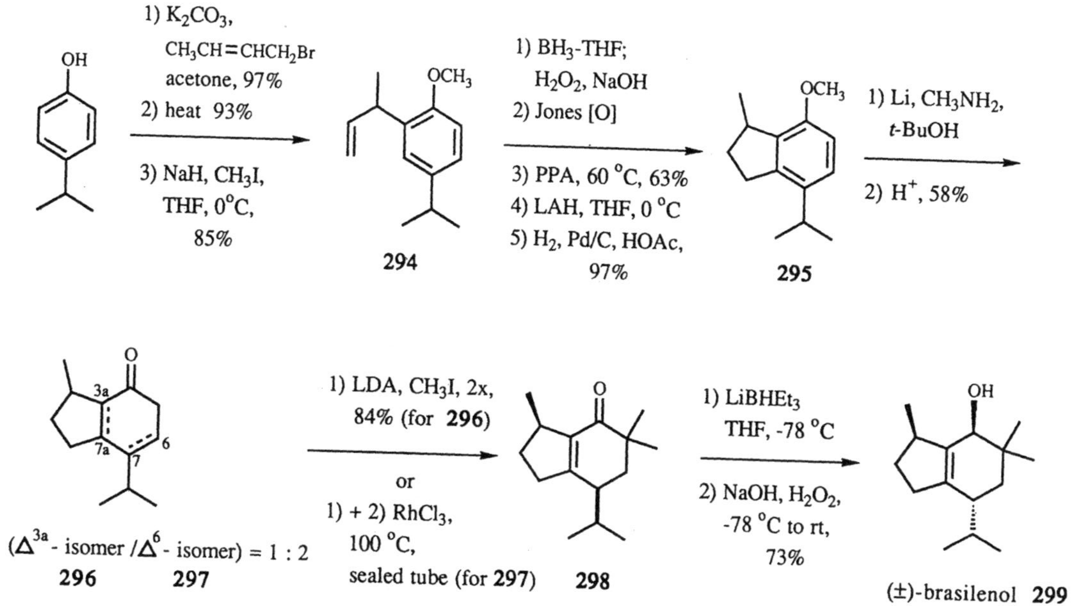

Scheme 50. Greene Synthesis of (±)-Brasilenol

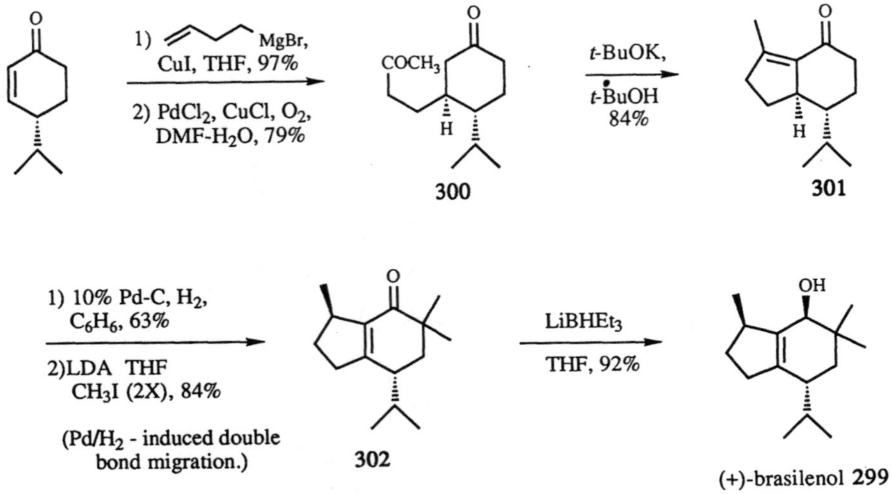

Scheme 51. Greene Synthesis of (+)-Brasilenol

2.2.4.2 *Oppositol*

Oppositol (**306**) is a brominated sesquiterpene, first isolated [91] from the red alga *Laurencia subopposita.* It is the first example of this skeletal class and possesses moderate activity *vs. Staphylococcus aureus.* The only synthesis of oppositol is that of Yamamura [92] illustrated in Scheme 52. Oppositol-like compounds have been obtained in acid-catalyzed reactions of epoxygermacrene-D [93]. Based upon this finding, and upon the results of molecular mechanics calculations, Yamamura has developed a biomimetic synthesis of **306**. Bromonium ion-induced cyclization of germacrene D provides bromoalcohol **304** as the major component of a mixture of seven products. Oxidative cleavage of the exocyclic methylene by successive treatment with osmium tetroxide and sodium periodate affords ketone **305**. Dehydration of **305** by reaction with phosphorus oxychloride and addition of methylmagnesium iodide gives oppositol (**306**) in 4 steps and 5.4% overall yield.

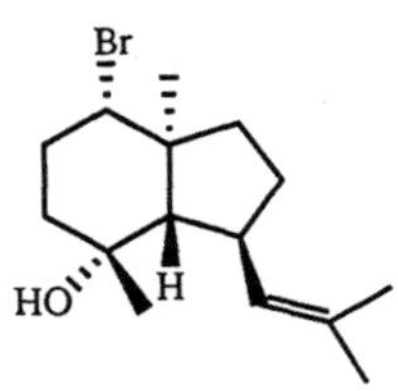

Oppositol 306

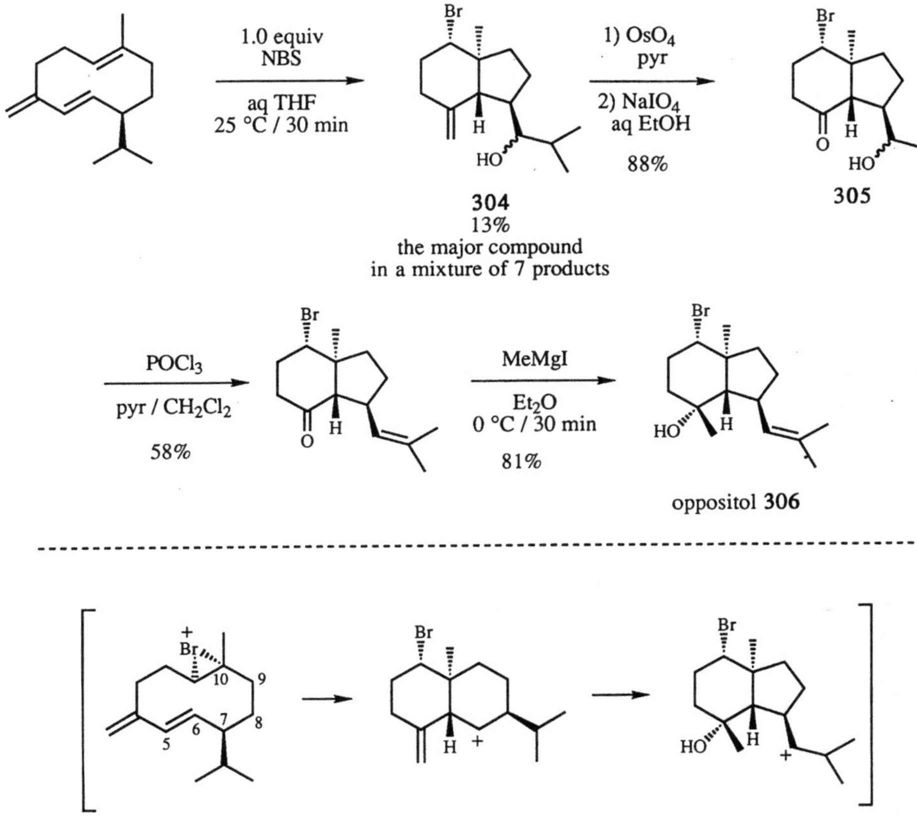

Scheme 52. Yamamura Synthesis of Oppositol from Germacrene D

2.2.4.3 *Pacifigorgiol*

Pacifigorgiol is an irregular terpenoid isolated [94] from the sea fan *Pacifigorgia adamsii* by Fenical in 1982. It shows moderate toxicity toward the reef-dwelling fish *Eupomacentrus leucostictus* [95]. Pacifigorgiol **(314)** possesses a *trans*-perhydroindane skeleton with five contiguous chiral centers and has been

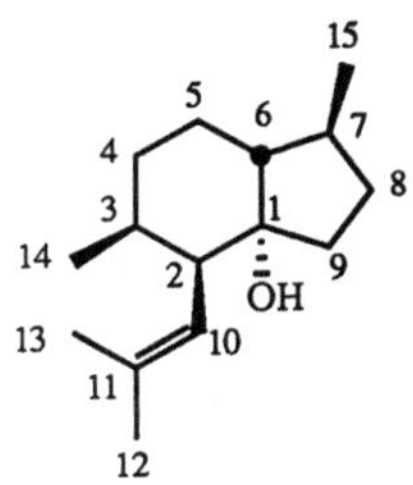

Pacifigorgiol **314**

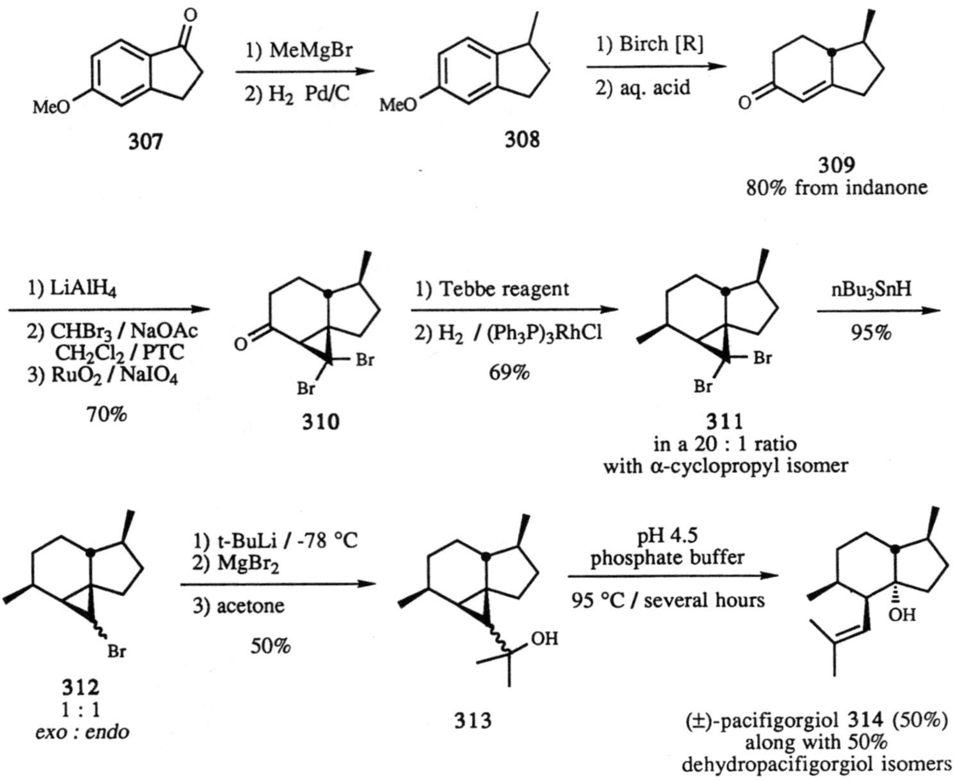

Scheme 53. Clardy Synthesis of (±)-Pacifigorgiol

prepared by Clardy [96] as shown in Scheme 53. Addition of methylmagnesium bromide to methoxyindanone **307** followed by hydrogenolysis affords indane **308**. Birch reduction followed by an acidic workup affords ketone **309** in 80% yield from methoxyindanone **307**, thus setting the correct relative stereochemistry at C-6 and C-7. LAH reduction of **309** to the allylic alcohol and addition of dibromocarbene produces a dibromocyclopropane, which is oxidized to ketone **310** under nonacidic conditions with ruthenium tetroxide and sodium periodate. Insertion of the C-3 methyl with correct stereochemistry occurs by methylenation with the Tebbe reagent and hydrogenation, directed by the bulky cyclopropyl group, to yield dibromocyclopropane **311**. Monodehalogenation of **311** with tri-*n*-butyltin hydride, produces a 1:1 mixture of diastereomeric bromides **312**. Metallation of **312** with *t*-butyllithium, transmetallation to afford the Grignard reagent and reaction with acetone gives tertiary alcohol **313**. Treatment of **313** with a phosphate buffer at pH 4.5 produces **314** in 50% yield along with 50% yield of dehydropacifigorgiol isomers. Compound **314** is produced in 11 steps and 9% overall yield.

2.2.5 Bicyclo[4.4.0] Ring Systems

2.2.5.1 Polygodial

Polygodial is a hot-tasting sesquiterpene first isolated [97] from the "water pepper" *Polygonum hydropiper*. It has also been isolated [98] from the East African medicinal trees *Warburgia ugandensis* and *W. stuhlmanii* along with the drimane sesquiterpenoids warburganal, muzigadial, cinnamolide and drimenin. Interest in its synthesis is due in part to its biological properties. Polygodial has been shown to possess antifeedant activity against African crop pests and it functions as a chemical defense substance in nudibranchs [99]. It possesses two vicinal formyl groups on a drimane skeleton and exhibits a small degree of instability.

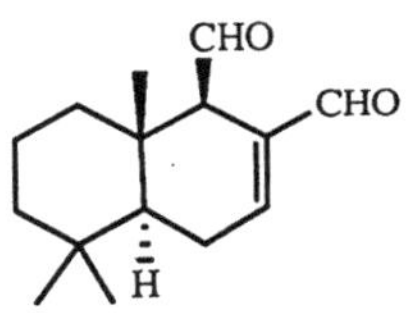

Polygodial **322**

The first synthesis of polygodial (**322**) is that of Kitahara in 1971 [100] (Scheme 54). Lewis acid-catalyzed cyclization of methyl monocyclofarnesate (**316**) followed by sensitized photo-oxidation yields allylic alcohol **317**. Acid-catalyzed rearrangement of allylic alcohol **317** and lactonization produces lactone **318**. LAH reduction of **318** followed by allylic oxidation transposes the carbonyl to provide (±)-cinnamolide (**319**). Lactone ring opening under basic conditions and reaction with diazomethane provides ester **320** as a mixture with unreacted starting material. Collins oxidation of the mixture and acetal formation provides a mixture of acetal **321** and the original lactone **319**, which are separable by chromatography on silica. Reduction of **321** with LAH and subsequent allylic oxidation provides an aldehyde acetal which is converted to polygodial (**322**) by treatment with aqueous oxalic acid. Polygodial is thus obtained in 12 steps and approximately 4% overall yield.

Nakanishi's synthesis of polygodial [101] is illustrated in Scheme 55. The general strategy involves formation of the *trans*-decalin system through a Diels-Alder reaction of dimethyl acetylenedicarboxylate. The vicinal carboxylate groups act as masked aldehyde equivalents so that completion of the synthesis only requires a slight amount of refunctionalization and oxidation state adjustment. The Diels-Alder reaction of β-cyclocitral-derived diene **323** with dimethyl acetylenedicarboxylate produces diene **324**. Allylic oxidation of **324** and catalytic hydrogenation produces ketone **325**, which is converted to alkene **326** by reduction, mesylation and elimination. Differentiation of the two hydroxyl groups by selective protection is acheived by reduction of the two carboxyl groups and sequential treatment with TBSCl and acetic anhydride to provide

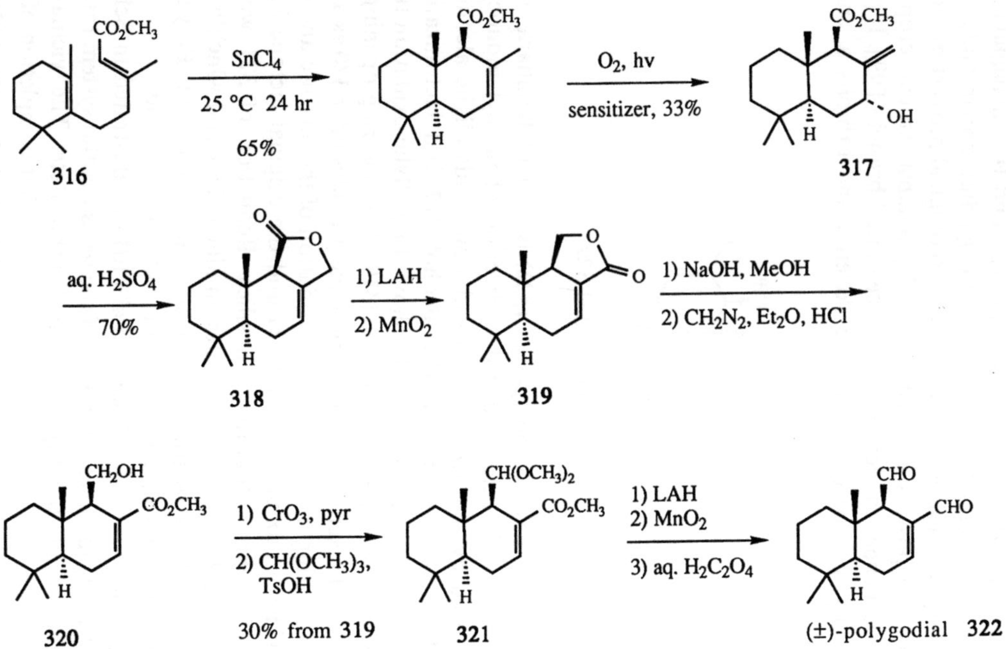

Scheme 54. Kitahara Synthesis of (±)-Polygodial

acetate **327**. Desilylation of **327**, allylic oxidation and protection of the resulting aldehyde gives acetal **328**. Saponification of acetate **328** followed by Collins oxidation and hydrolysis of the acetal affords polygodial after 15 steps and 22% overall yield.

Although Nakanishi's synthesis of polygodial (Scheme 55) proceeds in very good yield for a 15 step synthesis, the excessive functional group manipulations required in the latter stages decrease the yield and efficiency greatly. This problem was solved separately in an identical fashion by Ley [102] and Lallemand [103] in 1981. Both Ley's and Lallemand's syntheses of **322** begin with the Nakanishi common intermediate diester **324** obtained via the Diels-Alder reaction of β-cyclocitral-derived diene **323** with dimethyl acetylenedicarboxylate. In Ley's synthesis (Scheme 56) the diene diester is reduced under isomerizing conditions with hydrogen, palladium on carbon and a trace of HCl to afford the ester **325**. Reduction of **325** with LAH affords a diol which can be converted to polygodial (**322**) by careful Swern oxidation. An overall yield of 57% is obtained in 4 steps. The synthesis by Lallemand (Scheme 57) differs only in that diene **324** is isomerized to the conjugated diene **330** by treatment with LDA prior to hydrogenation. Polygodial (**322**) is obtained in 5 steps and approximately 31% yield.

The strategy of Ley and Lallemand has been modified by Mori [104] to achieve an enantiospecific synthesis of either isomer of polygodial (**322**) beginning from a common starting material (Scheme 58). Hydroxyketone **331** is obtained as a single enantiomer by reduction with baker's yeast. It can be converted to the required diene in five steps by silylation and alkylation to give trimethylated ketone **332** as a mixture of diastereomers. Addition of sodium acetylide to the ketone followed by elimination produces an enyne that is selectively reduced to the diene **333** under Lindlar conditions. Diels-Alder cycloaddition of diene **333** with dimethyl acetylenedicarboxylate proceeds without selectivity to give an equal mixture of diastereomers. The ready availability of both isomers allows for the preparation of either enantiomer of polygodial, as the inducing center is removed at a latter stage in the synthesis. It was found in the case of diene **334** that the reduction-isomerization procedure of Ley (Scheme 56) was not reliable. However, treatment of diene **334** with DBU causes equilibration to the conjugated diene, which is then selectively reduced to the alkene diester **335**. Conversion of alcohol **335** to the triflate gives elimination to produce diene **336**. Conversion of **336** to polygodial occurs by selective catalytic hydrogenation, reduction with LAH and Swern oxidation. Polygodial is obtained optically pure in 13 steps and 3% overall yield.

2.2.5.2 Zonarene

Zonarene is the major hydrocarbon component of the brown seaweed *Dictyopteris zonarioides* [105]. It has also been identified as a constituent in a variety of essential oils [106]. A very clever stereospecific synthesis of (−)-zonarene **341** has been developed by Williams [107] (Scheme 59). The synthesis is initiated by

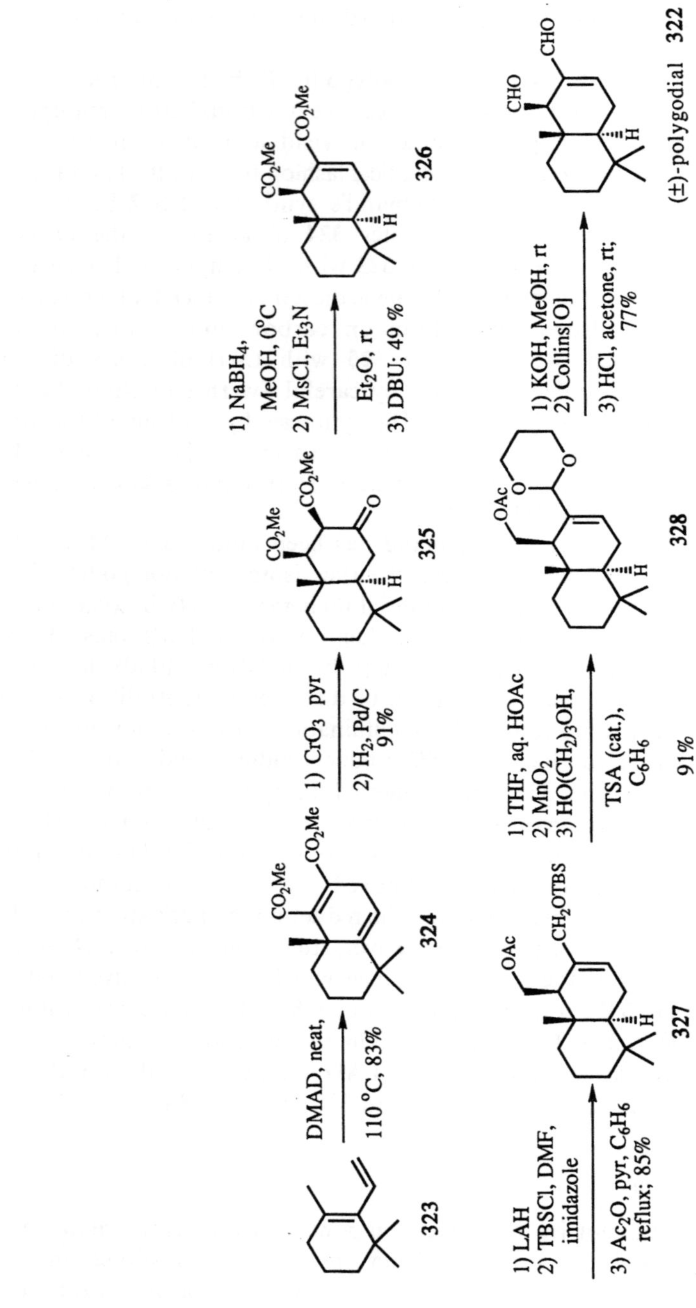

Scheme 55. Nakanishi Synthesis of (±)-Polygodial

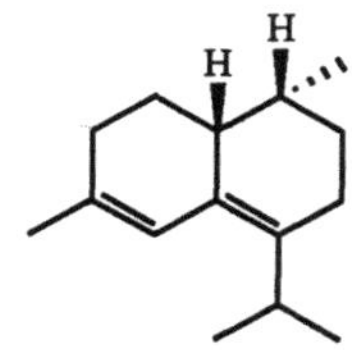

Zonarene **341**

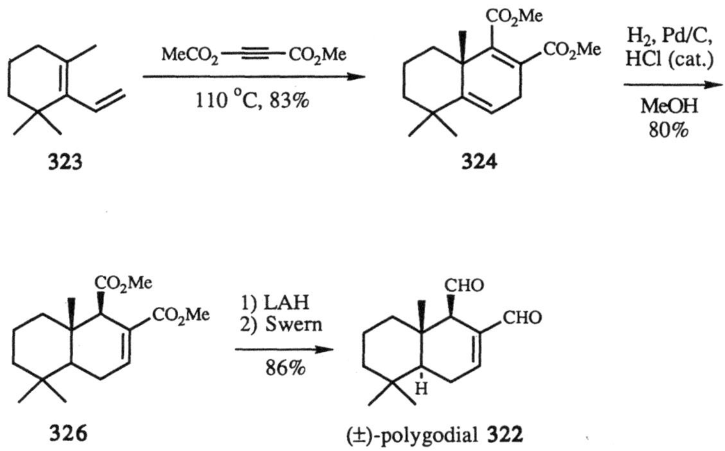

Scheme 56. Ley Synthesis of (±)-Polygodial

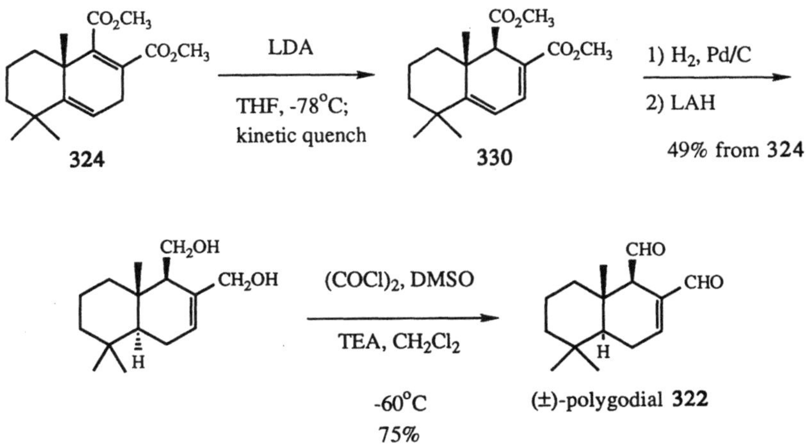

324 was synthesized by the method of:
Tanis SP, Nakanishi K (1979) J. Am. Chem. Soc. 101:4398

Scheme 57. Lallemand Synthesis of (±)-Polygodial

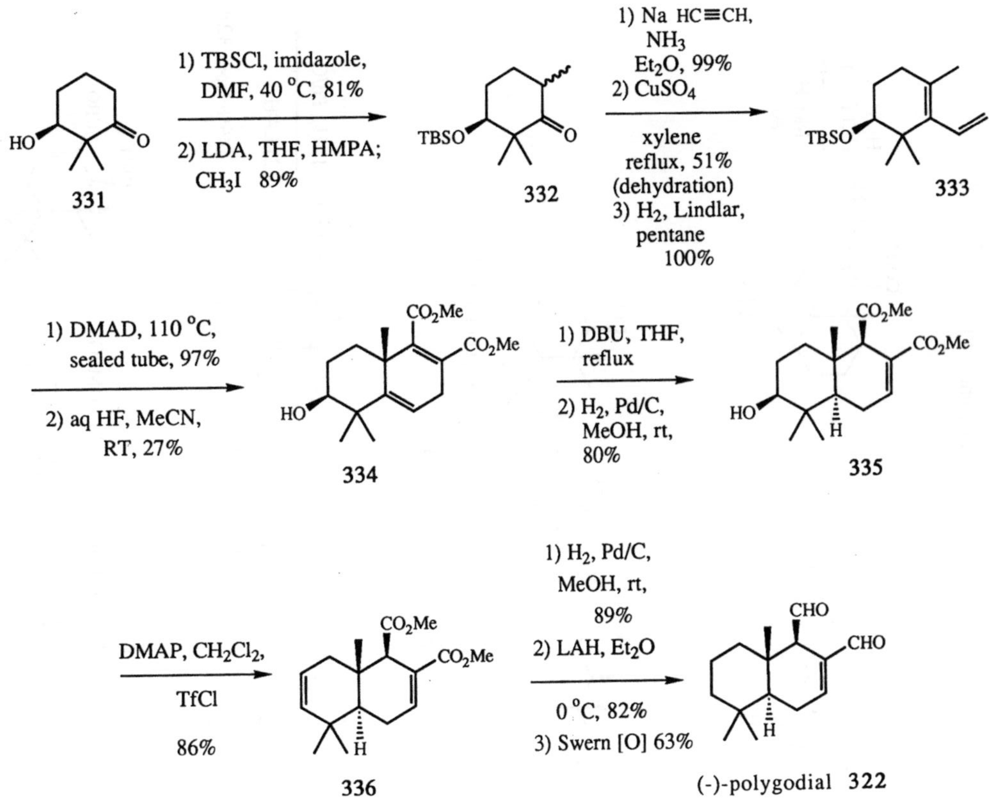

Scheme 58. Mori Synthesis of (−)-Polygodial

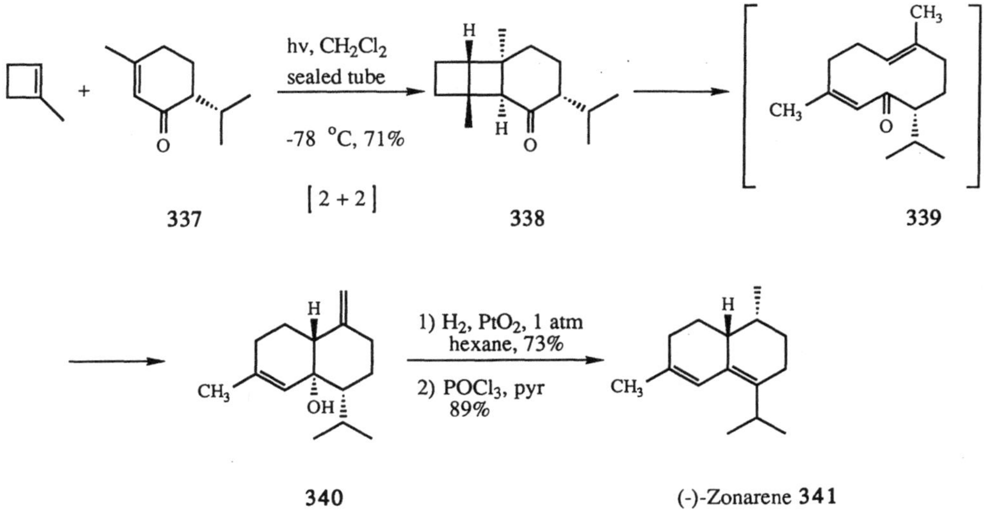

Scheme 59. Williams Synthesis of of (−)-Zonarene

the [2+2] photocycloaddition of 1-methylcyclobutene with (−)-piperitone (**337**) to afford the photoadduct **338** in a regio- and enantiospecific manner. Thermolysis of **338** proceeds via cycloreversion to give diene **339** as an intermediate. Under the reaction conditions, diene **339** undergoes a transannular ene cyclization reaction to provide *trans*-decalin **340**. Stereospecific catalytic hydrogenation of the exocyclic double bond, followed by elimination of the tertiary alcohol provides (−)-zonarene in 4 steps and 23% overall yield.

Zonarene has also been characterized as one of several products of the cyclization of farnesol (isomer mixture)-induced by $BF_3 \cdot Et_2O/CH_2Cl_2$ [108].

2.2.5.3 *Cycloeudesmol*

Cycloeudesmol **349** is a cyclopropane-containing sesquiterpene isolated from the marine alga, *Chondria oppositiclada* Dawson, by Fenical and Sims [109]. It exhibits antibiotic properties towards a variety of microbial organisms [110]. Although there was originally some controversy regarding the correct structure of cycloeudesmol, this was cleared up upon completion of the first synthesis of **349** by Chen [111] (Scheme 60). The synthetic strategy centers around a key

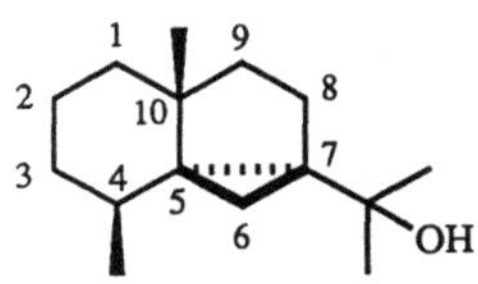

Cycloeudesmol **349**

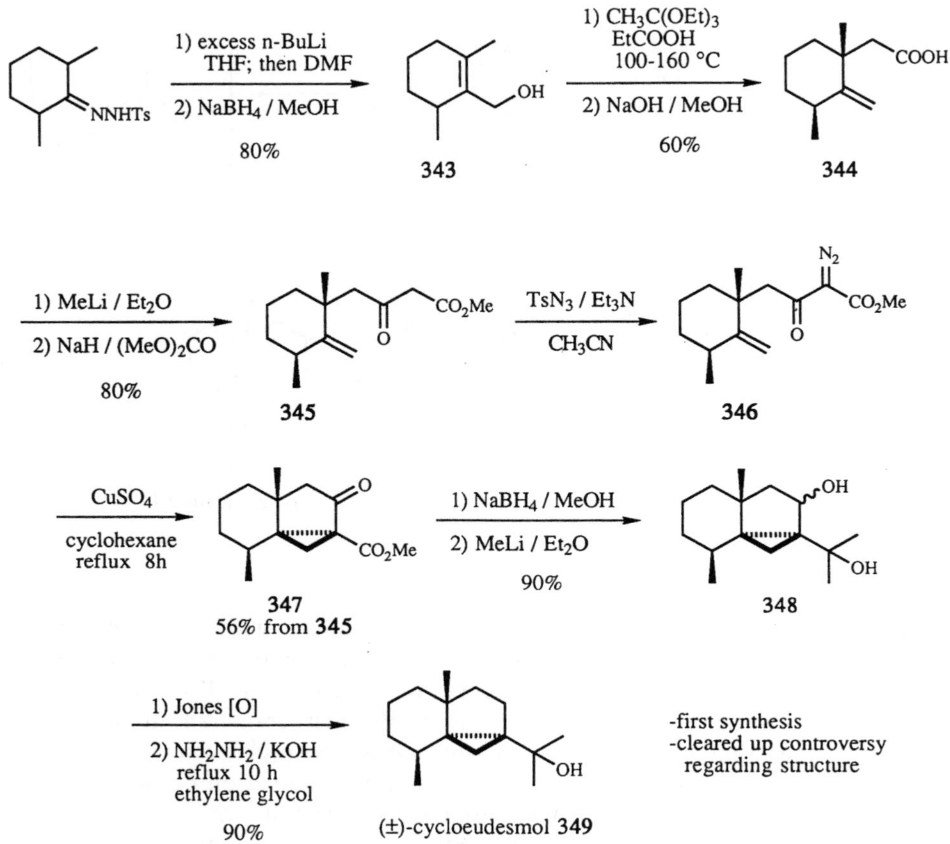

Scheme 60. Chen Synthesis of (±)-Cycloeudesmol

transformation, in which the relative stereochemistry at C5 and two of the three rings are constructed in a single step by intramolecular addition of a carbenoid species to an exocyclic double bond (**346** → **347**). The synthesis begins with alcohol **343**, available from 2,6-dimethylcyclohexanone in three steps. Claisen rearrangement of **343** and saponification provides acid **344** with the necessary *cis*-relationship between the two methyl groups. Conversion of acid **344** to the α-diazo keto ester **346** occurs via formation of the methyl ketone, methoxycarbonylation and diazotization with *p*-toluenesulfonyl azide. Thermolysis of **346** in refluxing cyclohexane with copper sulfate as a catalyst yields keto ester **347** as a single crystalline compound with the proper stereochemistry at all four stereocenters. Sodium borohydride reduction of the ketone **347** and addition of methyllithium produces the tertiary alcohol **348**. Deoxygenation of the secondary alcohol by sequential Jones oxidation and Wolff-Kishner reduction gives cycloeudesmol in 12 steps and 17% overall yield.

A different strategy developed for the preparation of cycloeudesmol **349** is that of Ando [112] (Scheme 61). The critical transformations in this approach are the stereoselective formation of the cyclopropane ring, via epoxy alcohol **353**, and the transformation of the vicinal *cis*-diol **354** to nitrile **355** with the correct relative stereochemistry at C-4. The synthesis begins with dienone **351**, available form enone **350** by DDQ oxidation. Hydrocyanation of dienone **351** followed by sodium borohydride reduction produces the β-alcohol **352** along with the corresponding α-alcohol. Sharpless epoxidation of **352** gives β-epoxy alcohol **353** which undergoes intramolecular ring opening upon treatment with LDA to afford the cycloeudesmol ring system **354**. Reaction of the vicinal diol **354** with *N*,*N*-dimethylformamide dimethyl acetal produces a cyclic orthoester which is eliminated by treatment with boiling acetic anhydride to give **355**. Hydroboration-oxidation of alkene **355** secures the correct relative stereochemistry at C-4. Deoxygenation of the secondary alcohol by treatment of the xanthate with tri-*n*-butyltin hydride affords nitrile **356**. DIBAL reduction of the nitrile and addition of methyllithium gives alcohol **357** which is converted to **349** by oxidation and addition of methyllithium. Cycloeudesmol is obtained in 14 steps and 9% overall yield.

2.2.5.4 *Lemnal-5a-en-2-one*

The only synthesis of the *Lemnalia* metabolite **363** is that of Wolf [113] (Scheme 62). Birch reduction of phenolic ether **359** followed by chelation-directed metalation with *n*-butyllithium and alkylation yields diene **360**. Hydrolysis of **360** to the ketone, reduction with LAH and methylation provides ether **361** in 31% overall yield. Deprotection of the THP ether and Jones oxidation produces acid **362**. Cation-olefin cyclization catalyzed by tin(IV) chloride sets the stereochemistry of four of the final five stereocenters and produces the lemnalone ring system in 40% yield. Demethylation of the methyl ether in 15% yield and elimination affords lemnal-5a-en-2-one in 12 steps and 1% overall yield.

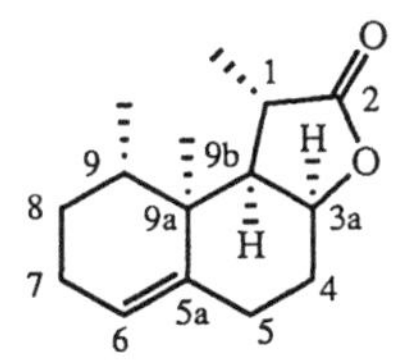

(±)-Lemnal-5a-en-2-one **363**

2.2.5.5 *Lemnalol*

Lemnalol (**368**) is a sesquiterpenoid antitumor agent isolated [114] from the Japanese soft coral *Lemnalia tenuis* Verseveldt. It is the first example of an oxygenated ylangene-type sesquiterpenoid, and the first ylangene-type sesquiterpenoid isolated from a marine source. Scheme 63 illustrates the synthesis of

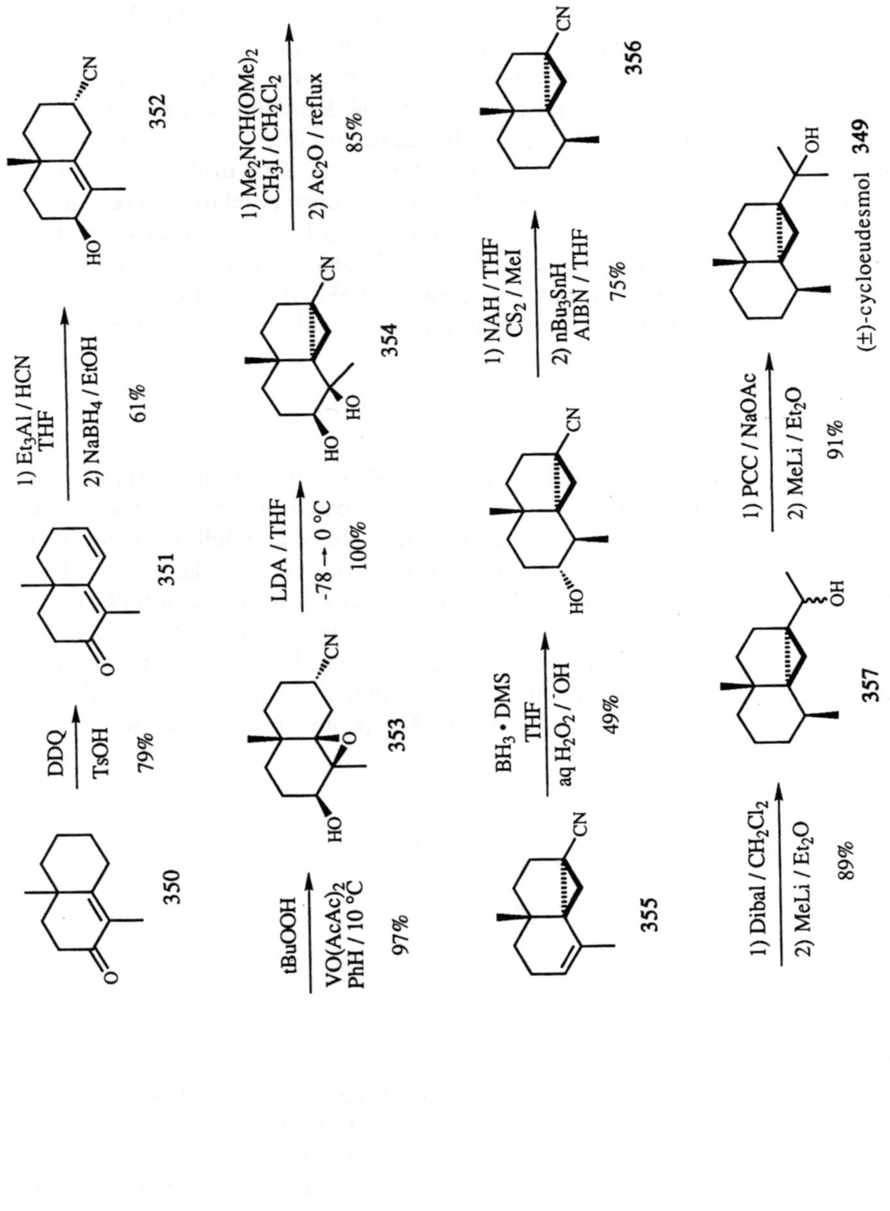

Scheme 61. Ando Synthesis of (±)-Cycloeudesmol

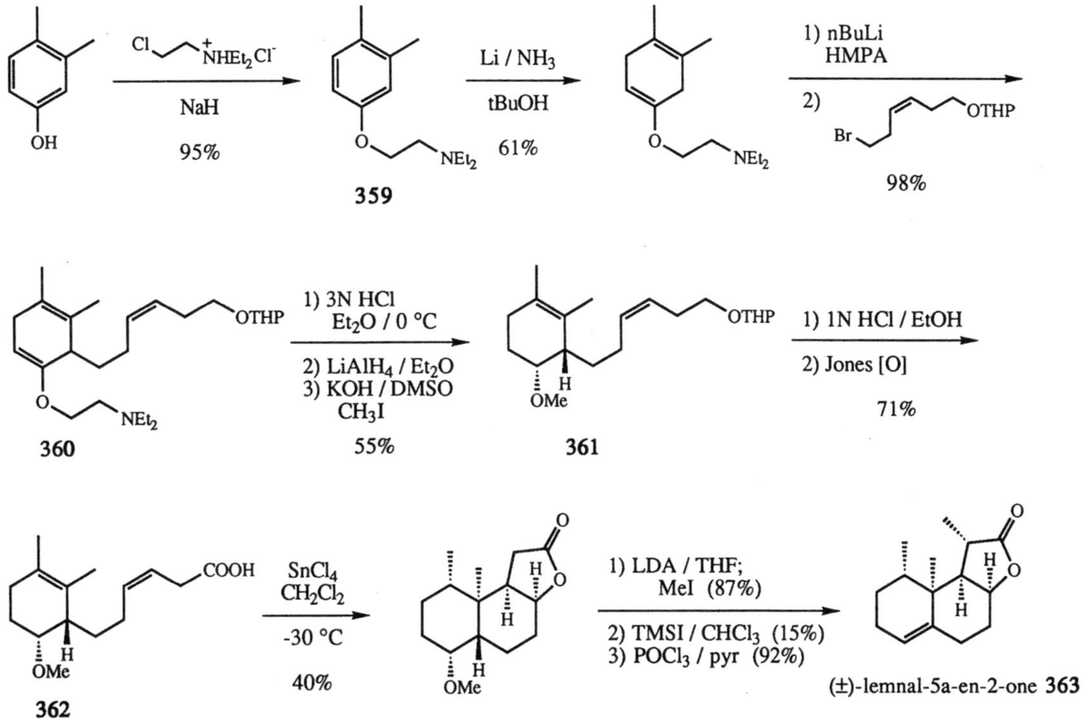

Scheme 62. Wolf Synthesis of *Lemnalia* Metabolite **363** [(±)-lemnal-5a-en-2-one]

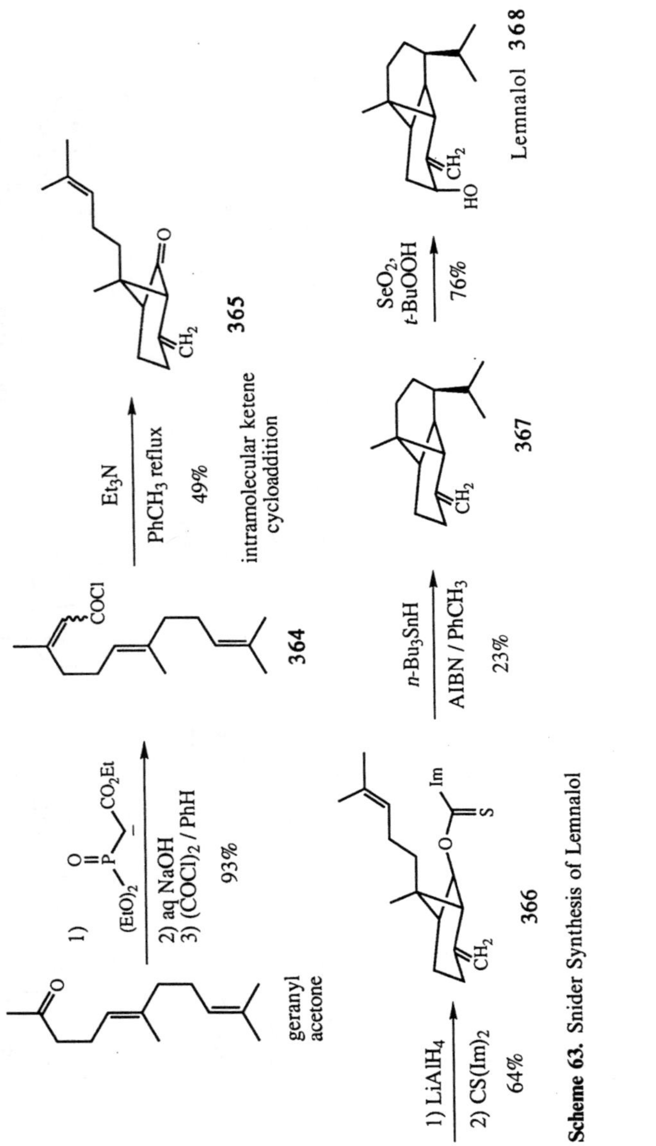

Scheme 63. Snider Synthesis of Lemnalol

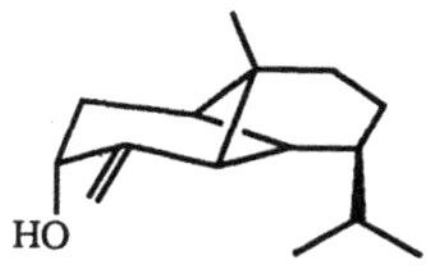

Lemnalol **368**

lemnalol by Snider [115]. The key step in the synthesis is an intramolecular ketene cycloaddition to produce a bicyclo[3.1.1] system with an appropriate sidechain that can be subsequently cyclized in an intramolecular fashion to provide the tricyclic ylangene ring system. The synthesis begins with the conversion of geranyl acetone to a mixture of farnesic acid isomers by Horner-Emmons olefination followed by saponification. Reaction of the acid with oxalyl chloride provides farnesyl chloride (**364**) as a mixture of isomers in 93% yield. Treatment of **364** with triethyl amine in refluxing toluene forms a vinyl ketene which undergoes intramolecular ketene cycloaddition in situ to afford ketone **365** containing an exocyclic methylene. Besides providing the exocyclic double bond needed in the final product, the vinyl substituent activates the ketene towards cycloaddition so as to make the entire transformation a viable synthetic process. Intramolecular cyclization onto the olefinic sidechain is achieved by a modification of the Barton-McCombie method for the deoxygenation of secondary alcohols [116]. Treatment of thiocarbonyl imidazolide **366** with AIBN and tri-*n*-butylin hydride generates a cyclobutyl radical which undergoes cyclization faster than reduction. β-Ylangene **367** is obtained in 23% yield, along with and equal amount of the *iso*-propyl isomer β-copaene. Oxidation of **367** with selenium dioxide and *t*-butylhydroperoxide produces the axial alcohol **368**, as is expected from related oxidations of β-pinene. Lemnalol (**368**) is obtained in 8 steps with an overall yield of 5%.

2.2.5.6 1,4-Epoxycadinane

The synthesis of the sesquiterpene 1,4-epoxycadinane (isolated [117] from the brown alga *Dilophus fasciola*) has been reported by Keay [118]. The synthetic strategy for the synthesis of **375** can be divided into three stages (Scheme 64). The first involves construction of an appropriately functionalized 2,5-disubstituted furan. The second stage, and the key step in the synthesis is an

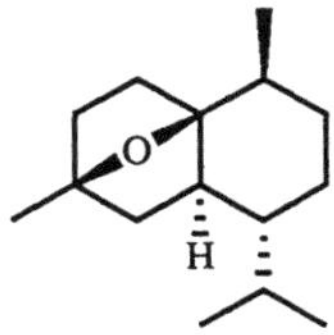

1,4-Epoxycadinane 375

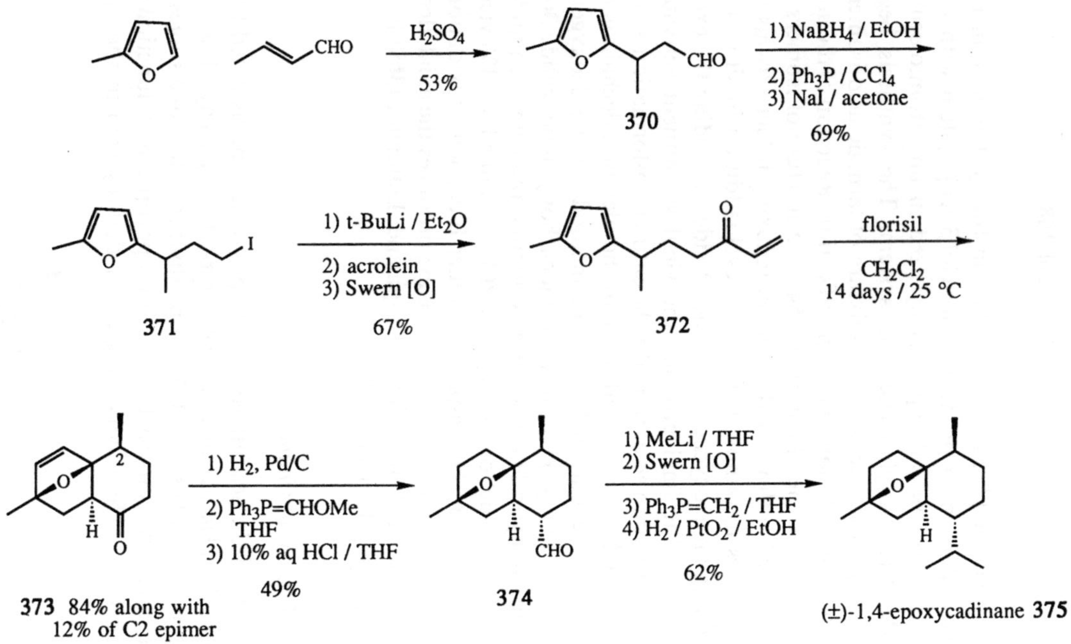

Scheme 64. Keay Synthesis of (±)-1,4-Epoxycadinane

intramolecular Diels-Alder reaction with the furan ring as the diene to set the relative stereochemistry of all four centers in a single step. Finally, all that is needed is minor refunctionalization to complete the preparation of 1,4-epoxycadinane. The appropriate furan **372** is obtained in a six-step sequence. Acid-catalyzed 1,4-addition of 2-methylfuran to crotonaldehyde affords the disubstituted furan **370**. Reduction of the aldehyde **370** to the alcohol and conversion to the iodide, by way of the chloride, produces iodide **371**. Metal-halogen exchange with *t*-butyllithium, addition to acrolein and Swern oxidation gives the key intermediate alkene **372**. After stirring for 12 hours in methylene chloride at room temperature the cycloaddition product **373** is obtained as a 1:1 mixture with its C2 epimer. After 14 days, the thermodynamic product mixture is obtained. The major product, in a 7:1 ratio, is the diastereoisomer **373** in which the C-2 methyl is equatorial. Hydrogenation, Wittig olefination and hydrolysis produces aldehyde **374** in 49% yield. The synthesis is completed by addition of methyllithium to aldehyde **374**, Swern oxidation, Wittig olefination and hydrogenation to secure the natural product in 62% yield. 1,4-Epoxycadinane is obtained in 15 steps and 6.2% overall yield.

2.2.5.7 *3β-Bromo-8-Epicaparrapi Oxide*

The bromine-containing sesquiterpene 3β-bromo-8-epicaparrapi oxide (**379**) has been isolated from *Laurencia obtusa* collected from the English Channel [119]. It has been prepared in seven steps from geranylacetone by Hoye [120] (Scheme 65). Reformatsky reaction of **376** with methyl bromoacetate provides the β-hydroxyester **377**. The ester is cyclized by treatment with mercury trifluoroacetate to give the cyclized organomercurial and brominated (SE_2) to give the cyclized bromoester as a 1:1 mixture of isomers at C-8. As this mixture was not easily separated, the ester was first reduced with DIBAL to give alcohol **378** in 22% yield. Conversion of alcohol **378** to the phenylselenide and selenoxide elimination affords the natural product **379** in 9% yield from hydroxyester **377**. The yield of the final transformation can be improved by direct conversion of the alcohol **378** to an arylselenide by reaction with *O*-nitrophenylselenocyanate and tri-*n*-butylphosphine.

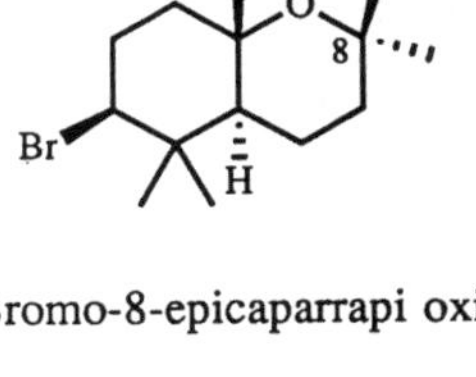

3β-Bromo-8-epicaparrapi oxide **379**

2.2.5.8 *β-Gorgonene*

The non-isoprenoid sesquiterpene β-gorgonene (**386**) was isolated from the sea fan *Pseudopterogorgia americana* in 1968 by Weinheimer [121]. In conjunction

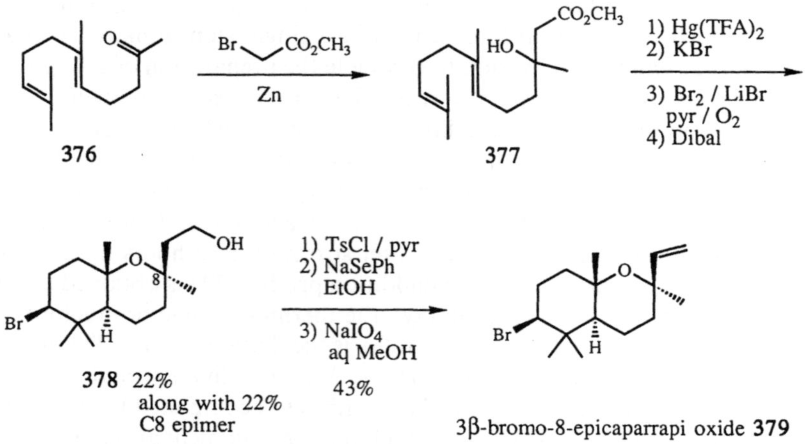

Scheme 65. Hoye Synthesis of 3β-Bromo-8-cpicaparrapi Oxide

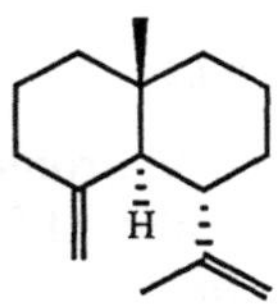

β-Gorgonene 386

with his studies on the interplay of stereoelectronic and steric factors in conjugate addition to α,β-unsaturated ketones, Boeckman [122] synthesized **386** as shown in Scheme 66. Decalone **381** is converted to α,β-unsaturated ketone **382** in 7 steps via standard transformations. Copper-catalyzed addition of isopropenylmagnesium bromide to **382** yields ketone **383** in 32% yield, along with 13% of two other diastereomers. Methylenation of **383** provides β-gorgonene in a total of 9 steps.

2.2.5.9 Albicanol and Albicanyl Acetate

The drimane-type sesquiterpenoid albicanol, along with its acetate, has been isolated [123] from the dorid nudibranch *Cadlina luteomarginata*. It exhibits potent fish antifeedant activity. It was originally isolated from the liverwort *Diplophyllum albicans* [124]. The first syntheses of **390** and **391** are those of Weiler [125] in 1986 (Scheme 67). Coupling of the (*Z*)-enol phosphate of ketoester **387** with trimethylsilylmethylmagnesium bromide in the presence of nickel(II) bisacetylacetonate affords silyl substituted unsaturated ester **388**. Treatment of **388** with mercuric trifluoroacetate in nitromethane initiates cation-olefin cyclization to afford exocyclic olefin **389** as a 3:1 (β:α) mixture of

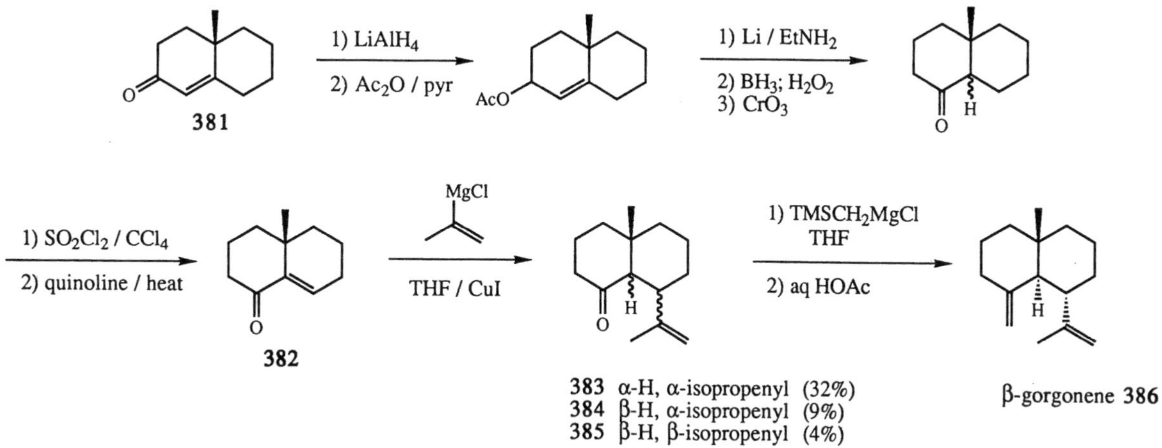

Scheme 66. Boeckman Synthesis of β-Gorgonene

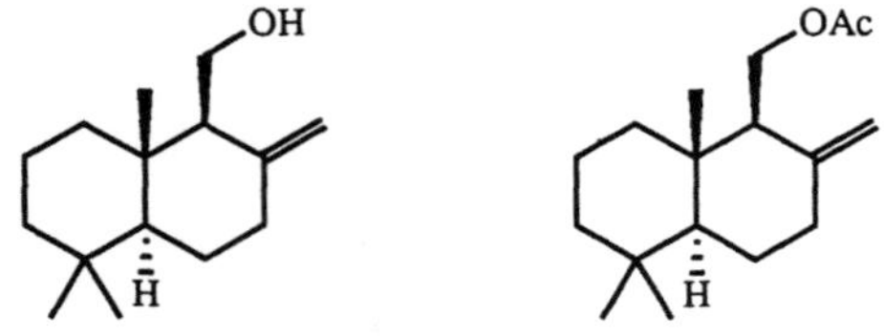

Albicanol **390** Albicanyl acetate **391**

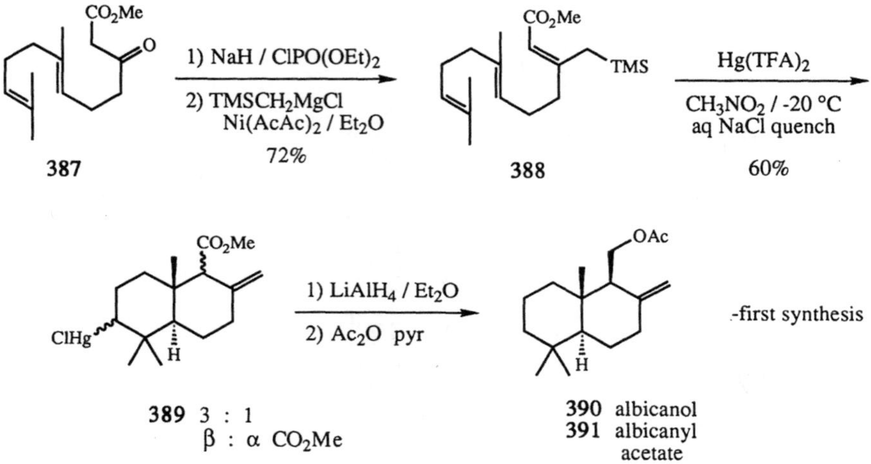

Scheme 67. Weiler Synthesis of Racemic Albicanol and Albicanyl acetate

isomers at the methoxycarbonyl. It was found that the cyclization was dependent on Lewis acid, solvent and temperature. Interestingly, the cyclization displays a lack of stereospecificity. Presumably, the α-epimer arises via a boat-chair conformation. Reduction of **389** with LAH provides albicanol in an overall 4 steps and approximately 32% overall yield. Albicanyl acetate (**391**) is obtained upon acetylation under standard conditions.

These metabolites have been prepared enantiospecifically by Fukumoto and Kametani [126] as shown in Scheme 68. The synthesis begins with ketone **393**, prepared in six steps from optically pure Wieland-Miescher ketone. Oxidation of **393** with LDA and 2-sulfonyloxaziridine **394** produces α-hydroxyketone **395** in 72% yield. Oxidative cleavage of **395** with lead tetraacetate, protection of the aldehyde and reduction-oxidation of the acid to aldehyde **396** proceeds in good yield. Wittig olefination of the free aldehyde, hydrolysis of the acetal and addition of hydroxylamine provides oxime **397** in 61% yield form the hydroxyketone **395**. Oxidation of **397** with sodium hypochlorite, and the ensuing intramolecular nitrile-oxide cycloaddition, provides isoxazoline **398** as a single

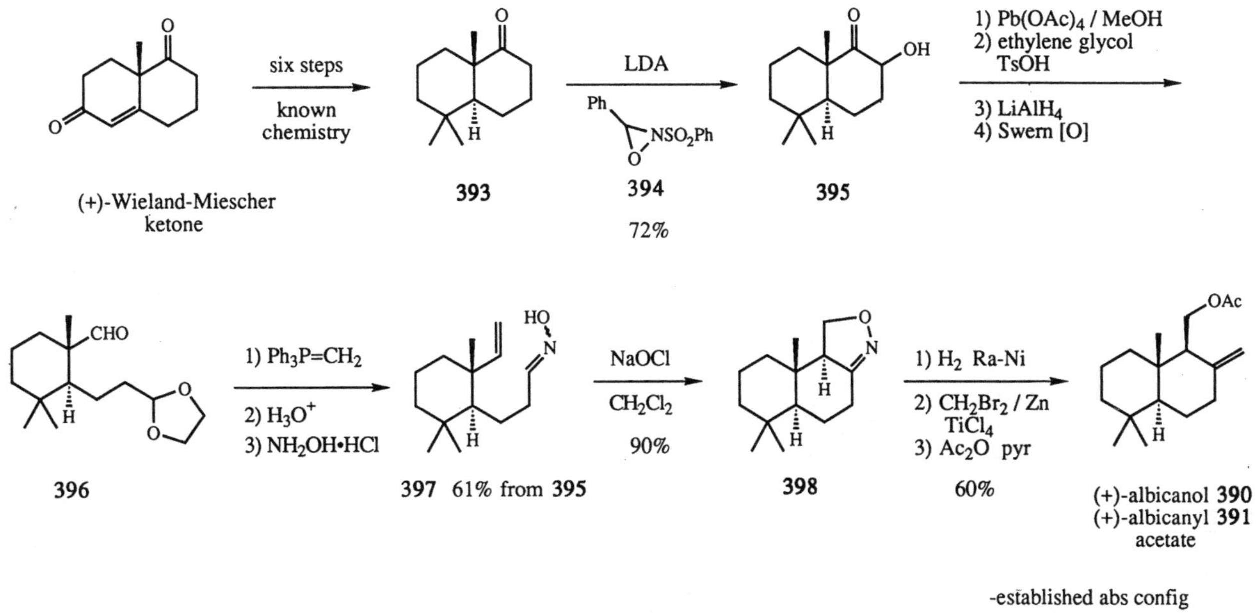

Scheme 68. Fulkumoto-Kametani Synthesis of (+)-Albicanol and Albicanyl acetate

isomer. Reductive hydrogenolysis of isoxazoline **398** and titanium mediated methylenation with zinc and methylene bromide provides (+)-albicanol **390** in 11 steps and 23% overall yield. As before treatment of albicanol with Ac_2O/pyridine produces albicanyl acetate in quantitative yield.

2.2.5.10 β-Dictyopterol

DeGroot's [127] synthesis of β-dictyopterol is outlined in Scheme 69. The synthesis is initiated by the MCPBA oxidation of dienol acetate **399** to afford diol **400** as a mixture of isomers. Selective oxidation of the allylic alcohol with PCC and reduction of the double bond with titanium trichloride in aqueous HCl provides dione **401** as a 2:1 *trans/cis* mixture of isomers. Utilization of the high degree of selectivity shown in the acetalization of hindered ketones with trimethyl orthoformate allows the less hindered carbonyl in **401** to be selectively converted to the dimethyl acetal. Epimerization of the remaining carbonyl under basic conditions produces diastereomerically pure **402**. Wittig olefination and hydrolysis affords ketone **403**. Reaction of ketone **403** with lithium reagent **404** to give an intermediate ketene *O,S*-acetal, followed by methanolysis provides ester **405** as a mixture of diastereomers. Protection of **405** as its THP ether, followed by addition of salt-free methylenetriphenylphosphorane and one equivalent of methyltriphenylphosphonium bromide, acting as a proton donor, provides the needed isopropenyl moiety [128]. Hydrolysis of the THP ether gives β-dictyopterol (**406**) in 14% yield, over 11 steps.

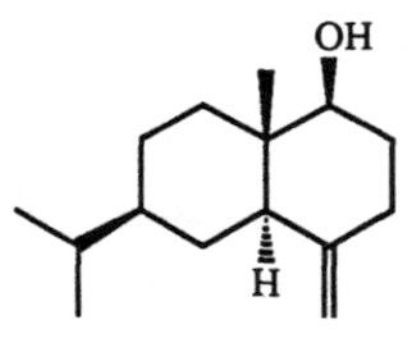

β-Dictyopterol **406**

2.2.6 Hydroazulenes

2.2.6.1 Africanol

Africanol (**418**) is a tricyclic sesquiterpene alcohol, first isolated in 1974 by Tursch [129] from the soft coral *Lemnalia africana*. Scheme 70 illustrates Paquette's synthesis [130] of africanol. The strategy utilizes the conformational rigidity of ketone **414** and the stereoselective reduction of unsaturated ketone **412** to obtain the correct relative stereochemistry, via Claisen rearrangement, at the 7–5 ring junction of **414**. However, the overall stereospecificity of the synthesis is lost in the later stages due to an inability to stereospecifically epoxidize an olefin. The synthesis begins with acylation of ketone **409** (readily available from 4,4-dimethylcyclohexanone (**408**), see Scheme 70) to provide ester

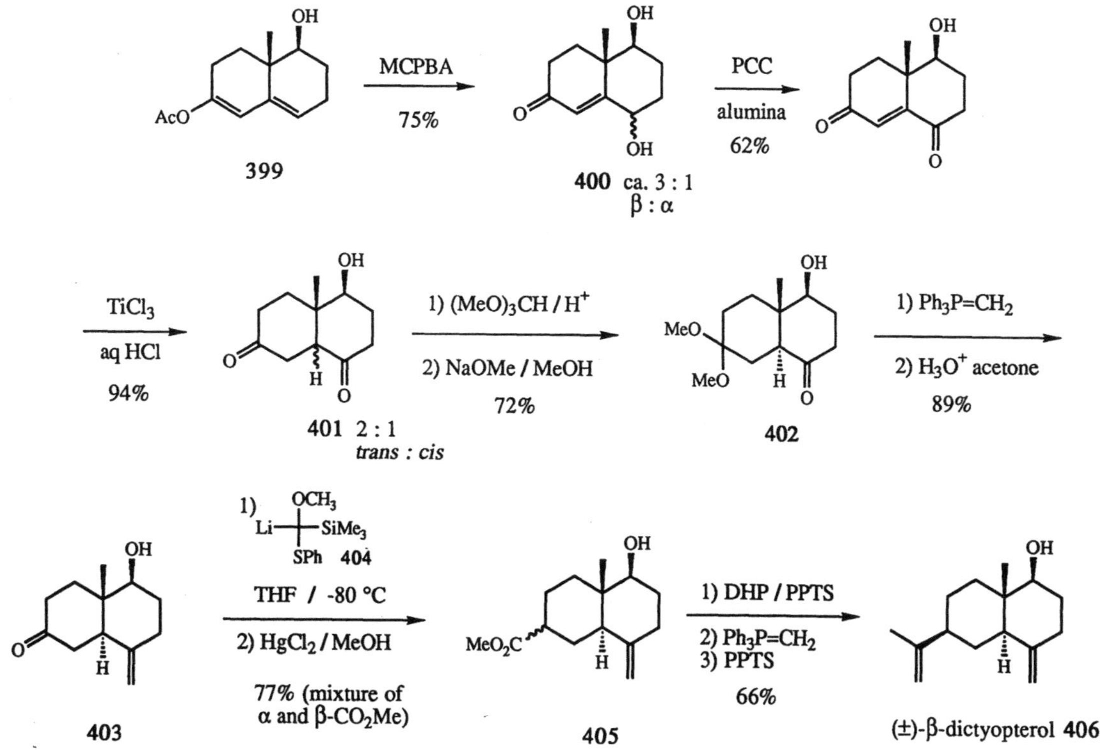

Scheme 69. de Groot Synthesis of (±)-β-Dictyopterol

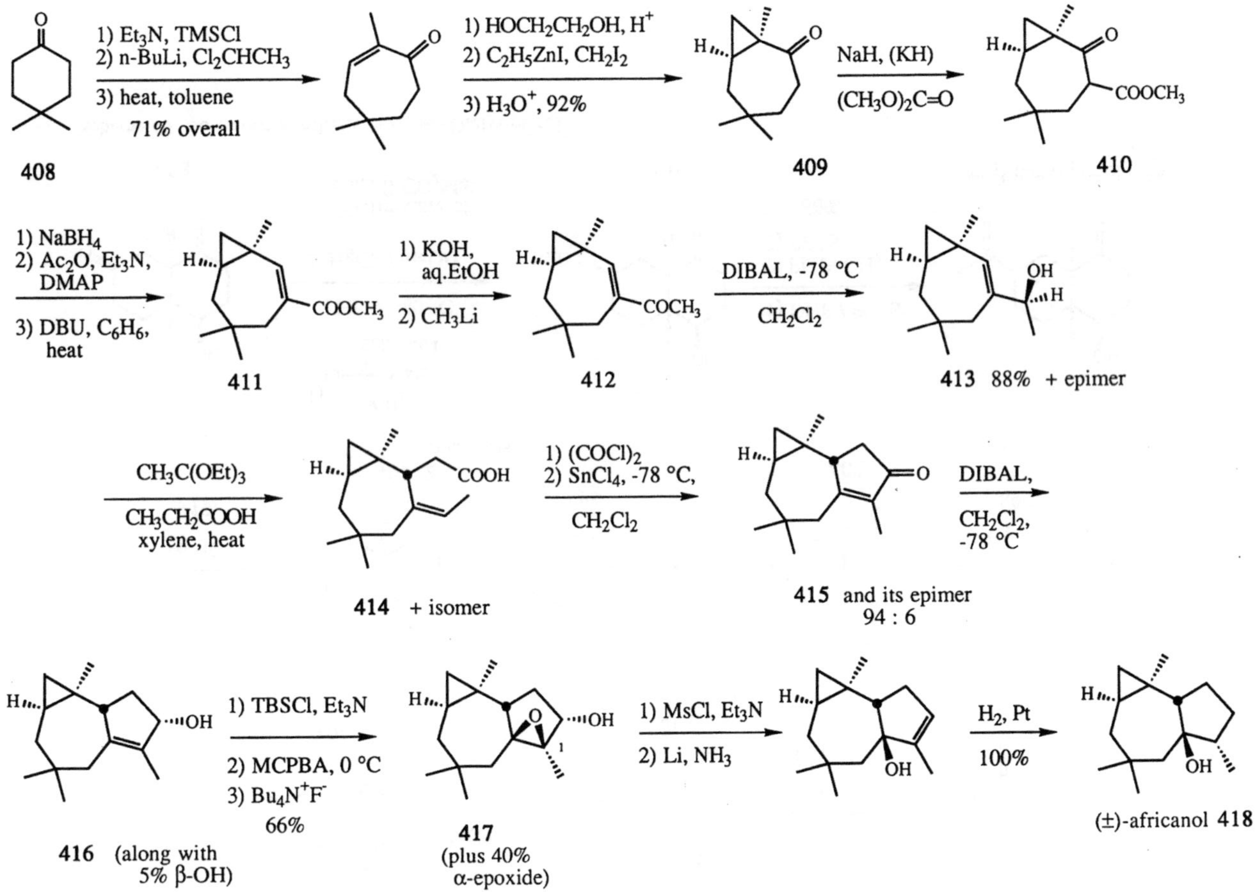

Scheme 70. Paquette Synthesis of (±)-Africanol

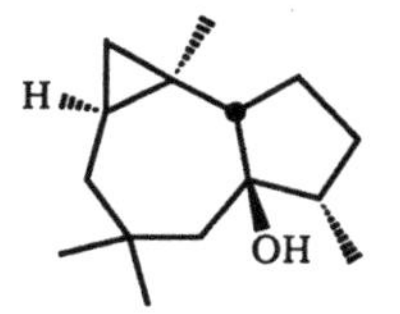

Africanol **418**

410. Reduction of the ketone, acylation and elimination yields unsaturated ester **411**. This is converted into an 88:12 mixture of secondary alcohols **413** by saponification of the ester, addition of methyllithium and stereoselective DIBAL reduction of the resulting ketone. Claisen rearrangement of the major isomer **413** stereospecifically produces acid **414**, thus providing the third stereocenter. Conversion of acid **414** to the acid chloride and Lewis acid-catalyzed ring closure provides ketone **415**. DIBAL reduction of ketone **415** gives alcohol **416** as a 19:1 mixture with the β-epimer. Silylation of the alcohol, epoxidation and desilylation produces a 3:2 mixture of epoxide **417** along with the α-epoxide. Mesylation of alcohol **417** and reductive elimination, with stereospecific opening of the epoxide, provides the fourth stereogenic center. Catalytic hydrogenation produces the fifth center and completes the synthesis of africanol in 22 steps.

Africanol has also been prepared from humulene-9,10-epoxide as an intermediate in Shirahama's synthesis of dactylol (see Scheme 75).

2.2.7 Cyclooctanoids

2.2.7.1 Precapnelladiene

Precapnelladiene (**425**), the presumed biogenetic precursor of the various capnellenes, was isolated [131] from the nonpolar extracts of the soft coral *Capnella imbricata*. The *cis* relationship of the two stereogenic centers in **425** was not confirmed until its first synthesis by Mehta [132] in 1984 (Scheme 71). The major problem in the synthesis of **425** is control of stereochemistry in the flexible 5–8 fused system. Mehta addresses this problem by setting the correct relative stereochemistry of the stereogenic centers in the more stereochemically rigid linear triquinane system. This occurs during the transformation of **419** to **422**. At this point the double bond of **422** can be cleaved to obtain the precapnelladiene

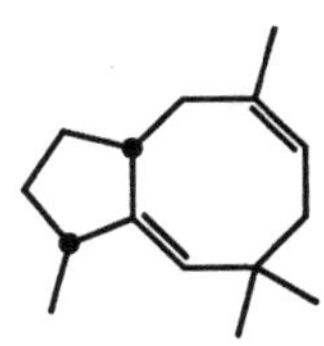

Precapnelladiene **425**

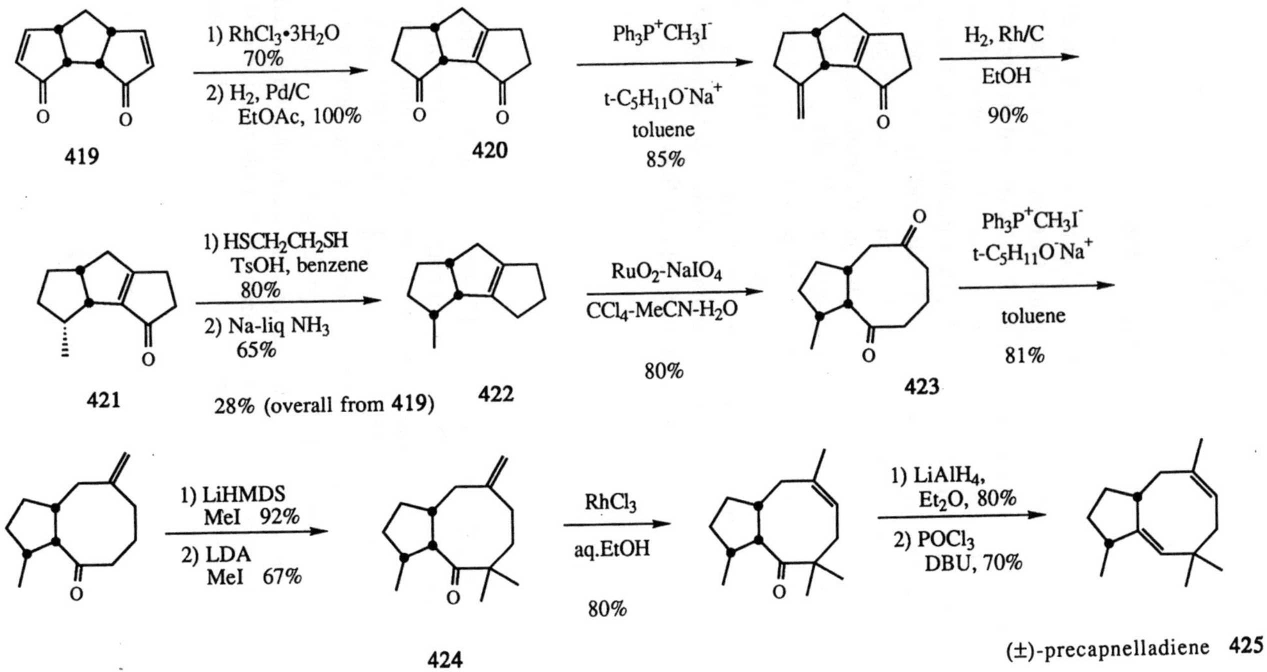

Scheme 71. First Mehta Synthesis of (±)-Precapnelladiene

ring system. The two carbonyl groups so obtained provide the necessary functionality for the introduction of the remaining three methyl groups and the generation of the 1,5-diene system. The synthesis begins with the readily available symmetrical *cis,cis* fused linear triquinane **419**. Isomerization of one of the double bonds to an internal position followed by partial reduction affords **420**. Selective Wittig olefination and reduction achieves the proper stereochemistry, giving unsaturated ketone **421**. Removal of the carbonyl by thioketalization-desulfurization and oxidative cleavage of the alkene causes ring expansion, thus achieving the desired 5–8 ring system. Wittig olefination of the less hindered carbonyl, followed by two regioselective kinetically-controlled methylations α- to the remaining carbonyl gives ketone **424**. Isomerization of the exocyclic olefin to an endocyclic position and reduction of the ketone followed by elimination of the subsequently formed alcohol affords **425**. Precapnelladiene is obtained in 13 steps and an overall yield of 5%.

In Paquette's approach [133] to **425** (Scheme 72), the relative stereochemistry of the two stereocenters that are carried on to the final product is set in the first step. The intramolecular ene reaction of unsaturated ketone **427** provides the proper relative stereochemistry in **428**. Baeyer-Villiger oxidation, methylation, basic opening of the lactone and PCC oxidation to afford ketone **429**, all proceed without loss of stereochemical integrity. Addition of vinyllithium reagent **430** to ketone **429** occurs with lactone formation to stereospecifically provide **431**. Methylenation with Tebbe reagent followed by heating effects Claisen rearrangement to give a mixture of epimers α- to the carbonyl in 87% yield. Formation of the tosylhydrazone and addition of *n*-BuLi leads to a mixture of 1,4-diene **432** and precapnelladiene **425**. Rhodium trichloride catalyzed isomerization of **432** → **425** increases the yield of precapnelladiene giving an overall yield of approximately 20% in 11 steps.

2.2.7.2 *Poitediol*

Poitediol (**445**) [134] isolated from the red seaweed *Laurencia poitei*, was described by Fenical in 1978. A nonstereoselective synthesis of poitediol (**445**) and 4-epipoitediol (**446**) by Gadwood appeared in 1984 [135]. The synthesis is initiated by the transformation of β-ketoester **433** into methyl ether **434** in five steps using standard methodology (Scheme 73). Ozonolysis of **434** and subsequent intramolecular aldol condensation affords methoxyketone **435**. Reduction and cyclopropanation of **435** yields **436** in 53% yield along with 7% of the

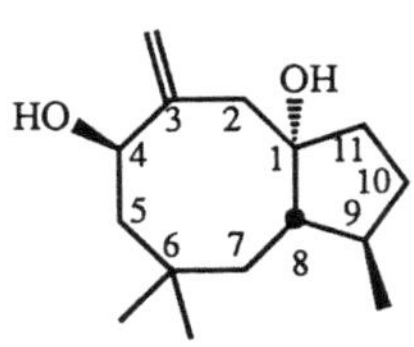

Poitediol **445**

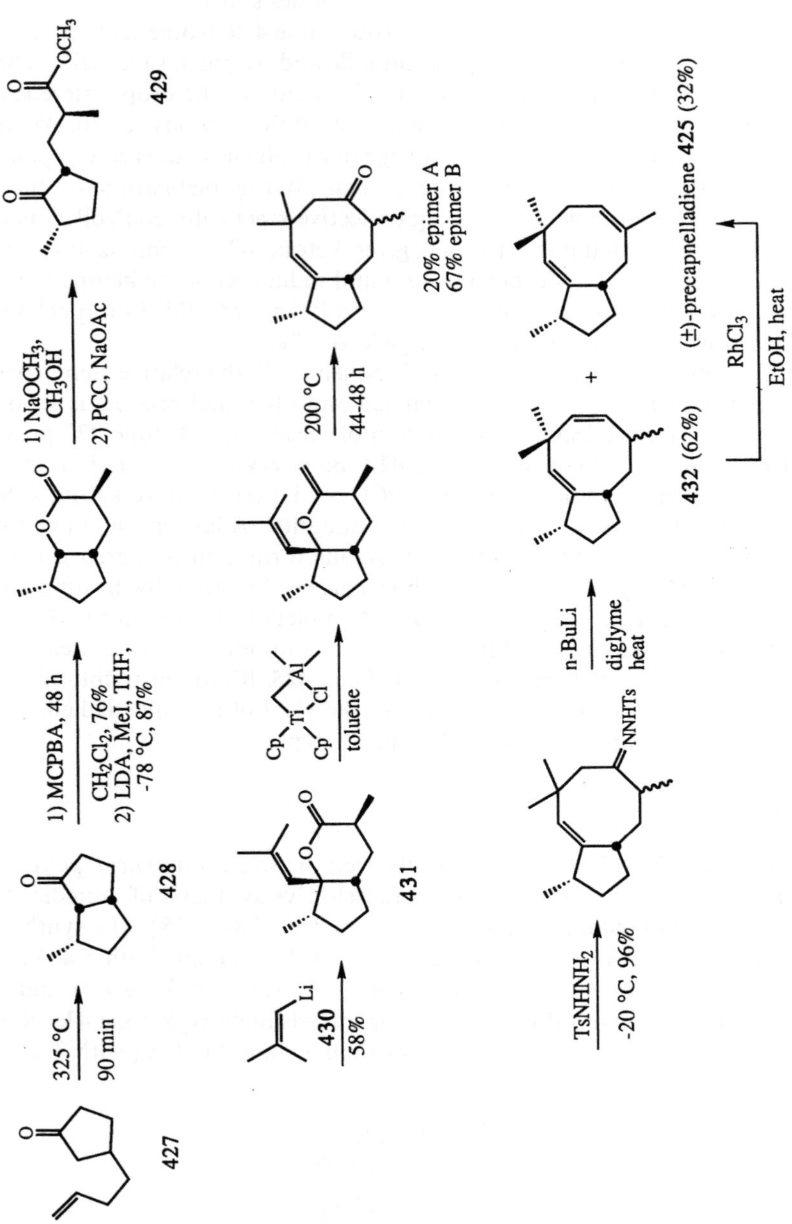

Scheme 72. Paquette Synthesis of (±)-Precapnelladiene

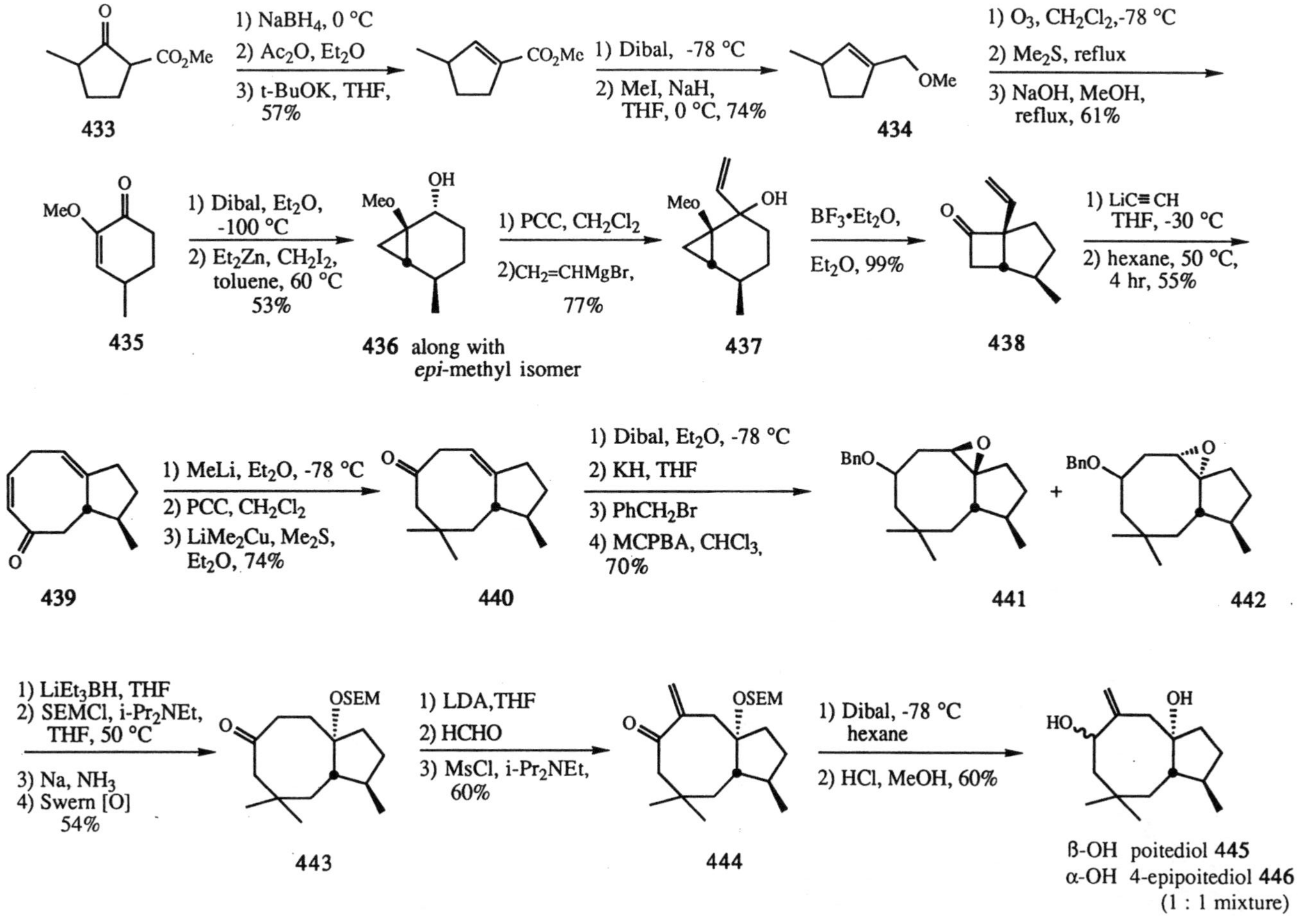

Scheme 73. Gadwood Synthesis of (±)-Poitediol and (±)-Epipoitediol

methyl epimer. PCC oxidation of **436**, addition of vinylmagnesium bromide and Lewis acid-catalyzed rearrangement transforms the 3–6 system to a 4–5 system **438**. Addition of lithium acetylide with warming effects anionic oxy-Cope rearrangement with ring expansion to give the bicyclo[6.3.0]undecane system **439**. Addition of methyllithium to ketone **439**, Dauben oxidative transposition with PCC and addition of lithium dimethylcuprate produces trimethylated ketone **440**. Reduction of the ketone, benzylation, and epoxydation gives a 3:2 mixture of epoxides **441** and **442**, which are separated after reduction to the alcohols. The minor alcohol isomer is protected, debenzylated and oxidized to give ketone **443**. Methylenation via a three step process produces **444**, which is converted to a 1:1 mixture of poitediol (**445**) and 4-epipoitediol (**446**) by DIBAL reduction and deprotection. Poitediol is obtained in 0.2% yield over 27 steps.

2.2.7.3 Dactylol

Dactylol is an irregular isoprenoid alcohol possessing the same bicyclo[6.3.0]undecane skeleton as precapnelladiene. It has been isolated [136] from the red seaweed *Laurencia poitei* and has also been found [137] in the Caribbean sea hare *Aplysia dactylomela*. The first synthesis of dactylol, by Gadwood [138] (Scheme 74), is a continuation of Gadwood's synthesis of

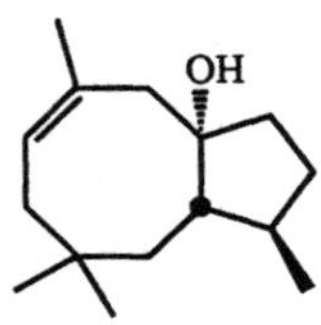

Dactylol **448**

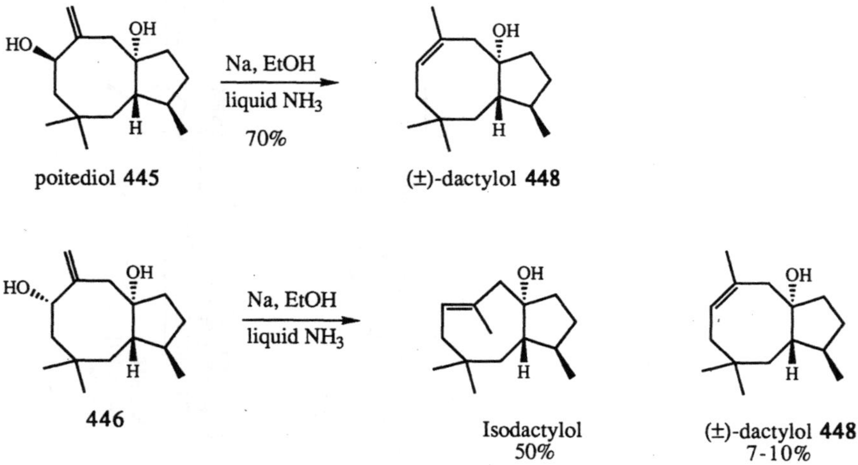

Scheme 74. Gadwood Synthesis of (±)-Dactylol and Isodactylol From Poietediol **445**

poitediol (**445**) (Scheme 73). Thus, treatment of poitediol (**445**) with sodium and ethanol in liquid ammonia results in the formation of dactylol (**448**) in 70% yield. Reductive elimination of 4-epipoitediol (**446**) under the same conditions gives only 10% yield of dactylol (**448**) along with isodactylol in 50% yield.

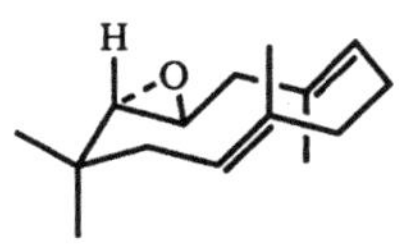

Humulene 9,10-epoxide
Conformer **450 A**

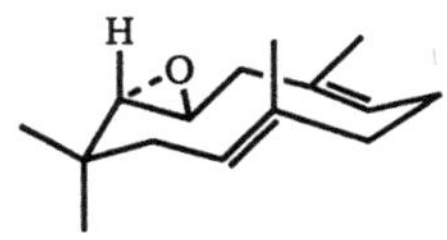

Humulene 9,10-epoxide
Conformer **450 B**

Biogenetically, dactylol **448** is thought to be derived from humulene **449**. A biomimetic synthesis of **448** beginning from humulene and proceeding via africanol (**418**) as a key intermediate has been developed by Shirahama [139] (Scheme 75). The synthetic strategy utilized by this group involves a conformationally selective transannular cyclization of humulene 9,10-epoxide. Two stable conformers of humulene are postulated with humulene 9,10-epoxide **450** thought to behave in a similar manner, giving rise to conformers **A** and **B**. Treatment of **450** with TMSOTf followed by desilylation gives alcohol **451** as a mixture of olefin isomers in 80% yield. Cyclization occurs preferentially via conformer **A**. Separation of the isomers on silver nitrate-impregnated silica affords **451A** in 27% yield. Conversion of **451A** to the bromide, and Birch reduction of the crude product affords tricyclic alkene **452** in 86% yield. Epoxidation of **452** gives epoxide **453** as the major isomer. Eliminative ring opening of the epoxide followed by hydrogenation of the newly formed allylic alcohol produces africanol **418** in 7 steps and 8% overall yield. Dehydration of africanol **418** produces a mixture of alkene isomers **454**, which, after epoxidation and separation give epoxide **455** in 42% yield. Treatment of **455** with five equivalents of boron trifluoride etherate initiates a "cyclopropane sliding rearrangement" with subsequent cyclopropane ring opening to provide alcohol **456** after further rearrangement (Scheme 76). Selective hydrogenation of **456** yields dactylol in 11 steps and 0.54% overall yield.

Paquette's formal synthesis [140] of dactylol **448** (Scheme 77) begins with ketone **458**, which is prepared in 54% yield from 4,4-dimethylcyclohexanone. Reaction of the lithium enolate of **458** with acetaldehyde followed by elimination affords unsaturated ketone **459**. Ketone **459** is reduced with sodium borohydride and subjected to an orthoester Claisen rearrangement to stereospecifically provide acid **460** after hydrolysis. The dactylol ring system is obtained as a mixture of three isomers by conversion of **460** to the acid chloride followed by Lewis-acid catalyzed cyclization. Under the conditions of thioketalization, the two isomers **461** and **462** are obtained in an 87:13 ratio. Raney nickel desulfurization, epoxidation and separation of isomers provides epoxide **464**.

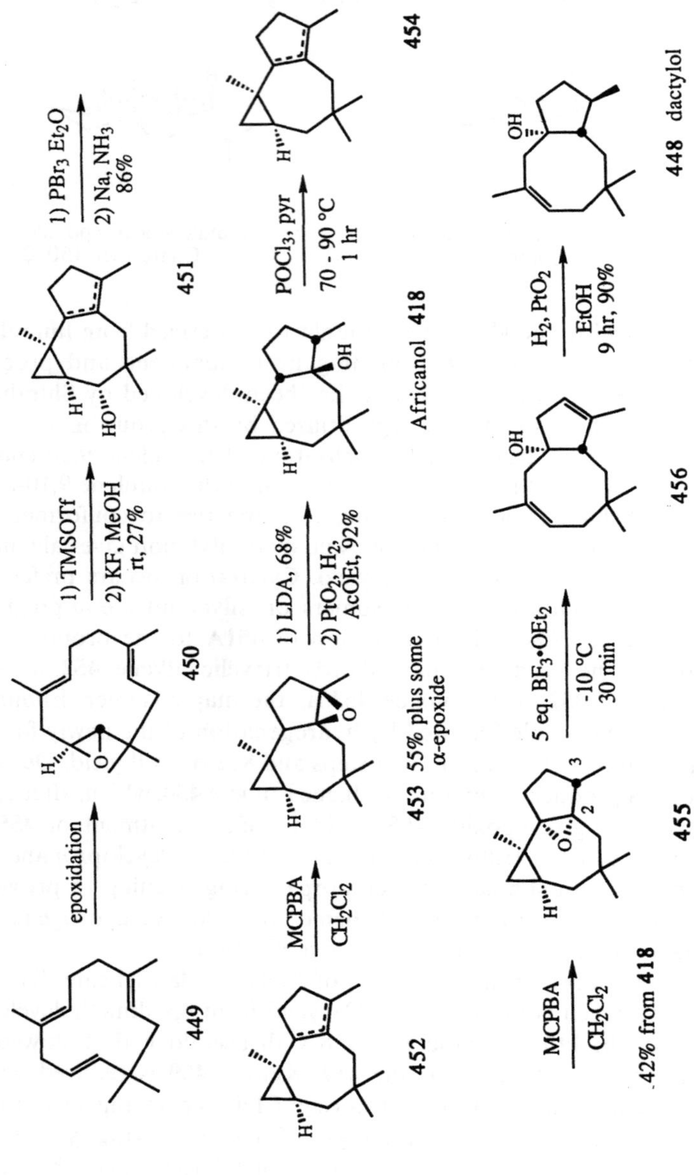

Scheme 75. Shirahama Conversion of Humulene to Dactylol

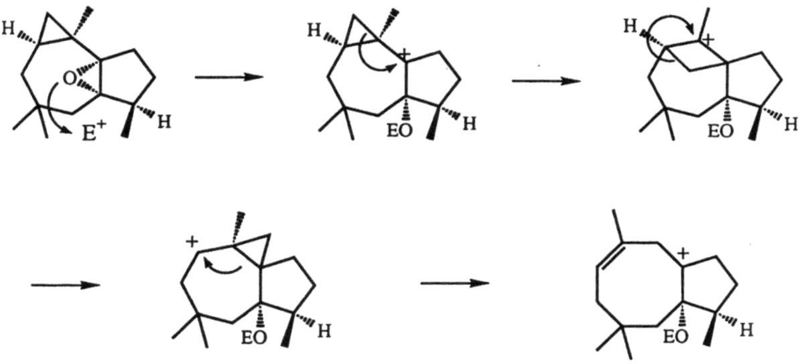

Scheme 76. "Cyclopropane Sliding Rearrangement" in Shirahama's Dactylol Synthesis

This completes the formal synthesis of dactylol, as **464** has previously been converted to dactylol **448** by Matsumoto.

The total synthesis of dactylol by Feldman [141] is shown in Scheme 78. The synthetic strategy can be divided into three distinct stages: preparation of the proper 2,4,7-trisubstituted tropone **468**, intramolecular tropone-alkene [6+2] photocycloaddition to obtain the bicyclo[6.3.0]undecane skeleton and re-functionalization of **470** to obtain dactylol **(448)**. The synthesis commences with the addition of sulfone-stabilized anion **466** to 4-methyltropone **(465)**, giving disubstituted tropone **467** in 51% yield along with 4% of the other regioisomer. Addition of the lithium anion of methyl phenyl sulfone to **467** affords trisubstituted tropone **468**. Irradiation of **468** gives an 11:1 mixture of stereoisomers at C9 **(469)** along with 12% of a tricyclic [8+2] adduct. The stereoselectivity of the photocycloaddition is postulated to occur during a stepwise reaction sequence involving a hydroxytropylium ion. Conversion of the tricyclic ketone **469** to the dactylol skeleton is achieved via Baeyer-Villiger oxidation of **469** with the strongly nucleophilic cumene hydroperoxide anion. The resulting lactone is reductively cleaved with LAH to diene-diol **470**. Selective hydrogenation of the disubstituted double bond, acylation of the primary alcohol and photochemical reduction of the resulting acetate provides dactylol **(448)** in 12 steps from 1,4-cyclohexadiene in 3% overall yield.

2.2.8 Triquinanes

2.2.8.1 $\Delta^{9(12)}$-Capnellene

The linear triquinane ($\pm$)-$\Delta^{9(12)}$-capnellene **(477)** is the parent hydrocarbon of a series of metabolites isolated from the soft coral *Capnella imbricata* [142]. This metabolite along with its oxygenated derivatives acts to inhibit the growth of microorganisms and the settlement of larvae. The parent hydrocarbon **477**

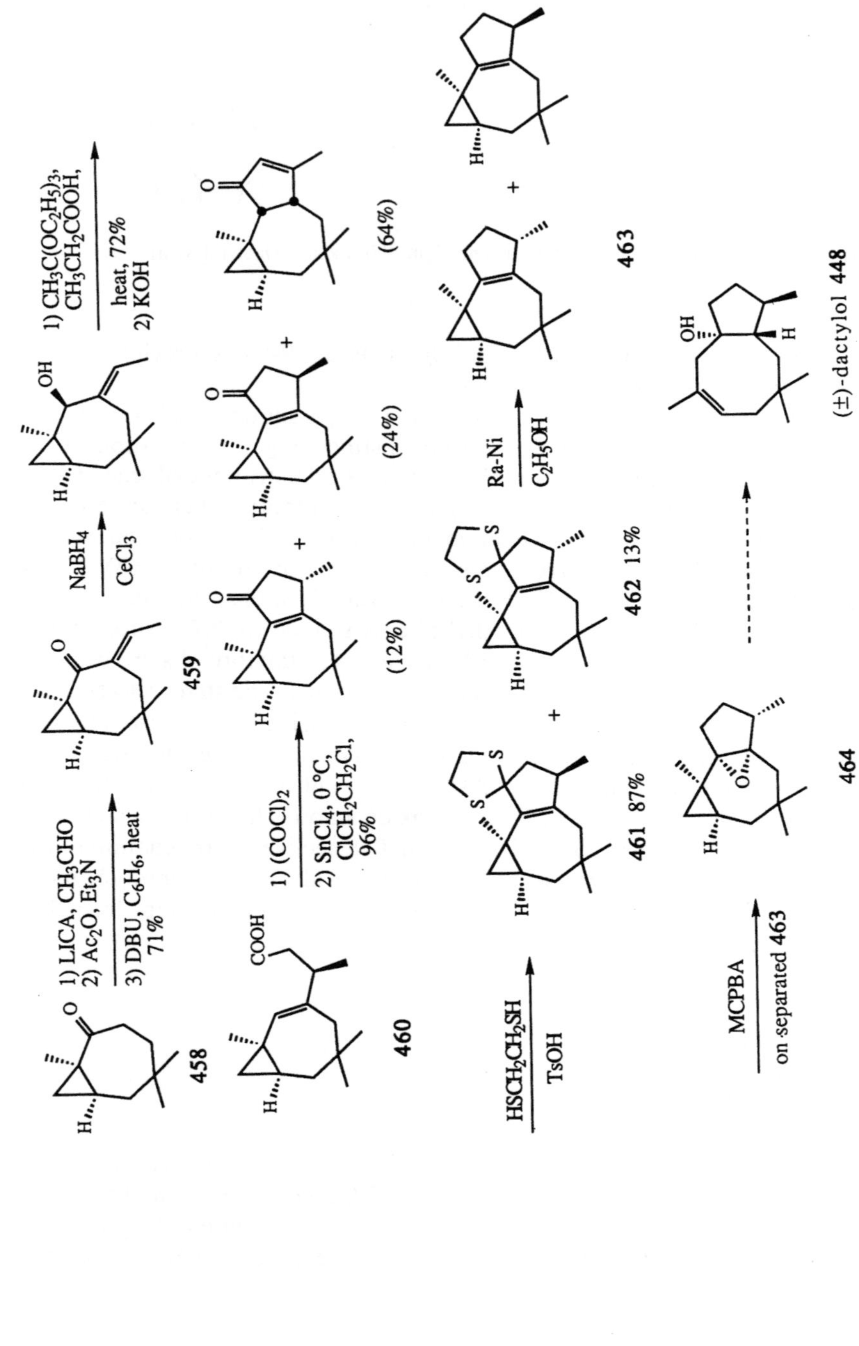

Scheme 77. Paquette Synthesis of (±)-Dactylol

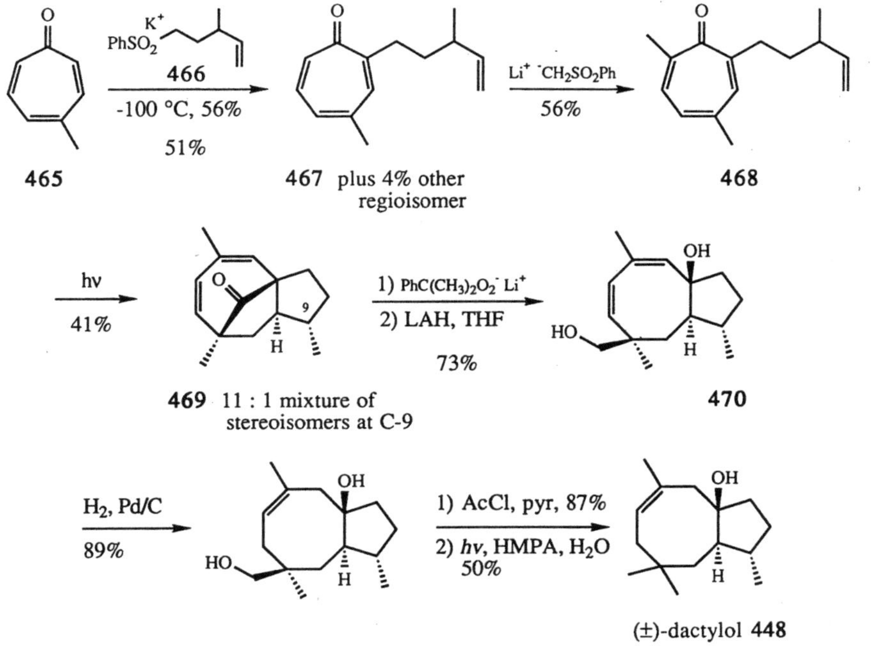

Scheme 78. Feldman Synthesis of (±)-Dactylol

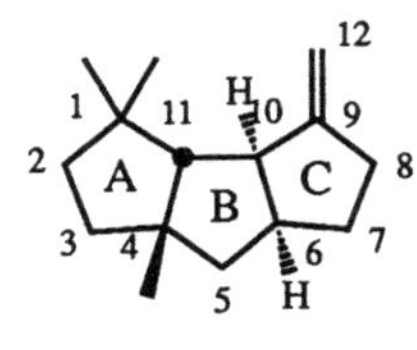

$\Delta^{9(12)}$-Capnellene **477**

having a tricyclo[6.3.0.0^{2,6}]undecane skeleton has been a particularly popular synthetic target for the development of five-membered ring annulation strategies. As the four chiral centers are present at the two ring fusions of the thermodynamically most stable *cis-anti-cis* configuration of the three rings, many synthetic strategies are simplified greatly. The relative stereochemistry of these centers will be a natural consequence of a variety of annulation tactics.

There have been thirteen total syntheses of **477** reported to-date. The strategies utilized for its synthesis fall into several general categories. The most common of these are sequential construction of the rings in a linear fashion (for example A → B → C) and construction of the rings outward from the central ring, as in the sequence B → AB → ABC. In both types of strategies are examples of both iterative processes and processes utilizing different types of methodology for each ring. There are also examples of both the simultaneous construction of all three rings and of the breaking down of more complex polycyclic systems to obtain the linear triquinane skeleton.

The Stille [143] synthesis of (±)-$\Delta^{9(12)}$-capnellene **(477)** (Scheme 79) is concise, giving **477** in only nine steps. The synthetic strategy involves application of an iterative three-carbon annulation process to a preformed A ring system to sequentially build up the B and C rings. This is accomplished by a palladium-catalyzed carbonylative coupling of vinyl triflate **472** with a vinyl stannane to provide divinyl ketone **473**, followed by Lewis acid-catalyzed silicon-directed Nazarov cyclization to give the A-B ring system **474** in 61% yield. After conjugate reduction and trapping to give vinyl triflate **475**, the annulation procedure is repeated to afford the *cis-anti-cis* linear quinane ring system **476** in 58% yield from **474**. Hydrogenation followed by Wittig olefination completes the synthesis to give (±)-$\Delta^{9(12)}$-capnellene in approximately 24% overall yield.

Oppolzer's [144] synthetic strategy for the preparation of (±)-$\Delta^{9(12)}$-capnellene **(477)**, shown in Scheme 80, follows an iterative strategy in which the A and B rings are constructed via sequential intramolecular "magnesium ene" reactions followed by an intramolecular aldol condensation to afford the C ring. This strategy is implemented by an intramolecular type-I-magnesium-ene reaction of allylic chloride **478**, followed by trapping of the δ,ε-unsaturated Grignard reagent with acrolein to provide allylic alcohol **479**. Chlorination with rearrangement gives allylic chloride **480**, which upon treatment with activated magnesium undergoes a second intramolecular magnesium-ene reaction to give **481** as a 3:2 mixture of diastereomers, thus forming the B ring. An oxidation-one carbon homologation-oxidation sequence followed by an intramolecular aldol reaction of the diastereomeric mixture of ketoaldehydes **482** gives the capnellene skeleton in the form of ketone **483** as a single diastereomer. The synthesis is completed by a selective hydrogenation-Wittig olefination sequence to give (±)-$\Delta^{9(12)}$-capnellene **(477)** in 10 steps and 4.8% overall yield.

Dreiding [145] (Scheme 81) takes advantage of the lack of functionality in **477** in devising an iterative strategy utilizing alkynone cyclizations via flash vacuum pyrolysis to prepare the B and C rings sequentially. The needed alkynone is prepared from **485** by conjugate reduction with "copper hydride" complex followed by hydrolysis and chlorination to give an acid chloride. Acylation of bis(trimethylsilyl)acetylene with the acid chloride derived from nitrile **486** accompanied by hydrodesilylation affords the desired alkynone in 60% yield from the unsaturated nitrile **485**. Flash vacuum pyrolysis of **487** (620 °C) results in formation of the A-B ring system in 78% yield. Standard chemistry led from **488** to **489**. Conversion to the alkynone as before followed by pyrolysis gives the linear quinane **490** in 33% yield along with 44% of the undesired angularly fused quinane. Hydrogenation and Wittig olefination of **490** completes the synthesis to give (±)-$\Delta^{9(12)}$-capnellene **(477)** in 0.68% yield over 17 steps.

Piers [146] utilizes a novel cuprate as a d^2, a^4 synthon in an iterative strategy involving a B → AB → ABC process for construction of the quinane ring system

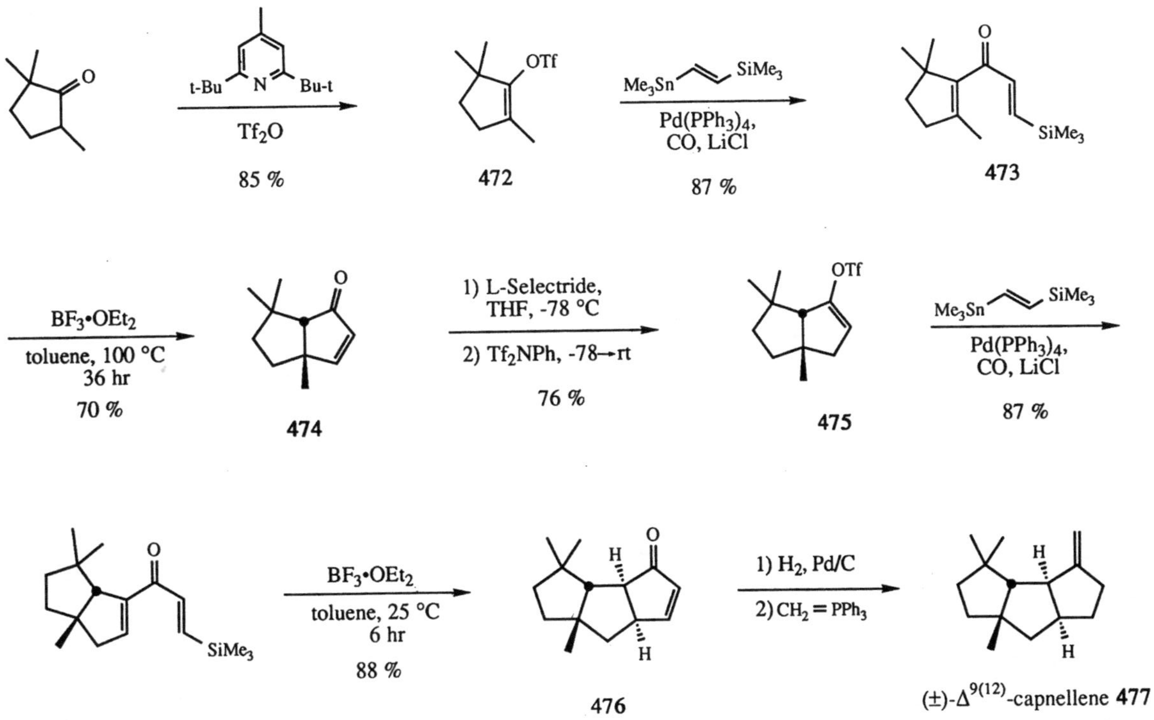

Scheme 79. Stille Synthesis of (±)-$\Delta^{9(12)}$-Capnellene

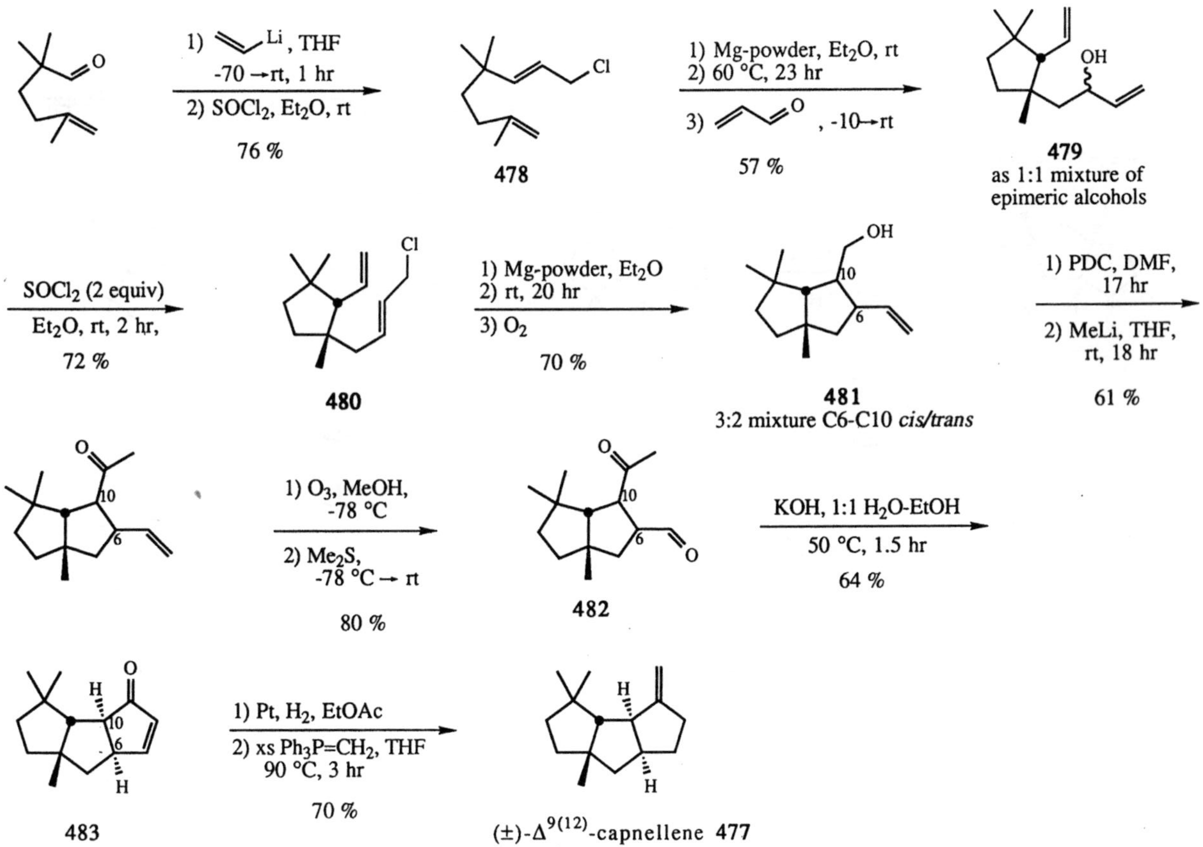

Scheme 80. Oppolzer Synthesis of (±)-$\Delta^{9(12)}$-Capnellene

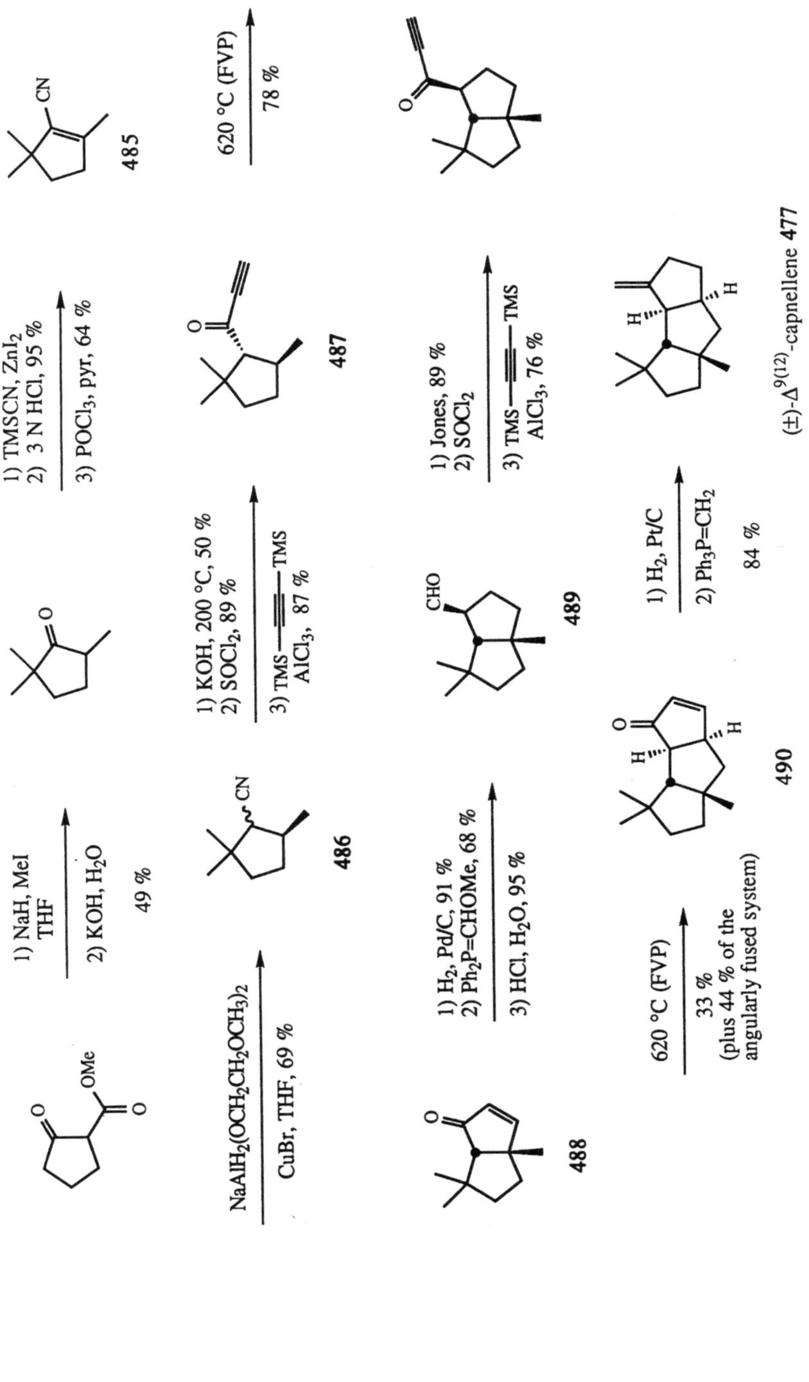

Scheme 81. Dreiding Synthesis of (±)-$\Delta^{9(12)}$-Capnellene

(Scheme 82). Lewis acid-catalyzed conjugate addition of the cuprate formed from **491** to methylcyclopentenone **(492)** followed by intramolecular alkylation yields the AB ring system in 60% yield. Formation of the gem-dimethyl moiety occurs by "protection" of the carbonyl as the alcohol, cyclopropanation and hydrogenolysis followed by oxidation to give ketone **493**. Conversion of **493** to the unsaturated ketone **494** allows the reapplication of the cuprate addition-intramolecular alkylation methodology to give **495** in 51% yield from ketone **494**. Removal of the carbonyl by reduction, xanthate formation and tri-*n*-butyltin hydride reduction to the hydrocarbon completes the synthesis to afford ($\pm$)-$\Delta^{9(12)}$-capnellene **(477)** in 12 steps with an overall yield of 10%.

The strategy followed by Paquette [147] involves stepwise construction of the B ring onto a pre-existing A ring via Nazarov cyclization followed by addition of the C ring by an intramolecular aldol condensation (Scheme 83). Addition of vinylmagnesium bromide to aldehyde **497** followed by allylic oxidation provides the divinylketone **498**. Treatment with acid initiates the Nazarov cyclization to give **499**, which upon reaction with lithium dimethylcuprate gives the A-B ring system in four steps. Two carbon homologation by addition of lithium acetylide followed by acid-catalyzed rearrangement of the acetylenic alcohol yields the α,β-unsaturated ketone **500**. The C ring is obtained by copper-catalyzed 1,4-addition of vinylmagnesium bromide to **500** followed by ozonolysis and intramolecular aldol condensation to give ketone **502**. Selective hydrogenation and Wittig olefination completes the synthesis to yield racemic **497** in 1.9% yield and 13 steps.

An alternative strategy to the iterative approaches involves adding the A and C rings to a preformed B ring. An interesting application of this strategy by Curran [148] involves a tandem radical cyclization to form both rings in a single step (Scheme 84). The preliminary steps in the synthesis involve formation of the central ring with proper stereochemistry. This is followed by refunctionalization of the side chains to prepare for cyclization and then tandem cyclization in the final step. The synthesis is initiated by $SN^{2'}$ anti opening of lactone **503** then reclosure via iodolactonization and elimination to give lactone **504**. Re-opening of **504** with Grignard reagent **505** gives the B ring with the needed relative stereochemistry. LAH reduction of the carboxyl functionality, conversion of the hydroxyl to the iodide and treatment with lithium acetylide affords the side chain needed for formation of the C ring in 34% yield. The *pro*-A ring sidechain is prepared via standard transformations in 63% yield from **506** to give bromide **507**. Treatment of **507** with tri-*n*-butyltin hydride initiates the radical cyclization process to produce **477** in 13 steps and 10% overall yield.

Another interesting approach to **477** is that of Mehta [149] (Scheme 85). The cornerstone of this strategy involves formation of all three rings simultaneously to give the linear quinane diketone **511** in three steps. At this point all that is needed is refunctionalization and isomerization to obtain **477**. This strategy is initiated by two successive pericyclic reactions. Diels-Alder cycloaddition of the methylcyclopentadiene isomers **509** with *p*-benzoquinone affords the tricyclic adduct **510** in 70% yield. Photochemical [2 + 2] intramolecular cycloaddition

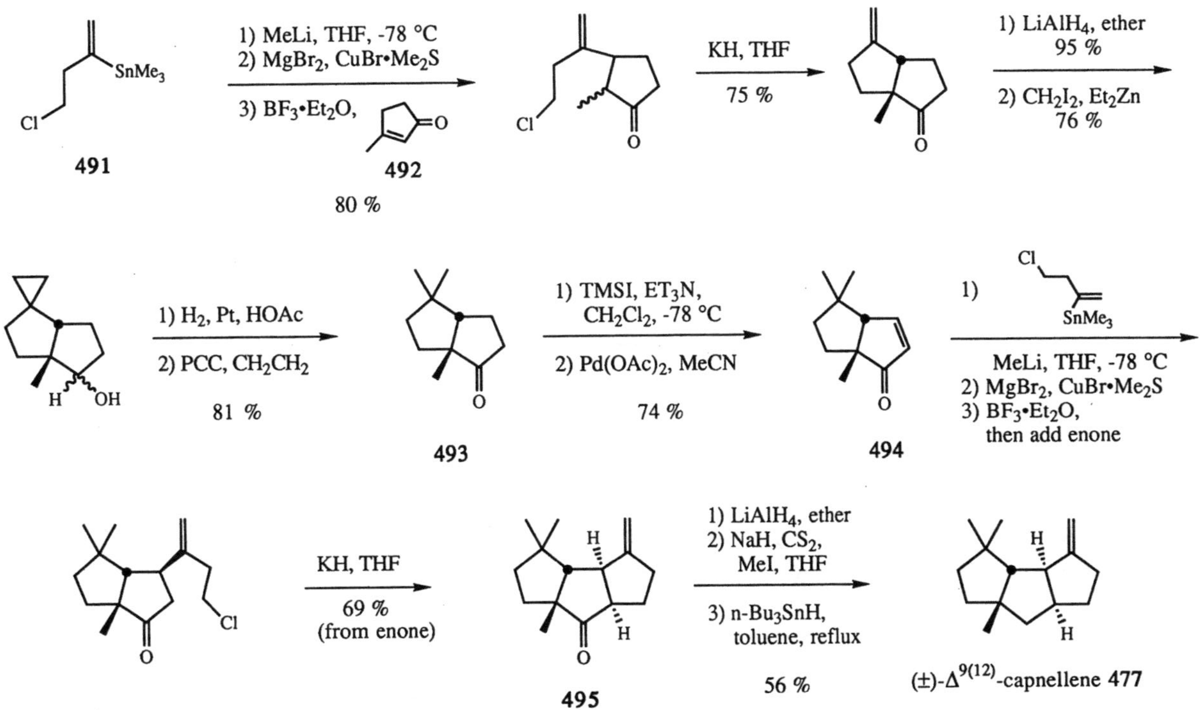

Scheme 82. Piers Synthesis of (±)-$\Delta^{9(12)}$-Capnellene

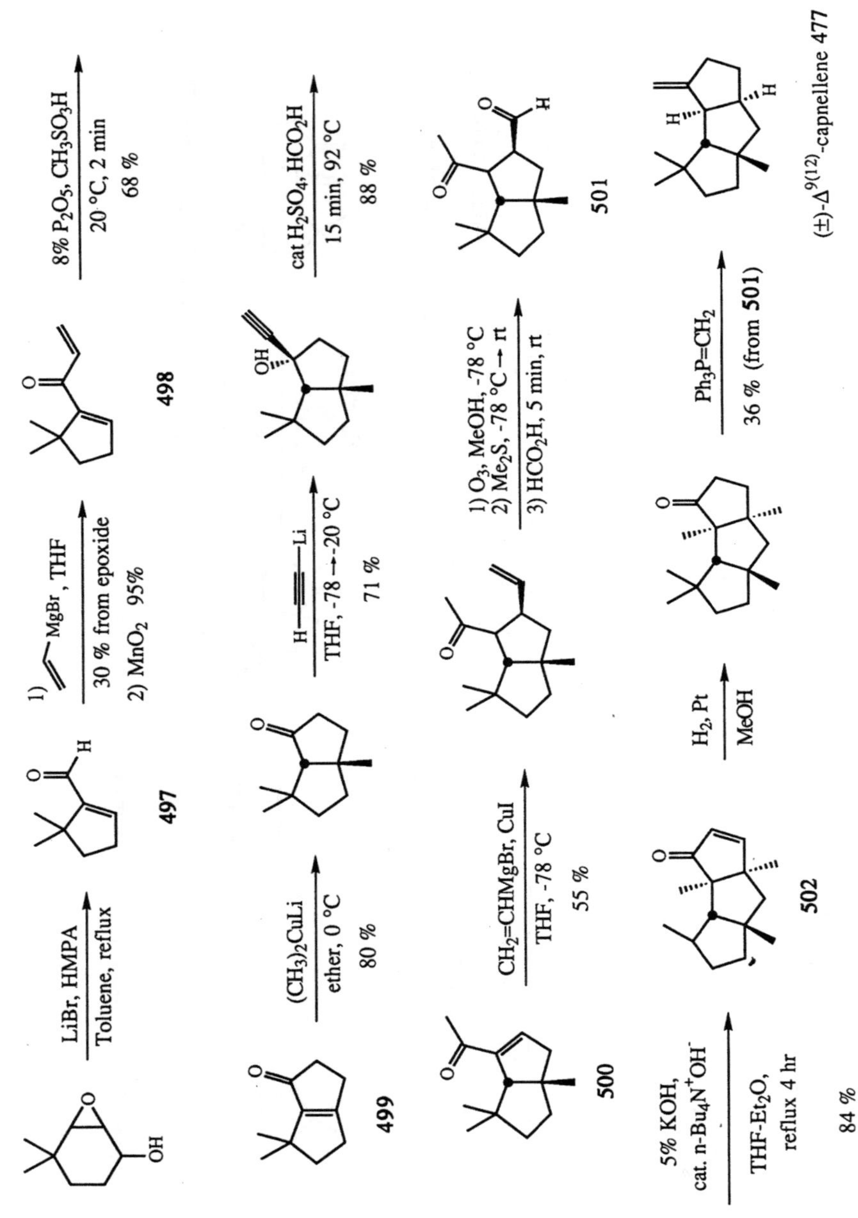

Scheme 83. Paquette Synthesis of (±)-$\Delta^{9(12)}$-Capnellene

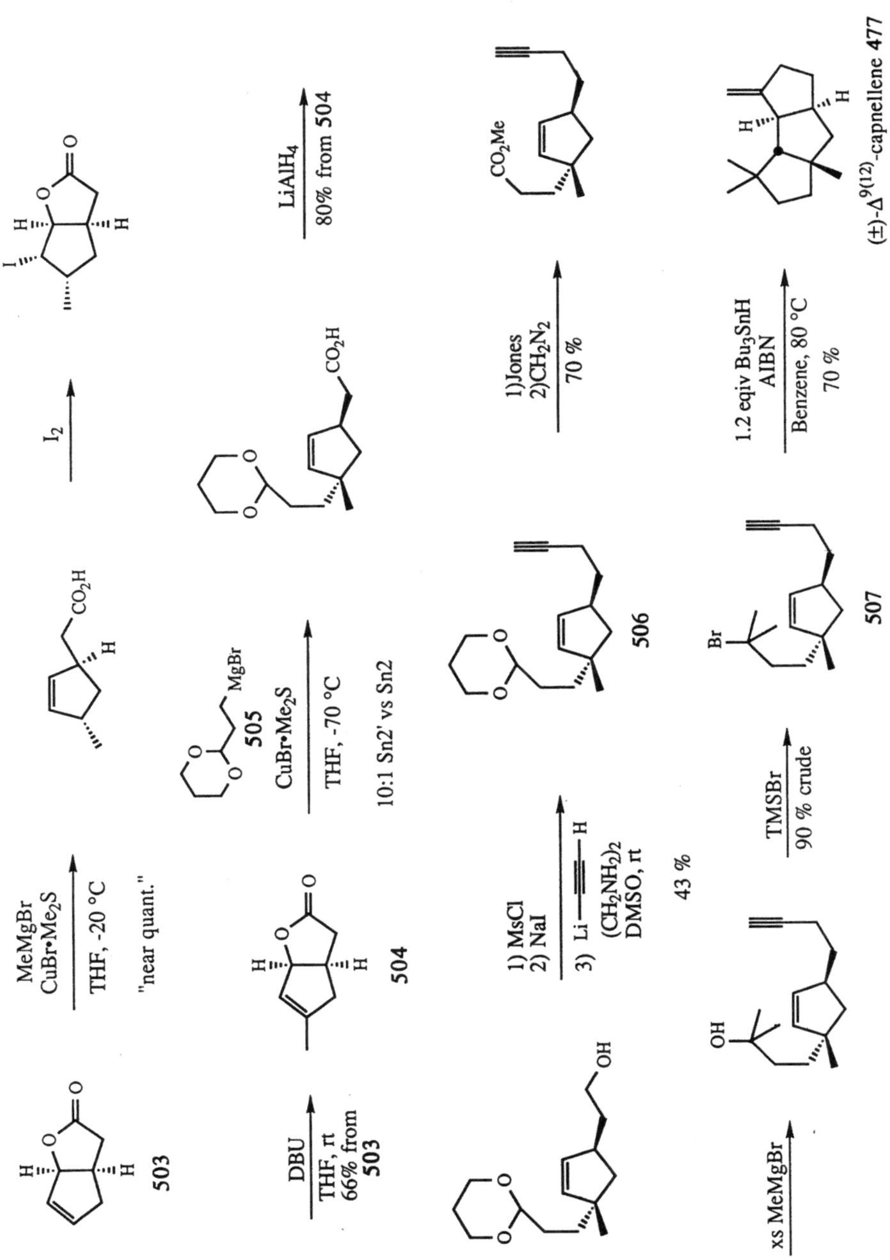

Scheme 84. Curran Synthesis of (±)-$\Delta^{9(12)}$-Capnellene

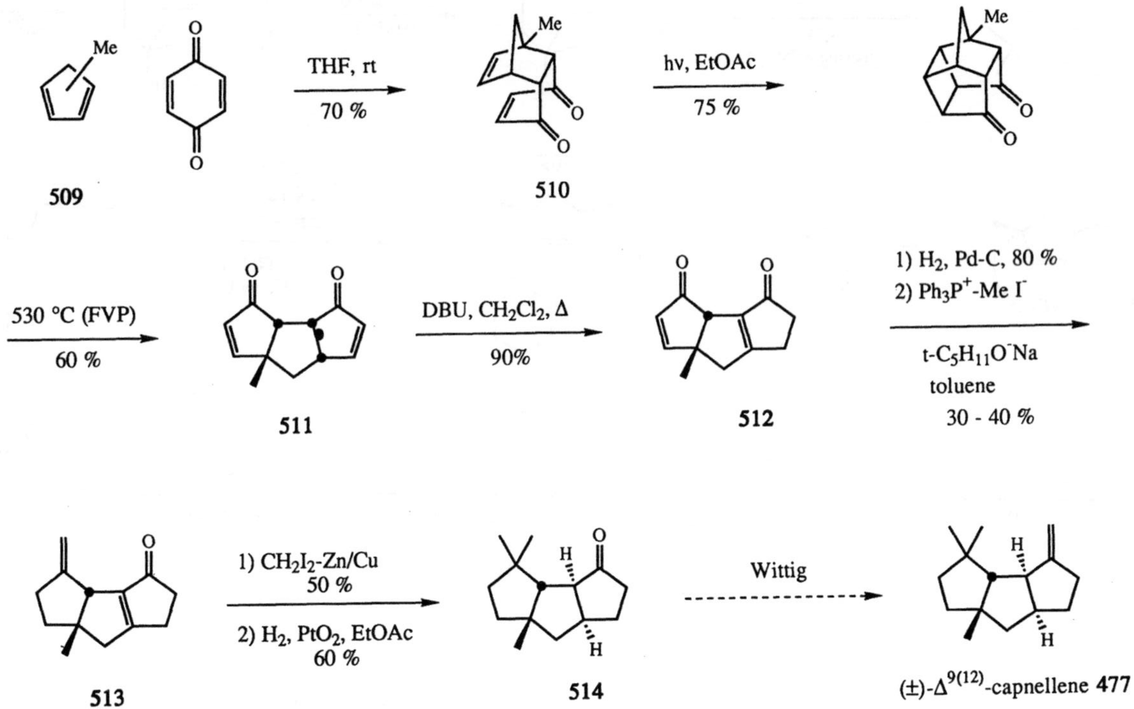

Scheme 85. Mehta Synthesis of $(\pm)$-$\Delta^{9(12)}$-Capnellene

followed by thermal cyclobutane fragmentation (530 °C) gives the *cis-syn-cis* triquinane **511** in 32% yield from **510**. Base catalyzed isomerization to give a more substituted double bond **512** followed by selective hydrogenation and Wittig olefination yields **513**. The gem-dimethyl functionality is introduced by cyclopropanation-hydrogenation to give ketone **514**, which has been converted to (±)-$\Delta^{9(12)}$-capnellene **(477)** in several previous syntheses. The entire process is completed in 9 steps.

In a second approach by Mehta [150] shown in Scheme 86, the intermediate **512** is converted to **477** by a different pathway. In this pathway the triquinane ring system **512** is refunctionalized using a different approach. The gem-dimethyl functionality is introduced by addition of methylmagnesium iodide followed by Dauben oxidative transposition to give rearranged diketone **515**. Controlled reduction with sodium borohydride results in initial 1,4-reduction of the tetrasubstituted olefin followed by reduction of the carbonyl to give alcohol **516**. Lewis acid-catalyzed cuprate addition to **516** followed by PCC oxidation and selective Wittig olefination of the less hindered ketone produces **517**. Lithium aluminium hydride reduction, xanthate formation and reduction with tri-*n*-butyltin hydride removes the carbonyl functionality to complete the synthesis of **477** in 12 steps and 1.5% yield.

Another approach that involves the simultaneous preparation of all three rings is that developed by Little [151] (Scheme 87) in the first synthesis of (±)-$\Delta^{9(12)}$-capnellene **(477)**. The key step involves an intramolecular trapping reaction of the 1,3-diyl **523** formed from **522** upon thermolysis (Scheme 88). Preparation of diazene **522** is initiated by a reduction-oxidation sequence of acid **519** to give the aldehyde followed by addition of cyclopentadienyllithium. Dehydration occurs *in situ* to afford fulvene **520**. Diels-Alder reaction of **520** with dimethyl azodicarboxylate followed by selective diimide reduction of the newly formed double bond gives **521**. Saponification and oxidation of **521** gives the needed diazene **522**. Thermolysis with intramolecular trapping of the 1,3-diyl **523** (Scheme 88) yields the linear triquinane ring system **524** (along with two other isomers). The synthesis is completed in short order by sequential hydroboration-double oxidation to give ketone **525**, and Wittig olefination resulting in (±)-$\Delta^{9(12)}$-capnellene **(477)** in 10 steps and 3.2% yield.

A formal synthesis of **477** by Oda [152] involves rapid initial construction of the A-B ring system followed by attachment of the C ring onto the A-B ring system (Scheme 89). The A-B ring system is obtained by a ring expansion/ring contraction process via rearrangement and concomitant deoxygenation of the bicyclo[4.2.0]octane skeleton **526** with TMSI to afford the α,β-unsaturated ketone **527**. Activation of **527** by conversion to the β-diketone **528** facilitates Michael addition to ethyl acrylate to give **529**. Cuprate addition followed by titanium induced intramolecular coupling of the ketone and ester carbonyls leads to the formation of ketone **530**, which has been converted to (±)-$\Delta^{9(12)}$-capnellene **(477)** in several previous syntheses. Most noteworthy is the fact that the natural product is obtained in only 7 steps from commercially available starting materials.

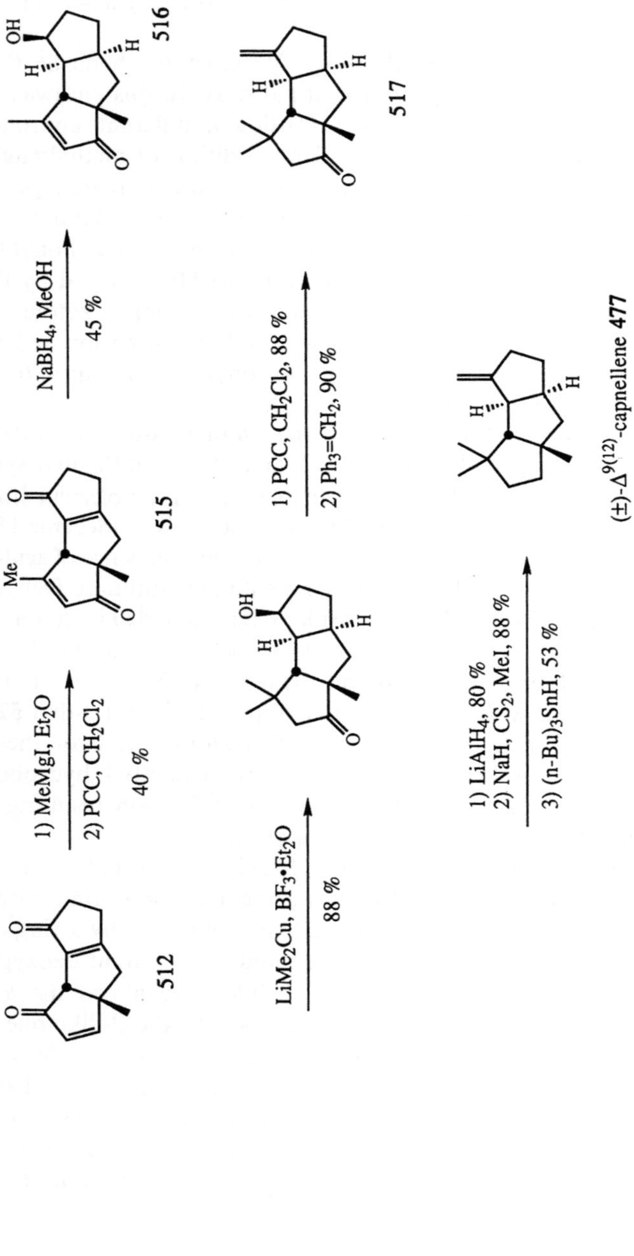

Scheme 86. Second Mehta Synthesis of (±)-$\Delta^{9(12)}$-Capnellene

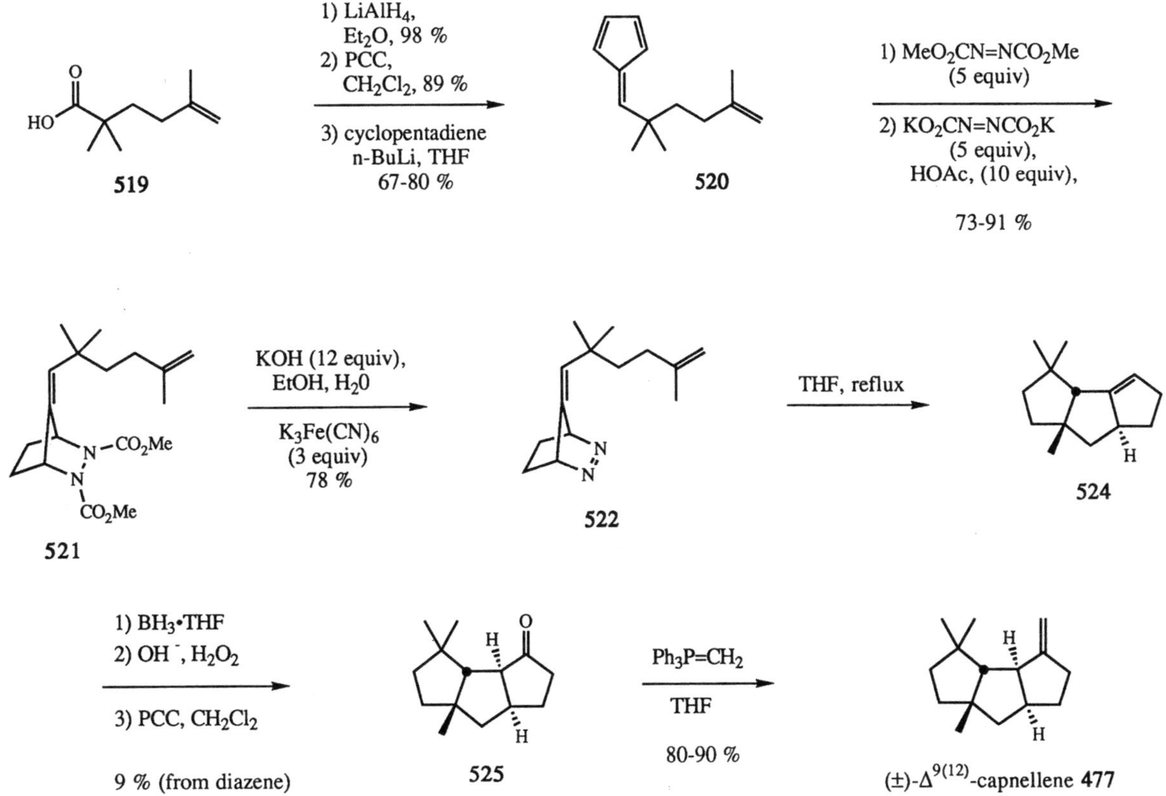

Scheme 87. Little Synthesis of $(\pm)$-$\Delta^{9(12)}$-Capnellene

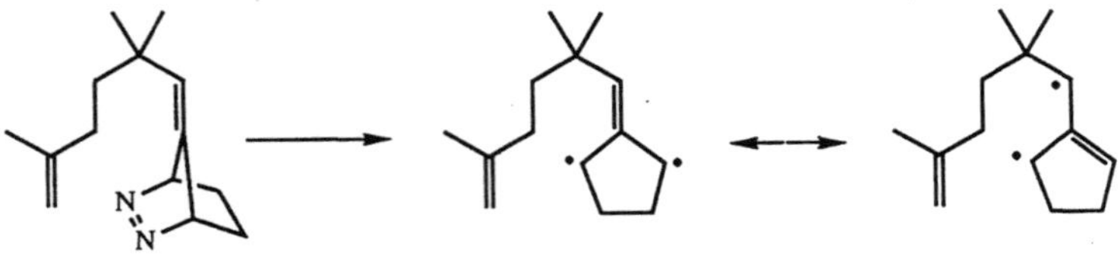
522 523

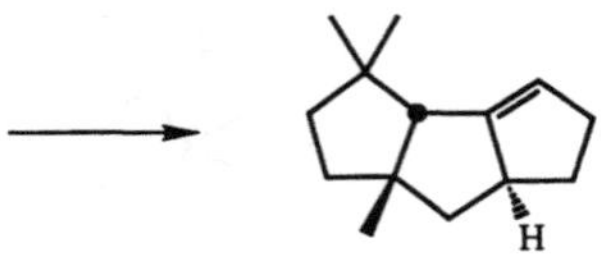
524

Scheme 88. Mechanism of Diyl Trapping

A somewhat longer synthesis of **477** by Liu is shown in Scheme 90 [153]. The overall strategy involves building the B ring onto the C ring via a ring expansion process. The A ring is appended to the B-C system as a six-membered ring followed by a formal contraction process. The synthesis is initiated by a [2+2] photocycloaddition to give **532**. Conversion of **532** to **533** followed by treatment with ethyl diazoacetate in the presence of $BF_3 \cdot Et_2O$ preferentially forms the ring expanded keto-ester **534** as the major product. Formation of the unsaturated keto-ester followed by Lewis acid-catalyzed Diels-Alder reaction with isoprene forms the single adduct **535**. The required methyl group at the A-B ring juncture is inserted by reduction of the ester with RED-A1, formation of the monothiocarbonate and treatment with tri-*n*-butyltin hydride to afford **536**. Deoxygenation of alcohol **536** to give **537** followed by ozonolysis yields **538** where cyclization has occurred during the reductive work-up. The gem dimethyl functionality is inserted by Beckmann rearrangement to give ketone **539**. Wittig olefination and cyclopropanation provides **540** which is converted to **541** in 36% yield by hydrogenation, deprotection and oxidation. Wittig olefination completes the synthesis of **477** in 21 steps, giving an overall yield of 0.85%.

The strategy developed by Grubbs [154] involves formation of the A-B ring system, and all four stereocenters in a single step in the form of a bridged tricyclic system (Scheme 91). The C ring is obtained via rearrangement through the intermediacy of a tetracyclic metallacycle. The synthesis is initiated by treatment of the lactol of **542** with a Wadsworth-Emmons reagent resulting in condensation with the aldehyde followed by intramolecular conjugate addition to give **543**. The ether is opened and tosylated to give the unsaturated ester **544**. Displacement of the tosylate with cyclopentadienylmagnesium chloride followed by intramolecular Diels-Alder reaction gives **545** in which all four stereocenters of the final product are set in a single step. Reaction of **545** with the Tebbe reagent produces titanacyclobutane **546**. Upon heating, **546** ring-opens to

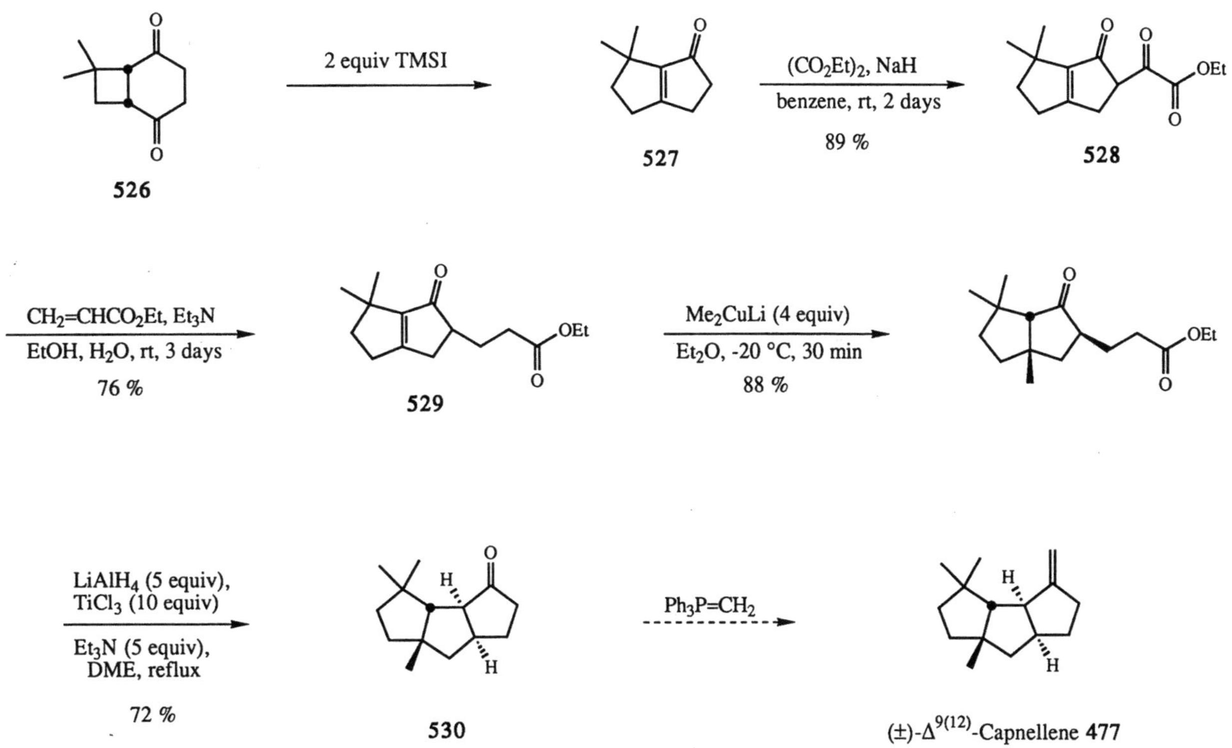

Scheme 89. Oda Formal Synthesis of (±)-$\Delta^{9(12)}$-Capnellene

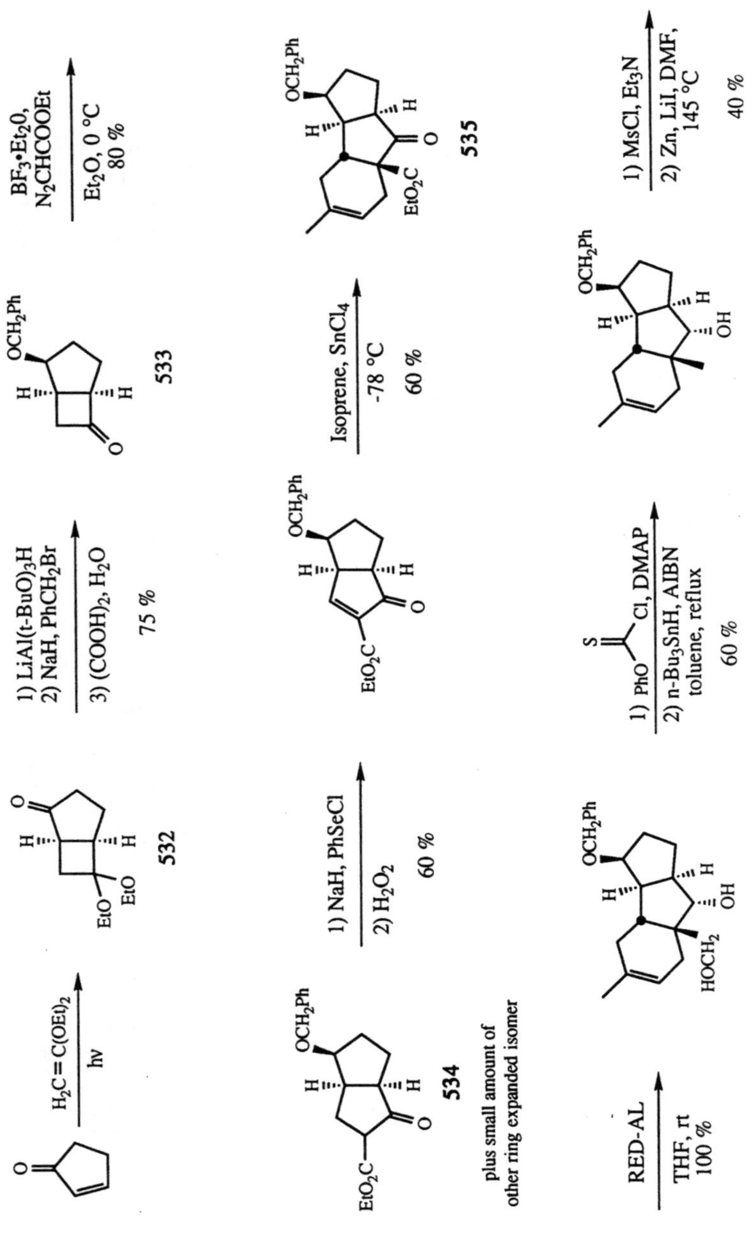
H2C=C(OEt)2
hv
EtO
EtO
H
H
O
532
1) LiAl(t-BuO)3H
2) NaH, PhCH2Br
3) (COOH)2, H2O
75 %
OCH2Ph
533
BF3•Et20,
N2CHCOOEt
Et2O, 0 °C
80 %
EtO2C
534
plus small amount of
other ring expanded isomer
1) NaH, PhSeCl
2) H2O2
60 %
Isoprene, SnCl4
-78 °C
60 %
535
RED-AL
THF, rt
100 %
HOCH2
OH
1) PhO Cl, DMAP
S
2) n-Bu3SnH, AIBN
toluene, reflux
60 %
536
1) MsCl, Et3N
2) Zn, LiI, DMF,
145 °C
40 %

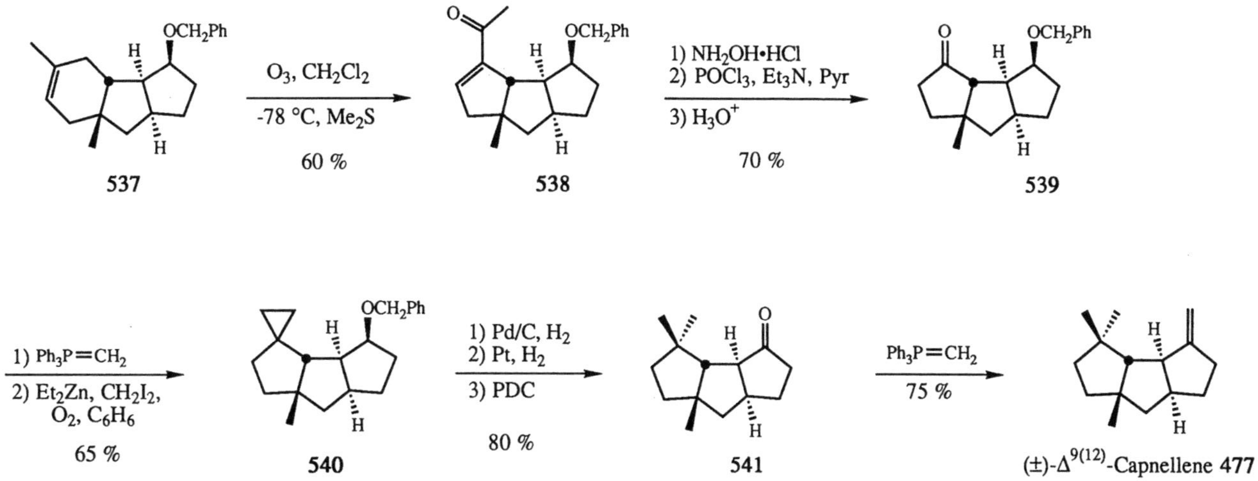

Scheme 90. Liu Synthesis of (±)-$\Delta^{9(12)}$-Capnellene

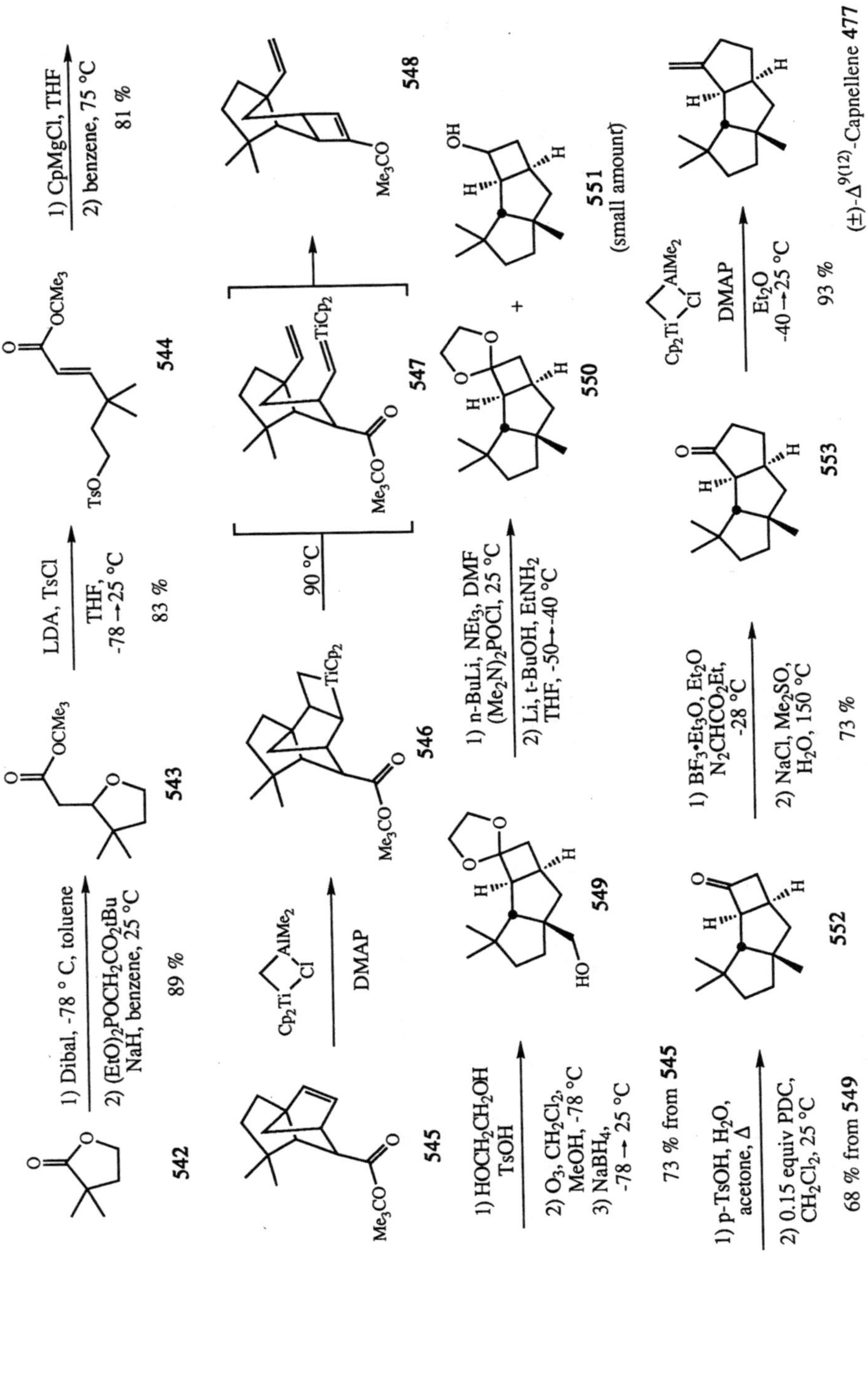

Scheme 91. Grubbs Synthesis of $(\pm)$-$\Delta^{9(12)}$-Capnellene

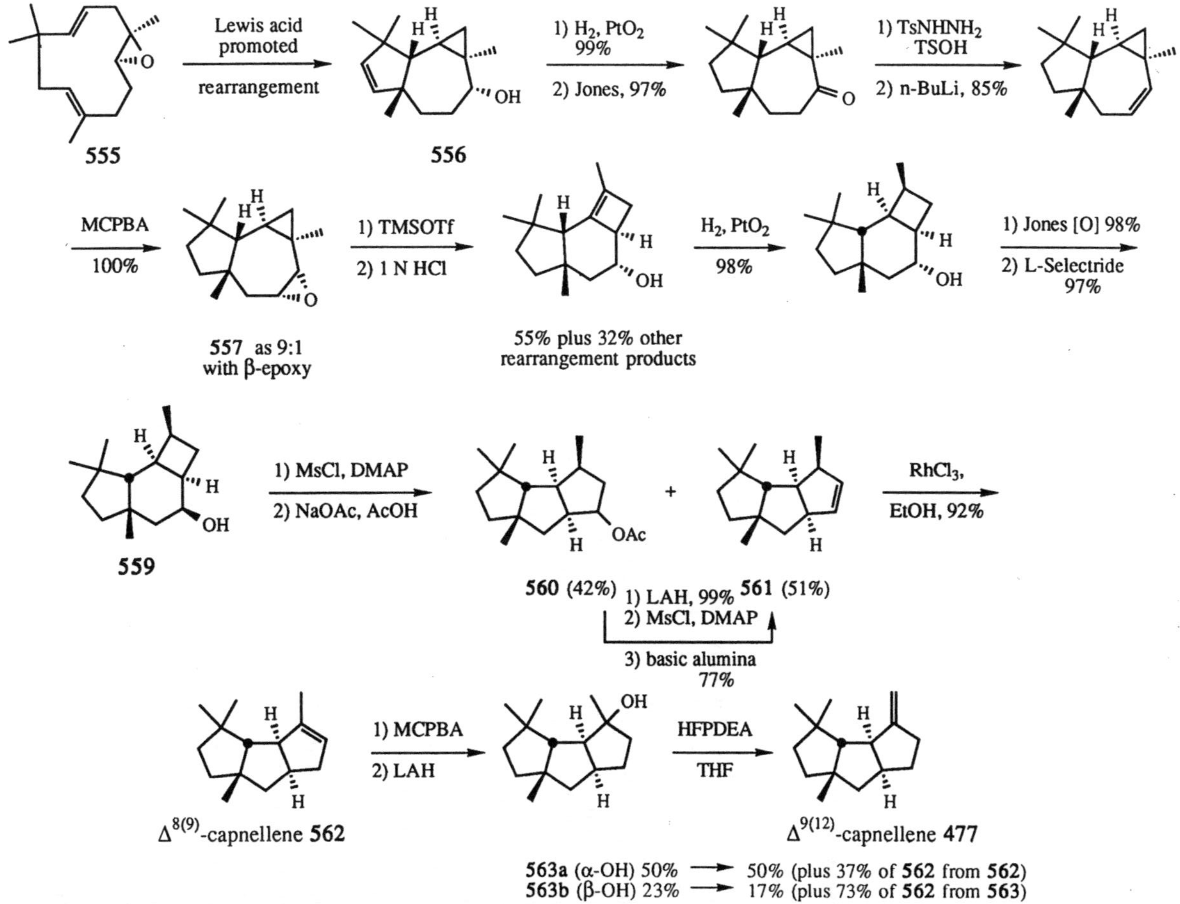

Scheme 92. Shirahama Conversion of Humulene $\Delta^{9(12)}$-Capnellene

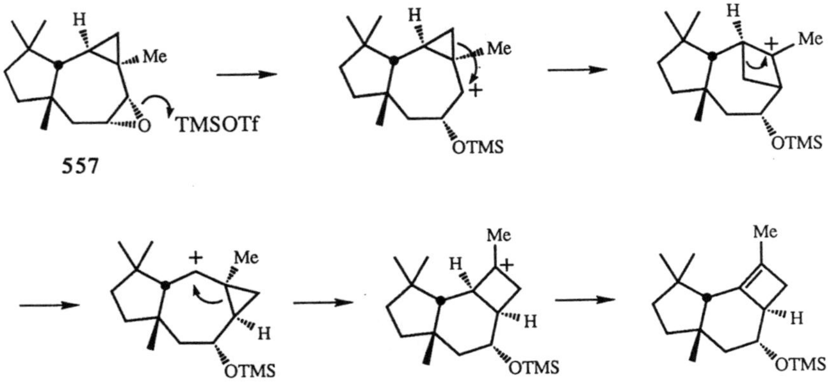

Scheme 93. Rearrangement of **557** to **558** via Cyclopropane "Sliding" Reaction

the alkylidene **547** which is trapped in an intramolecular fashion to provide the cyclobutenyl ether **548**. Hydrolysis and ketalization yields a ketal. The methyl at the A-B ring fusion is obtained via cleavage of the olefin and reduction to the alcohol **549** followed by dissolving metal reduction of the tetramethylphosphorodiamidate ester of the alcohol to afford **550** along with a small amount of alcohol **551**. Ring expansion of **552** by treatment of with ethyl diazoacetate in the presence of $BF_3 \cdot Et_2O$ produces ketone **553** which is converted to (±)-$\Delta^{9(12)}$-capnellene **(477)** by treatment with the Tebbe reagent. An overall yield of approximately 20% is obtained over 14 steps.

Shirahama's strategy [155] for the construction of $\Delta^{9(12)}$-capnellene **(477)** begins with the 11-membered ring of humulene 6,7-epoxide **(555)** and gives the A-ring via rearrangement to a fused 5-7-3 tricyclic ring system **556** (Scheme 92). Further rearrangements convert **556** to a 5-6-4 system via a cyclopropane sliding reaction, and then finally to the linear triquinane system. Lewis acid promoted rearrangement of humulene 6,7-epoxide affords alcohol **556**. Conversion of **556** to epoxide **557** proceeds in very good yield using standard transformations. In a key rearrangement, involving a cyclopropane sliding reaction (Scheme 93), **557** is transformed into cyclobutene **558**. A final rearrangement involving alcohol **559** gives the linear triquinane system as a mixture of acetate **560** and alkene **561**. Conversion of the acetate to the alkene gives a total yield of 83% for the transformation of **559** to **561**. Isomerization of the double bond to the exocyclic position occurs in four steps. The final transformation involves the dehydration of the two hydroxyl diastereomers **563a** and **563b** by treatment with 1:1 mixture of 1,1,2,3,3,3-hexafluoropropyl diethyl amine and 1,2,3,3,3-pentafluoropropenyl diethyl amine to give a combined yield of 67% of $\Delta^{9(12)}$-capnellene **(474)**. A yield of approximately 14% is obtained over 14 steps.

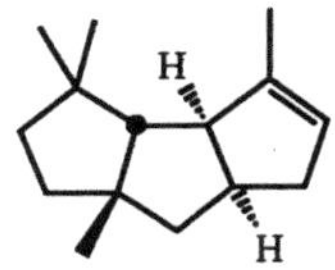

$\Delta^{8(9)}$-Capnellene **568**

2.2.8.2 $\Delta^{8(9)}$-Capnellene and $\Delta^{9(12)}$ capnellenes

The only other synthesis of a non-oxygenated capnellene is Pattenden's synthesis [156] of ($\pm$)-$\Delta^{8(9)}$-capnellene (Scheme 94). The general strategy differs from those normally used for the capnellenes in that the C ring is formed first, followed by the simultaneous formation of the A-B rings. This is accomplished by the formation of an angularly fused tricyclic 6,4,5 ring system which is subjected to ring expansion, functionaliztion and ring contraction. The tricyclic ketone **564** is obtained through an intramolecular photochemical [2+2] cyclization. Dimethylation provides the gem-dimethyl moiety at C1 of what will eventually be the A ring. Basic hydrolysis of the ester **565** initiates a retro-aldolization to give the C ring fused to an 8-membered diketone. Protection of the less hindered ketone followed by reduction, elimination and deprotection gives **566** in good yield. Wittig olefination of the remaining ketone followed by $RhCl_3$ isomerization of the exocyclic olefin to an endocyclic position forms *epi*-precapnelladiene **567**. Treatment of **567** with $BF_3 \cdot Et_2O$ results in a biogenetically patterned transannular cyclization to provide **568** in 13 steps and 3% overall yield.

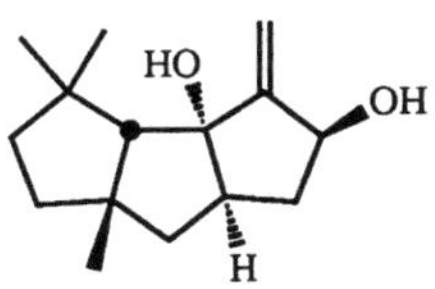

($\pm$)-$\Delta^{9(12)}$-Capnellene-8β,10α-diol **577**

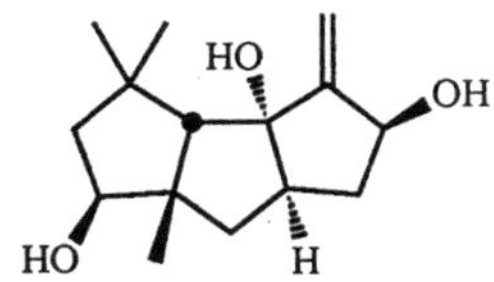

($\pm$)-$\Delta^{9(12)}$-Capnellene-3β,8β,10α-triol **583**

The strategies followed for the syntheses of the oxygenated capnellenes are similar to those used in the preparations of the parent hydrocarbon. Slight modifications are usually required to allow for the introduction of the hydroxyl functionalities. The oxygenated capnellenes that have been synthesized are ($\pm$)-$\Delta^{9(12)}$-capnellene-8β, 10α-diol **(577)** and ($\pm$)-$\Delta^{9(12)}$-capnellene-3β,8β,10α-triol **(583)**.

The pathway followed by Shibasaki [157] in the synthesis of racemic **577** and **583** is flexible in that a slight modification at an early stage of the synthesis allows for the preparation of both metabolites (Schemes 95 and 96). The general strategy involves building the rings in a sequential fashion in the order B → A-B → A-B-C. Oxygenation at C8 is inserted with construction of the C

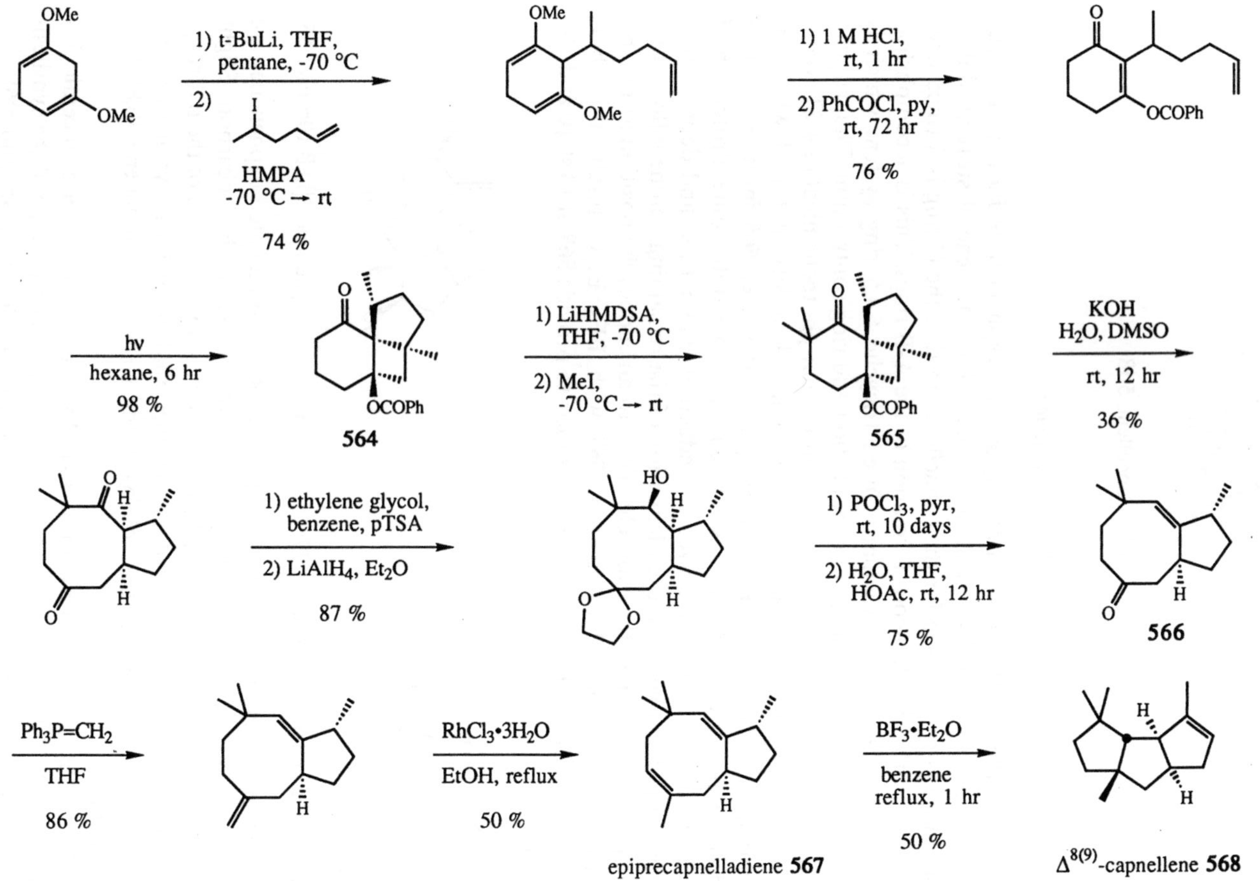

Scheme 94. Pattenden Synthesis of (±)-$\Delta^{8(9)}$-Capnellene

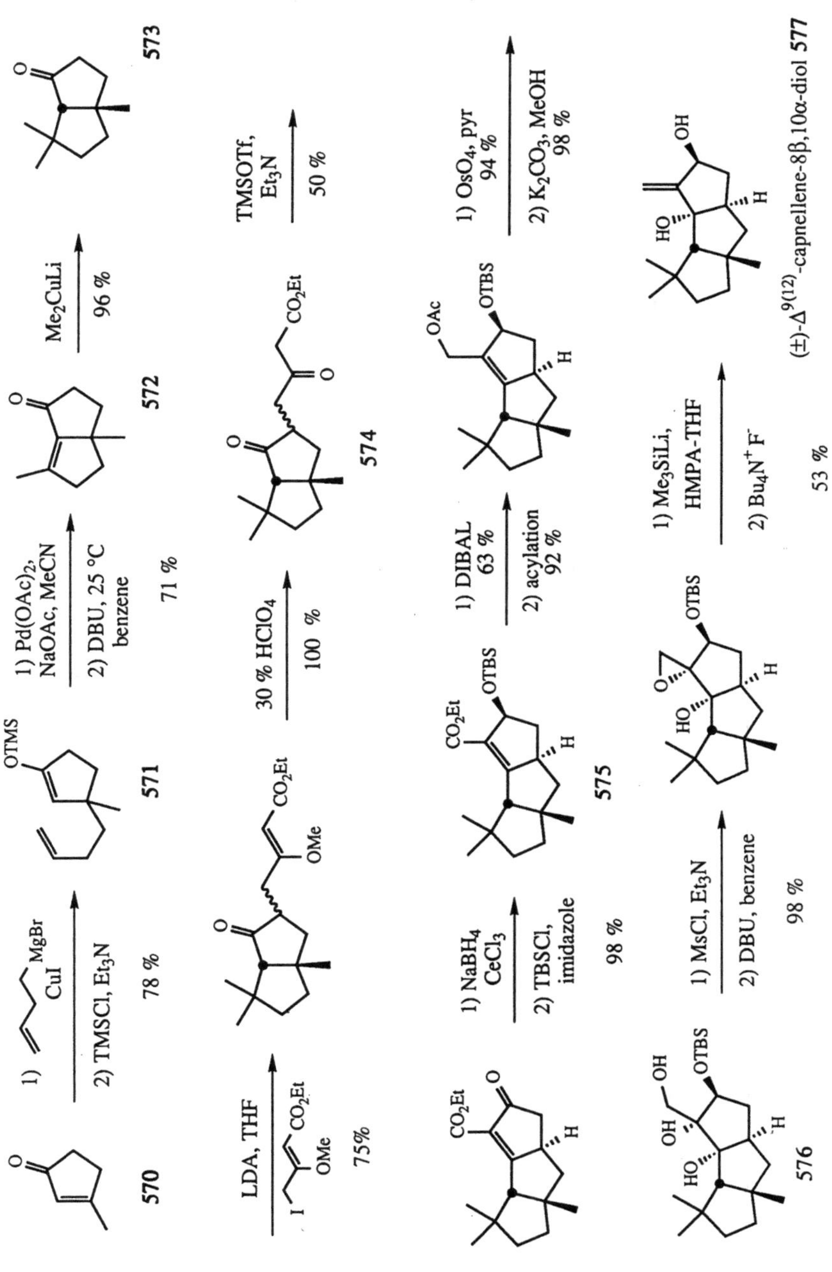

Scheme 95. Shibasaki Synthesis of (±)-$\Delta^{9(12)}$-Capnellene-β,10α-diol

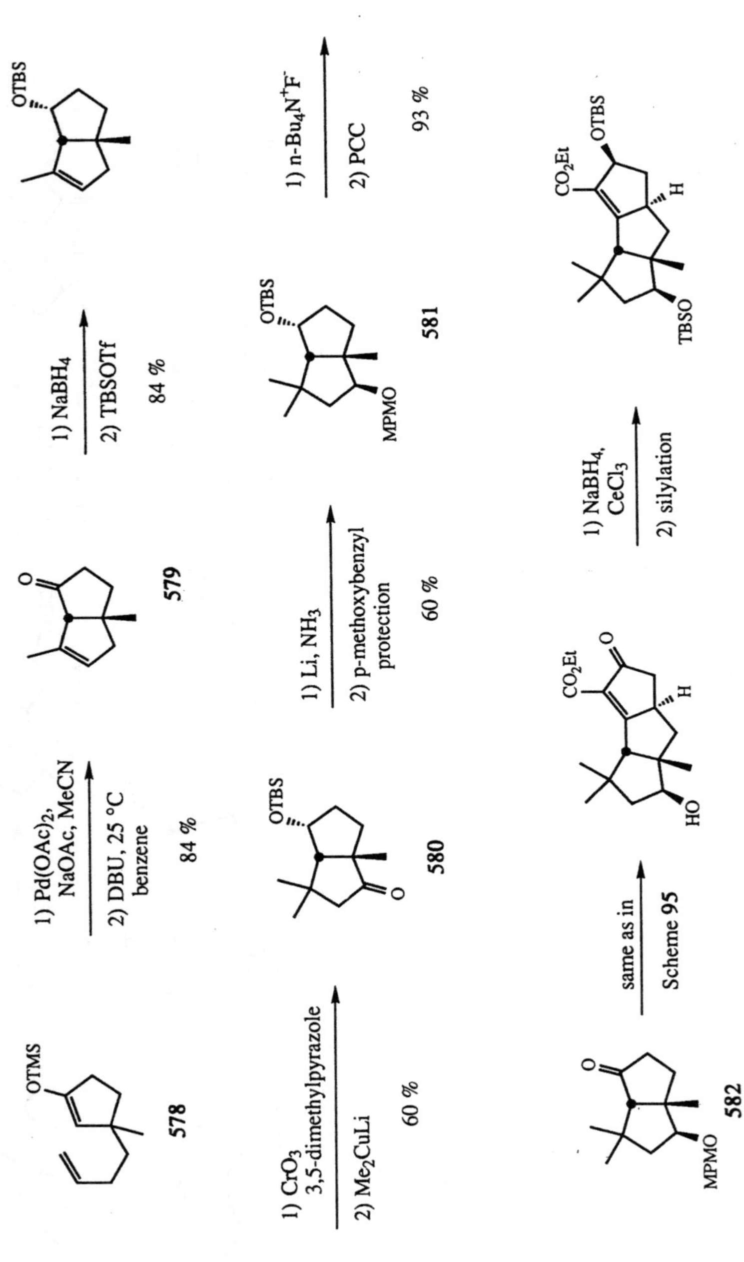
OTMS
578
1) Pd(OAc)2,
NaOAc, MeCN
2) DBU, 25 °C
benzene
84 %
O
579
1) NaBH4
2) TBSOTf
84 %
OTBS
1) CrO3
3,5-dimethylpyrazole
2) Me2CuLi
60 %
OTBS
O
580
1) Li, NH3
2) p-methoxybenzyl
protection
60 %
OTBS
MPMO
581
1) n-Bu4N+F-
2) PCC
93 %
O
MPMO
582
same as in
Scheme 95
CO2Et
O
H
HO
1) NaBH4,
CeCl3
2) silylation
CO2Et
OTBS
H
TBSO

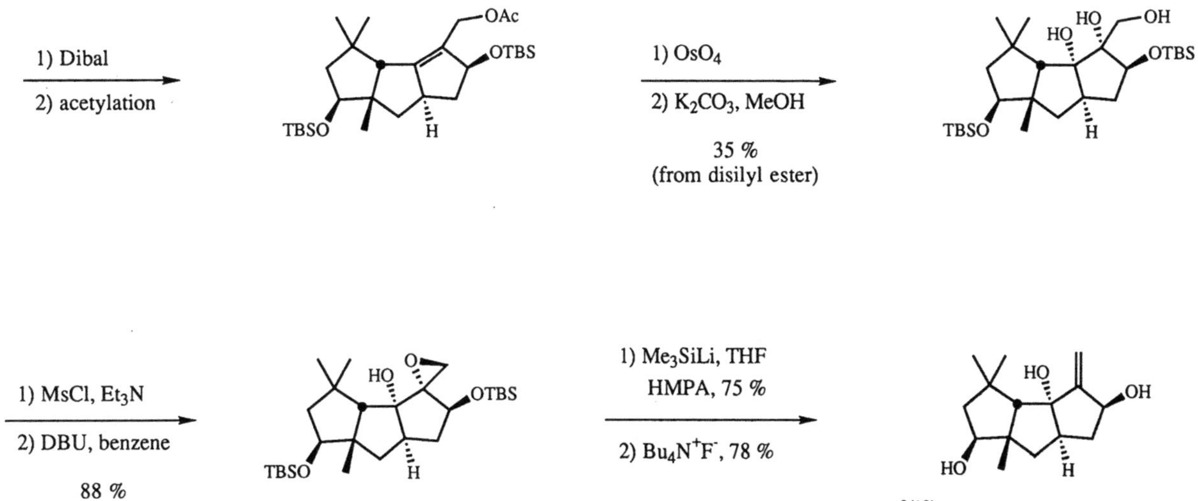

(±)-$\Delta^{9(12)}$-capnellene-3β,8β,10α-triol **583**

Scheme 96. Shibasaki Synthesis of (±)-$\Delta^{9(12)}$-Capnellene-3β,8β,10α-triol

ring. The 10α-hydroxyl is added to the completed quinane ring system. Synthesis of **577** is initiated by a palladium mediated cyclization of diene **571** to give bicyclic ketone **572** as a mixture of olefin isomers. Equilibration of the mixture with DBU at room temperature gives **572** in 71% yield. Addition of lithium dimethylcuprate followed by alkylation of **573** with hydrolysis affords the gem-dimethyl functionality and secures the needed functionality for annulation to form the C ring. Cyclization of **574** is effected by treatment with TMSOTf to drive the aldolization via irreversible dehydration. Reduction and protection of the ketone provides the oxygen functionality at C8 **(575)**, while ester reduction and protection followed by osmylation of the 9,10-double bond provides the C10 hydroxyl **576**. Epoxide formation followed by deoxygenation finishes the synthesis to give (±)-$\Delta^{9(12)}$-capnellene-8β, 10α-diol **(577)** in 15 steps and 0.6% yield.

The same strategy, with slight modifications, is utilized by Shibasaki to prepare (±)-$\Delta^{9(12)}$-capnellene-3β,8β,10α-triol **583**. As shown in Scheme 96 the synthesis begins with the palladium mediated cyclization of **578** to form a mixture of bicyclic ketones as double bond positional isomers. Treatment of the mixture with DBU in refluxing benzene results in the formation of **579** in 84% yield. The carbonyl in **579** is protected and C3 oxygenation is obtained through treatment with chromium trioxide-3,5-dimethylpyrazole complex, treatment of which with lithium dimethylcuprate produces the gem-dimethyl functionality **580**. Reduction and protection of the carbonyl gives the protected C-3 hydroxyl **581**, while deprotection and oxidation regenerates the C9 carbonyl **582**. Ketone **582** is strictly analogous to bicyclic ketone **573** (Scheme 95) and from this point the synthesis is carried forward in an identical fashion as the synthesis of (±)-$\Delta^{9(12)}$-capnellene-8β,10α-diol **(577)** (Scheme 95) to produce (±)-$\Delta^{9(12)}$-capnellene-3β,8β,10α-triol **(583)** in 22 steps.

Pattenden [158] utilizes a strategy similar to that of Shibasaki in a synthesis of (±)-$\Delta^{9(12)}$-capnellene-8β,10α-diol **(577)** (Scheme 97). The general strategy involves building the rings in a sequential fashion in the order B → A-B → A-B-C. Michael addition of an unsaturated cuprate to **570** followed by cation-olefin cyclization gives the A-B ring system in the form of bicyclic ketone **585**. The functionality needed to prepare the C ring is obtained by alkylation of **585** to give the chloro-substituted γ,δ-unsaturated ketone **586** followed by protection of the ketone by reduction to the alcohol **587**. Formation of the alkyne, oxidation and intramolecular reductive radical cyclization produces the A-B-C ring system **(588)** along with the C-10 hydroxyl. Addition of the C-8 hydroxyl by catalytic selenium dioxide oxidation followed by inversion to obtain the correct stereochemistry results in the formation of **577** in 10 steps and 0.6% yield.

2.2.8.3 Subergorgic Acid

Subergorgic acid **(596)** is an angular triquinane (tricyclo[6.3.0.0^{1,5}]undecane), isolated [159] from the Pacific gorgonian coral *Subergorgia suberosa*. The basic ring system is identical to several other sesquiterpenes, including isocomene

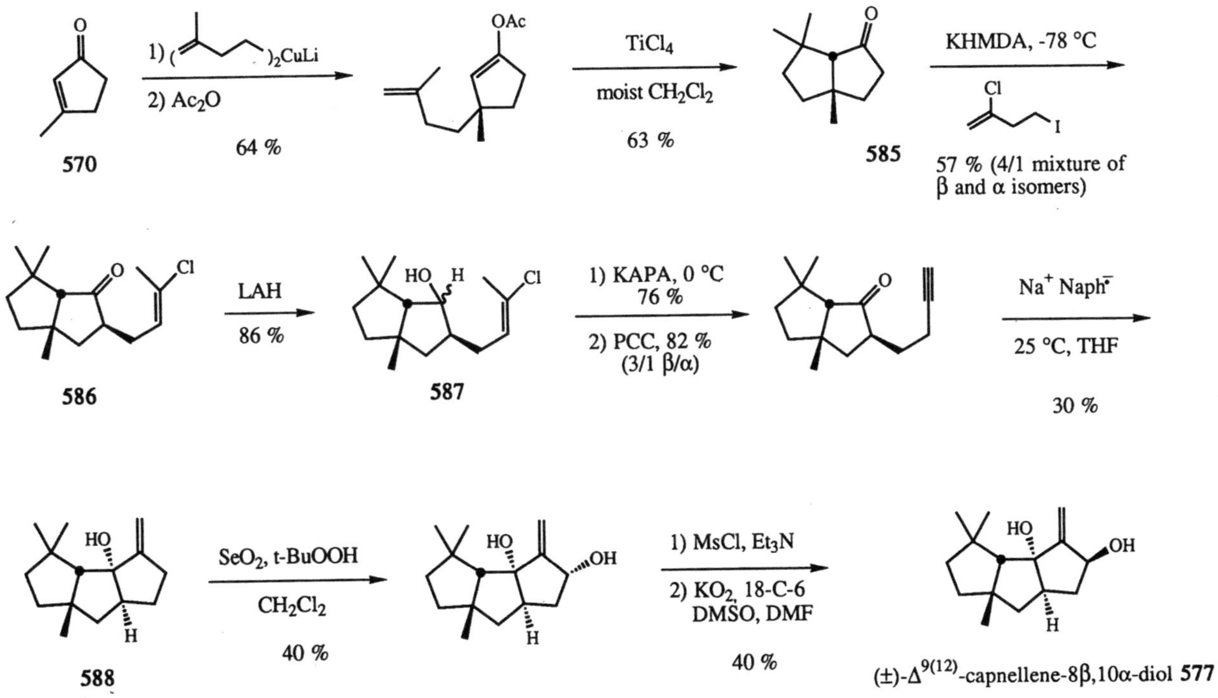

Scheme 97. Pattenden Synthesis of $(\pm)$-$\Delta^{9(12)}$-Capnellene-8β,10α-diol

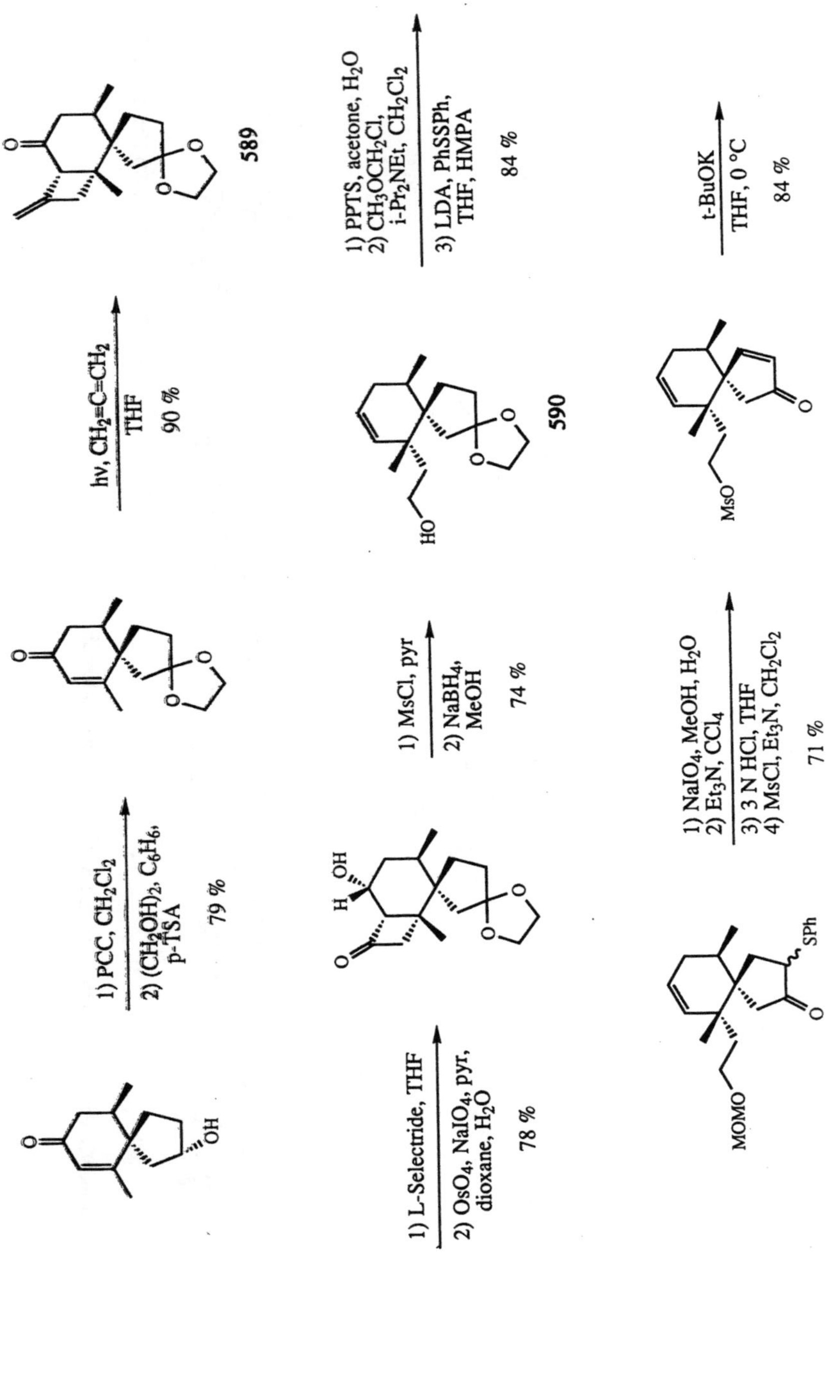

1) PCC, CH_2Cl_2
2) $(CH_2OH)_2$, C_6H_6, p-TSA
79 %
hν, $CH_2{=}C{=}CH_2$
THF
90 %
589
1) L-Selectride, THF
2) OsO_4, $NaIO_4$, pyr, dioxane, H_2O
78 %
1) MsCl, pyr
2) $NaBH_4$, MeOH
74 %
HO
590
1) PPTS, acetone, H_2O
2) CH_3OCH_2Cl, i-Pr_2NEt, CH_2Cl_2
3) LDA, PhSSPh, THF, HMPA
84 %
MOMO
SPh
1) $NaIO_4$, MeOH, H_2O
2) Et_3N, CCl_4
3) 3 N HCl, THF
4) MsCl, Et_3N, CH_2Cl_2
71 %
MsO
591
t-BuOK
THF, 0 °C
84 %

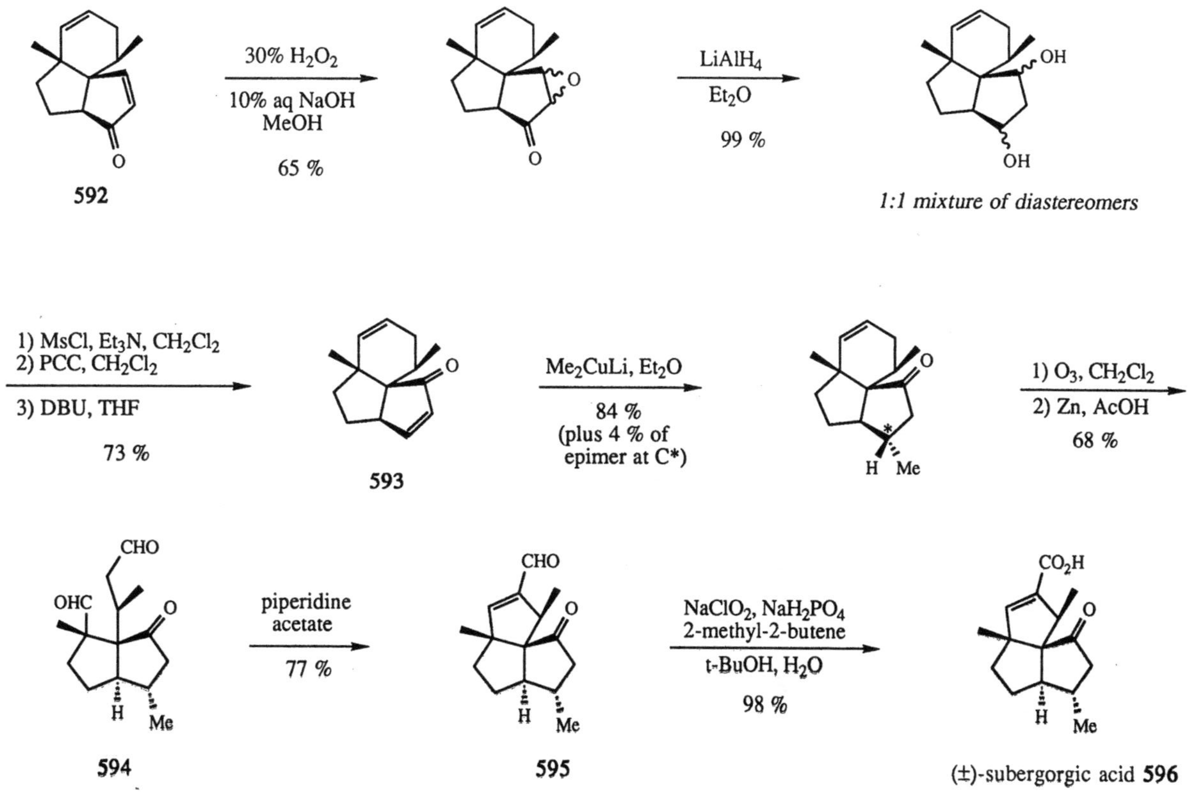

Scheme 98. Iwata Synthesis of (±)-Subergorgic Acid

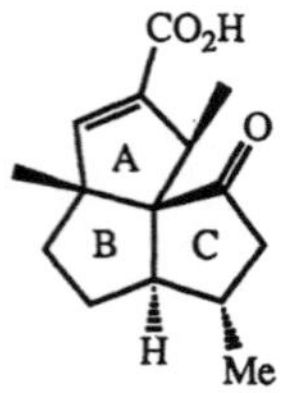

(±)-Subergorgic acid **596**

[160] and silphinene [161]. However, the methyl substitution pattern is novel and the name "subergane" has been suggested for the new skeleton. Unexpectedly, **596** exhibits cardiotoxic properties and inhibits neuromuscular transmission. The approach utilized by Iwata [162] for its synthesis (Scheme 98) involves construction of the B ring onto a spiro[5.4] system already containing the C ring, followed by ring contraction of the 6,5,5-fused ring system to give the A ring and the completed angular quinane ring system. Preparation of the B ring is initiated by a [2+2] photocycloaddition to afford **589** followed by cleavage of the 4-membered ring to produce a net two carbon homologation **590**. The cleavage process involves a facile Grob-type fragmentation via an unusual *cis*-elimination process. Obtaining the B ring requires a series of protection-deprotection steps and functional group interconversions to develop unsaturation in the 3,4-position and prepare for intramolecular alkylation via **591**. Treatment of **591** with potassium *tert*-butoxide yields the B-C ring system in 14 steps. Addition of the final methyl group involves a five step "inversion" of the α,β-unsaturated ketone **592** to **593** followed by cuprate addition. The entire process can be accomplished in three steps with a lower degree of diastereoselectivity. The synthesis is completed by ozonolysis, to form the dialdehyde **594**, aldol condensation with elimination, to form the A ring **595** and oxidation to afford (±)-subergorgic acid **(596)** in 23 steps and 4% yield.

2.2.9 Sinularenes

2.2.9.1 Sinularene

The sesquiterpene (−)-sinularene was isolated [163] from the soft coral *Sinularia mayi*. Subsequently both (+)- and (−)-12-acetoxysinularene were obtained [164] from *Clavularia inflata*. The tricyclo[4.4.0.0^{2,8}]decane skeleton is present in other sesquiterpenes such as sativene, copacamphene, ylangocamphor and copacamphor. The arrangement of substituents in (−)-sinularene is novel, and the normal head-to-tail linkage of isoprene units is not present. To date there have been four total syntheses of (±)-sinularene by Collins, Oppolzer, Fallis and Piers. The related (±)-12-acetoxysinularene has been prepared by Oppolzer.

The strategy followed by Oppolzer [165] in the synthesis of **605** is guided by the use of an intramolecular type I magnesium ene reaction (see Scheme 80) to

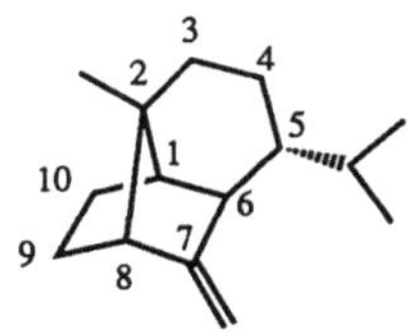

Sinularene **605**

efficiently construct the sinularene skeleton (Scheme 99). The known norbornene tosylate (**598**) is converted to the iodide which, surprisingly, alkylates the dienolate **599** regiospecifically at the γ-position. Conversion of **600** to the allylic chloride **601**, followed by the intramolecular magnesium ene reaction affords a cyclized Grignard reagent which can be trapped with carbon dioxide to give the completed sinularene skeleton **602** with the stereochemistry inverted at C5. Reduction of **602** followed by methylation gives methyl ether **603**. Proper stereochemistry at C5 is obtained by ozonolysis of the isopropenyl moiety to afford the methyl ketone, base catalyzed epimerization to the more stable isomer and Wittig olefination to regenerate the isopropenyl functionality. Hydrogenation and demethylation to give alcohol **604** followed by pyrolysis of the derived acetate results in (±)-sinularene in 15 steps and 3.5% overall yield.

As shown in Scheme 100, the approach utilized by Collins [166] also involves construction of the six-membered ring of **605** from a preformed norbornane skeleton. The key step involves intramolecular epoxide opening with an enolate to produce the sinularene framework. Formation of the needed keto-epoxide **609** proceeds from the known norbornane lactone **606** to the iodoketal **607** in five steps via standard transformations. Treatment of the iodide **607** with an excess of the allylnickel complex **608** followed by deketalization and MCPBA oxidation secures the keto-epoxide **609** as a 1:1 mixture of diastereomers. Treatment of the mixture of epoxides with base followed by separation and acylation gives acetate **610** in 21% yield from **609**. Ester pyrolysis and hydrogenation gives **611**, which is converted to **605** by addition of MeLi, acylation and a second ester pyrolysis. (±)-Sinularene is obtained in 15 steps and 1% yield.

A completely different strategy is utilized by Piers [167] in the synthesis of **605** (Scheme 101). The cornerstone of this synthesis involves Cope rearrangement of a highly functionalized bicyclo[3.1.0]hex-2-ene followed by an intramolecular alkylation to afford the sinularene ring system late in the synthesis. The early stages of the synthesis involve construction of the proper Cope substrate with the correct functionality and relative stereochemistry. The synthesis is initiated by triethyl orthoacetate-based Claisen rearrangement to stereospecifically give **613**, which is converted by standard methodology to the diazoketone **614**. Copper acetate-catalyzed cyclization of **614** yields **615** which is reduced under Lindlar conditions to afford the *cis*-alkene **616**. Insertion of unsaturation via the procedure of Saegusa, vinyl cuprate addition and silyl enol

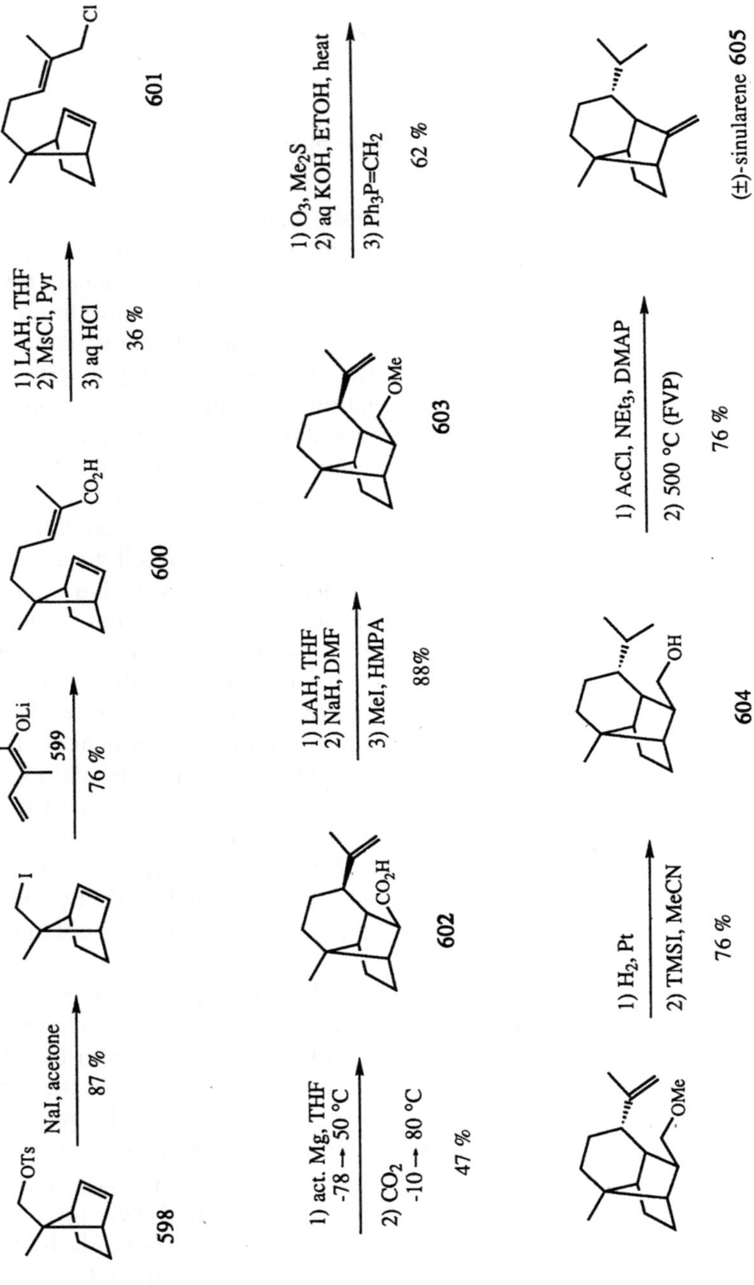

Scheme 99. Oppolzer Synthesis of (±)-Sinularene

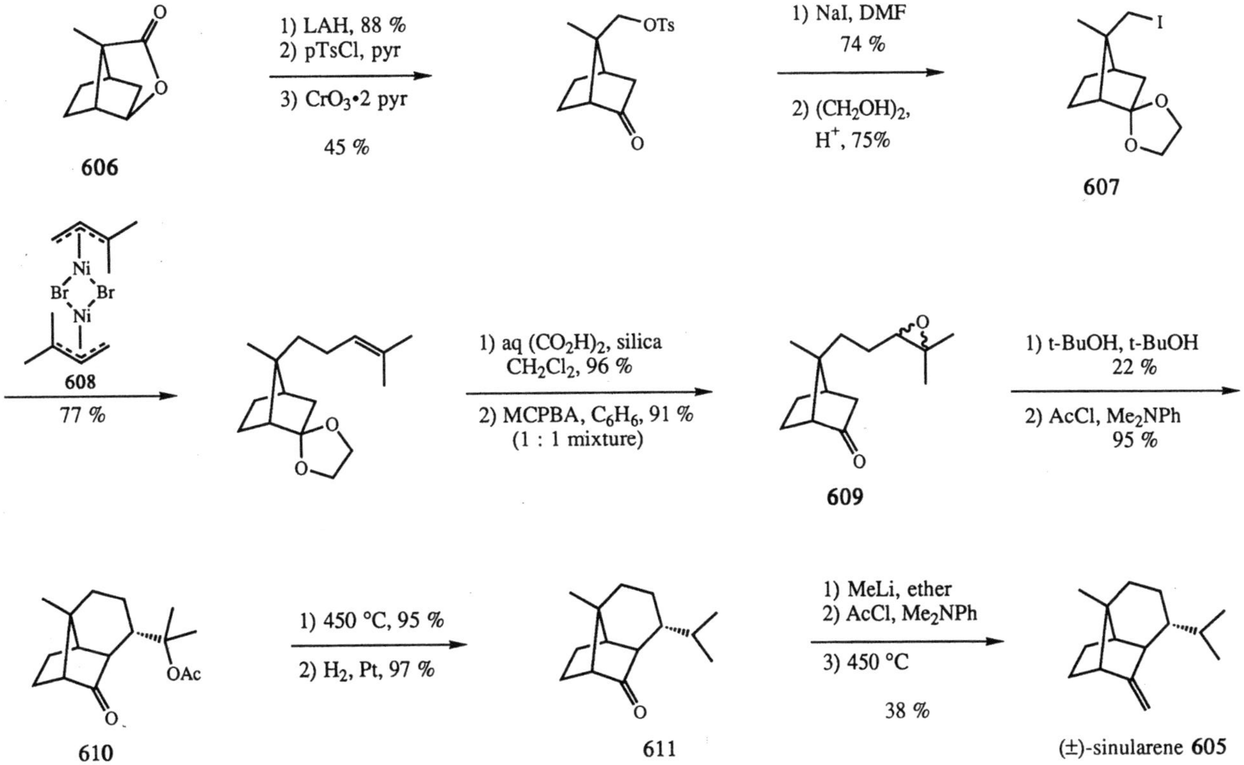

Scheme 100. Collins Synthesis of (±)-Sinularene

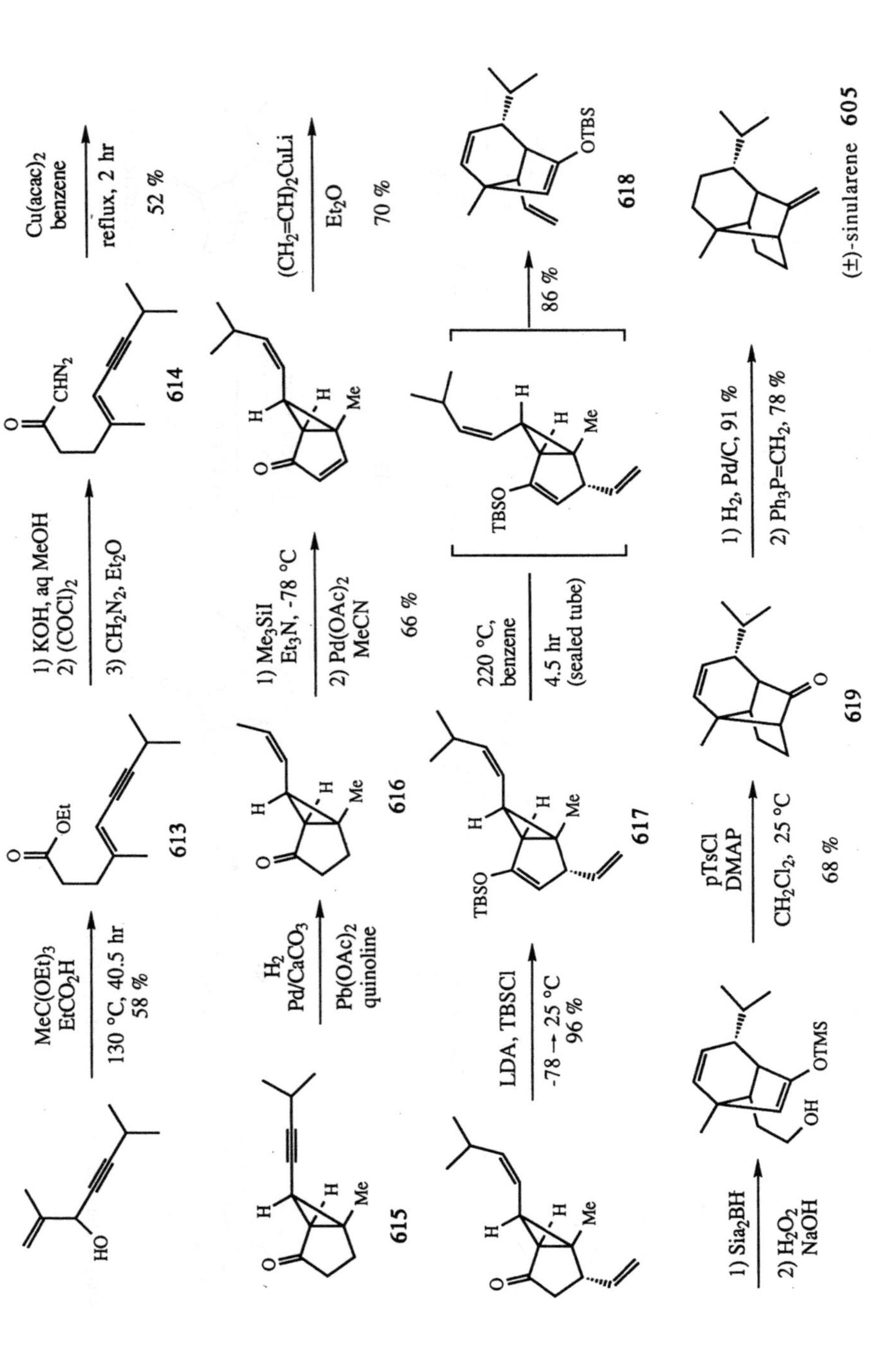

Scheme 101. Piers Synthesis of (±)-Sinularene

ether formation produces the Cope substrate **617**, with inverted stereochemistry at the point of isopentenyl attachment. This did not matter in that **617** presumably rearranges upon heating to **618** (possibly via a diradical process) thus allowing a facile Cope rearrangement to give **618** in good yield. Hydroboration-oxidation of **618** followed by tosylation and intramolecular alkylation of the silyl enol ether provides ketone **619**. Hydrogenation and Wittig olefination converts **619** to **605** in 15 steps and approximately 5% overall yield.

The Fallis synthesis [168] of (±)-sinularene **(605)** (Scheme 102) involves an intramolecular Diels-Alder reaction to construct two of the three rings of the tricyclic sinularene skeleton. This strategy is made feasible by blocking of the facile 1,5-sigmatropic rearrangements that the required substituted cyclopentadienes are prone to undergo. The cyclopropyl moiety in **624** blocks the rearrangement processes and acts as a latent form of the C-2 methyl. Interestingly, it was found that oxygenation at C5 is required for the intramolecular cycloaddition to proceed. The triene **623** is assembled by two condensation processes. Allylic oxidation of **620** followed by addition of the ester enolate of **621** yields **622**. Protection of the secondary alcohol followed by a reduction-oxidation sequence produces an aldehyde, which is subjected to Wittig olefination to afford triene **623**. Warming of **623** in THF initiates the cycloaddition giving **624** in 95% yield. The latent methyl group is revealed and the alcohol deprotected by hydrogenolysis. Formation of the chloride of **625** under Appel conditions results in elimination to afford **626**. Double reduction of **626** followed by acylation and ester pyrolysis gives (±)-sinularene **605** in 13 steps and 8.5% yield.

2.2.9.2 12-Acetoxysinularene

The only synthesis of an acetoxy sinularene is that of Oppolzer [169], shown in Scheme 103. The strategic approach is similar to that utilized in Oppolzer's synthesis of **605**. An intramolecular magnesium-ene reaction in a norbornane system is used to construct the six-membered ring portion of the target. Although the positions of the reaction partners have been switched (the magnesium-ene unit is now part of the norbornane system), the wrong stereochemistry is still obtained at C5. This is less of a problem here in that most of the refunctionalization needed to complete the synthesis takes place at the C5 side chain. The synthesis is initiated by copper catalyzed coupling of an allylic Grignard reagent with iodoacetal **607**. The unsaturated ketone **627** thus obtained is converted to a vinyl anion, formylated and reduced to provide the allylic alcohol **628**. Conversion to the allylic chloride **629** followed by reaction with activated magnesium powder provides the cyclized Grignard reagent. Oxidation with MoOPH affords alcohol **630**, which is converted to the acid **631** by Jones oxidation. Epimerization of the C5 center was achieved by conversion of **631** to the methyl ketone followed by equilibration on a base-washed alumina column results in **632** after two cycles. Addition of methyllithium followed by acetylation with elimination yields (±)-12-acetoxysinularene **633** in 12 steps and 5% yield.

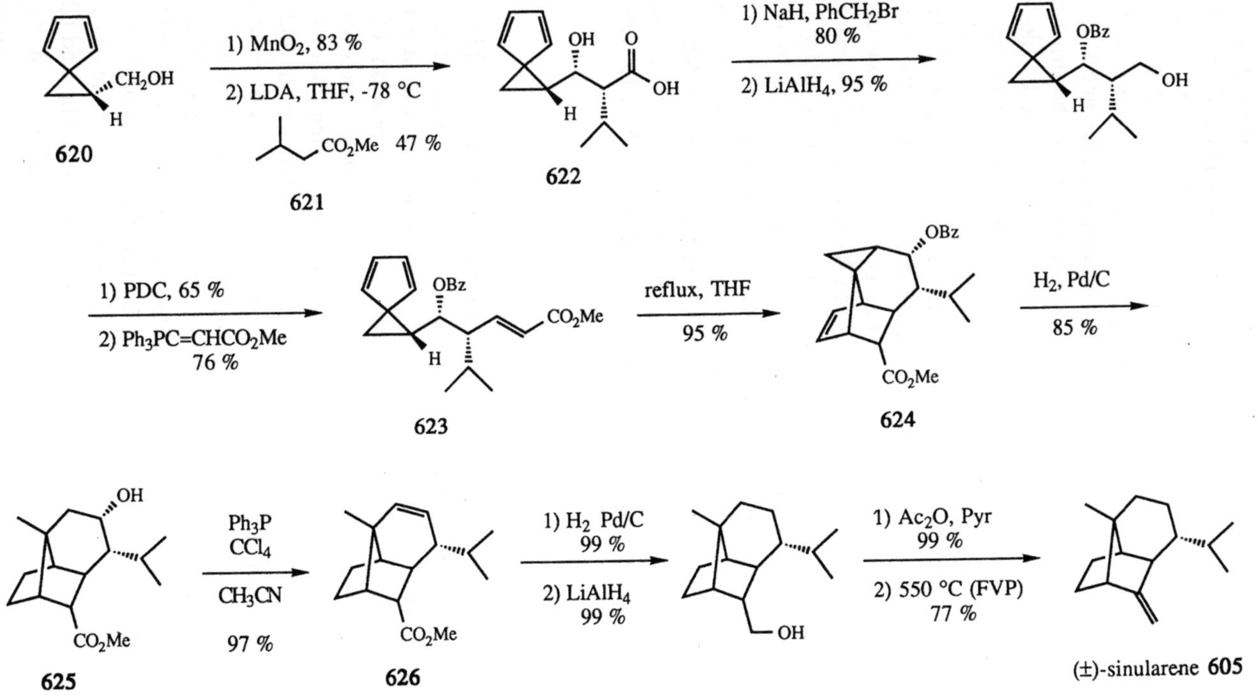

Scheme 102. Fallis Synthesis of (±)-Sinularene

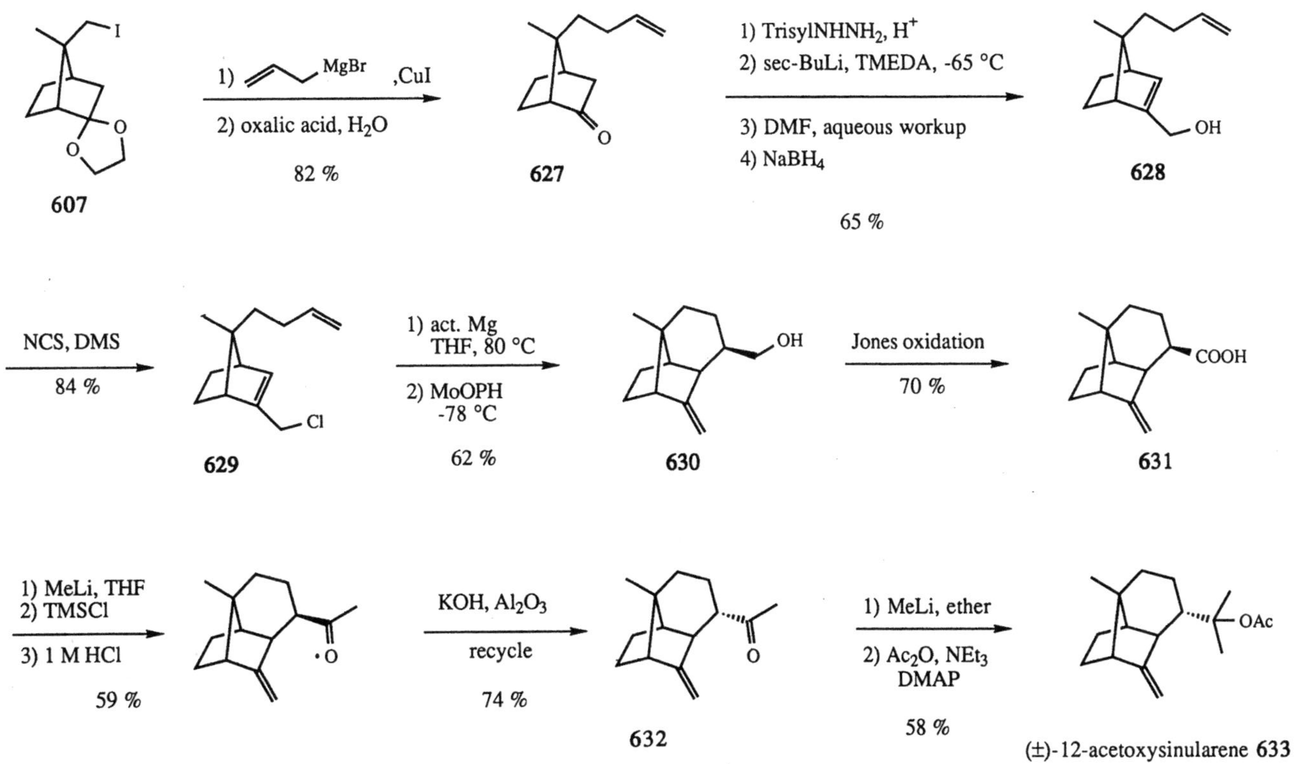

Scheme 103. Oppolzer Synthesis of (±)-12-Acetoxysinularene

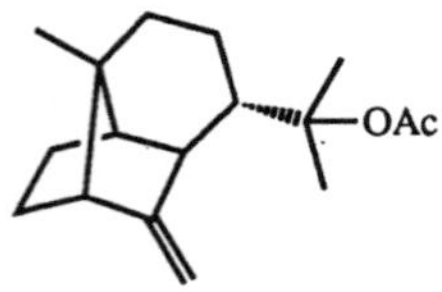

12-Acetoxysinularene **633**

2.2.10 Caespitanes

2.2.10.1 8-Desoxyiso- and Isocaespitol

The caespitanes are a series of polyhalogenated bisabolene-type sesquiterpenes isolated [170] from the marine alga *Laurencia caespitosa*. Of these metabolites, isocaespitol **(647)** and 8-desoxyisocaespitol **(638)** have been subjected to synthesis. Confirmation of the structure of 8-desoxyisocaespitol **(638)** was obtained through a non-regiocontrolled 3 step synthesis. Its synthesis, by Gonzalez [171] is shown in Scheme 104. Oxidation of farnesol acetate with NBS in aqueous THF affords the terminal bromohydrin **634**. Treatment of **634** with lithium perchlorate and aqueous acetic acid gives a mixture of products, including **635** and **636**, from which the desired bicycle **637** is isolated in 9% yield. Reaction of **637** with the mixed halogen BrCl produces 8-desoxyisocaespitol **(638)** along with an isomer **639** in a 1:2 ratio A yield of approximately 1.6% is obtained for **638**.

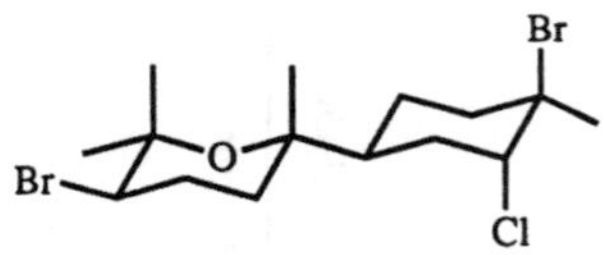

8-Desoxy-isocaespitol **638**

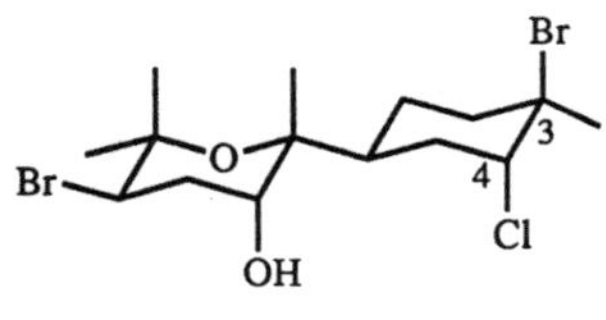

Isocaespitol **647**

A non-stereocontrolled, non-regiocontrolled, synthesis of the related isocaespitol **647** has also been developed by Gonzalez [172] (Scheme 105). Ozonolysis of *trans*-3-terpineol **(641)**, available from (+)-limonene, followed by olefination and LAH reduction of the resulting ester provides allylic alcohol **642** as a 1:1 mixture of isomers. Conversion of **642** to the tertiary monoacetate in three steps, formation of the allylic bromide and displacement with the anion of ethyl acetoacetate yields ketoester **643** in 76% yield. Bromination of the ketoester and deacylation gives bromoester **644**. Addition of a deficiency of methylmagnesium bromide to **644** and removal of unreacted starting material by chromatography produces bromohydrin **645**. Epoxidation of **645** and cyclization on acid-washed alumina affords didehalocaespitol **646** along with its Δ^2-isomer. Addition of BrCl gives isocaespitol **647** as the minor product in a 1:3 mixture with the 3-

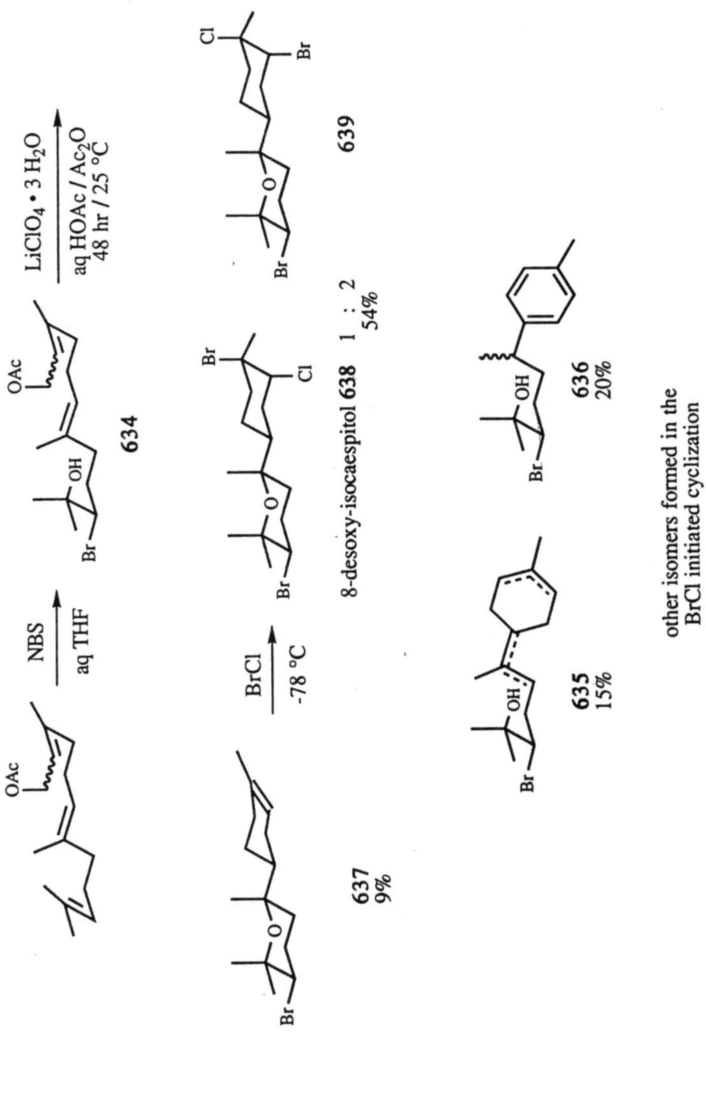

Scheme 104. Gonzalez Synthesis of 8-Desoxyisocaespitol

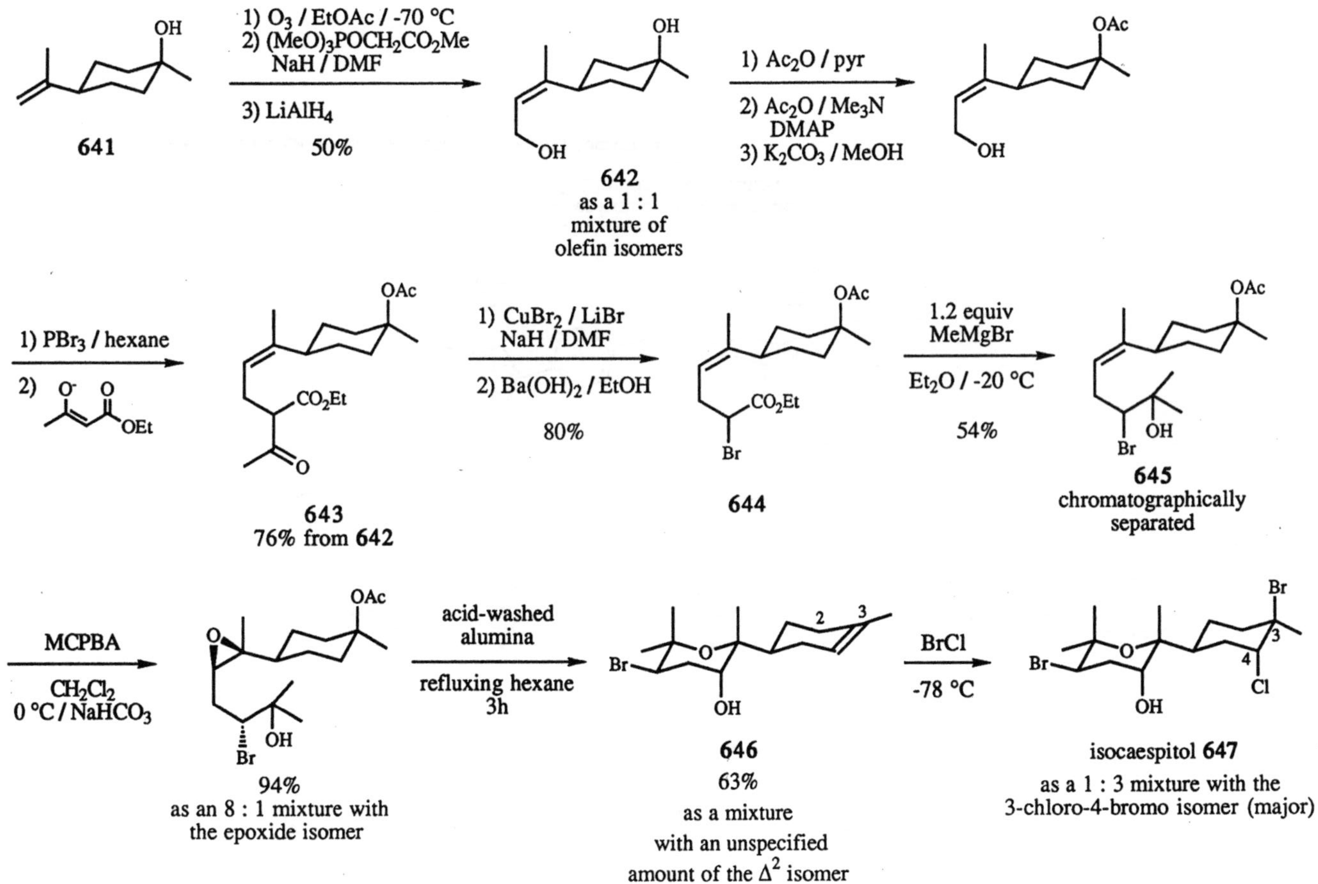

Scheme 105. Gonzalez Synthesis of Racemic Isocaespitol

chloro-4-bromo isomer. Isocaespitol is obtained in 14 steps and approximately 1% overall yield.

2.2.10.2 Obtusenol

Obtusenol (**649**) is a dibrominated monocyclic sesquiterpene isolated [173]. from the red alga *Laurencia obtusa* and has been prepared in a biomimetic synthesis by Gonzalez [174] (Scheme 106). Bromination of *trans,trans*-farnesol acetate with NBS in aqueous acetone yields a mixture of products, from which dibromide **648** is obtained in 14% yield. Dibromide **648** is hydrolysis, converted to the tribromide and treated with wet silica to afford obtusenol (**649**) in 4 steps and 8% overall yield.

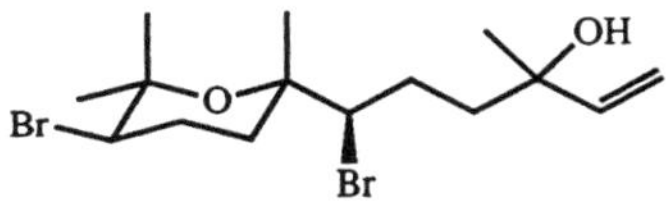

Obtusenol **649**

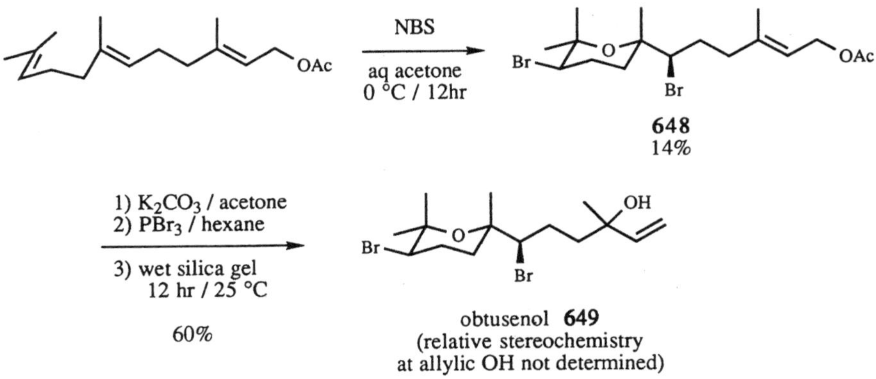

Scheme 106. Gonzalez Synthesis of Racemic Obtusenol

2.2.11 Isonitriles and Related Metabolites

A curious class of marine metabolites are the terpene isonitriles. The compounds in this series which have been synthesized are shown in Scheme 107. The biosynthetic origin of the atoms of the isonitrile groups in these metabolites has been the subject of a variety of studies. Biosynthetic investigations on sponge-derived isonitriles are rare due to the difficulty of keeping these organisms alive in the laboratory. However, there has been at least one ingenious solution to this problem [175]. It now appears that the isonitrile carbon in a terrestrial blue-green algae metabolite originates from a C1 donor related to tetrahydrofolate

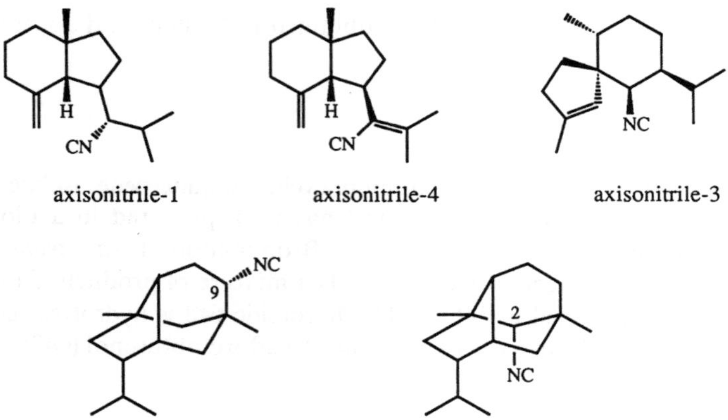

axisonitrile-1 axisonitrile-4 axisonitrile-3

9-isocyanopupukeanane 2-isocyanopupukeanane

Scheme 107. Naturally Occurring Sesquiterpene Isonitriles

metabolism, although the biosynthesis of the isonitrile group may differ from organism to organism [176]. Radiotracer experiments suggest cyanide ion as a source of at least the isonitrile carbon atom, although the exact role of cyanide in the biosynthesis of isonitrile-containing metabolites is unclear.

2.2.11.1 Axisonitriles-1, -3, -4 and Axamide

Axisonitriles-1 and -3 are perhydroindanone derivatives isolated from the sponge *Axinella cannabina.* Frequently, marine isonitriles are accompanied by the corresponding amide and isocyanate in the organism and are related biogenetically. Since amines can readily be converted to isonitriles, formamides and isocyanates, an entire series of metabolites may be formed from a common amine precursor. Piers has synthesized axisonitrile-1 (**657**) and its corresponding formamide axamide-1 (**656**) via such an approach (Scheme 108) [177]. Methylene cyclohexane annulation of the cyclopentenone **650**, as previously described (Scheme 82) yielded ketone **651**. Introduction of the conjugated double bond was accomplished leading to enone **652**. Conjugate addition of isovaleric acid via the silyl ketene acetal **653** catalyzed by $TiCl_4$ gave rise to only the desired stereo-

Axamide-1 **656** Axisonitrile-1 **657**

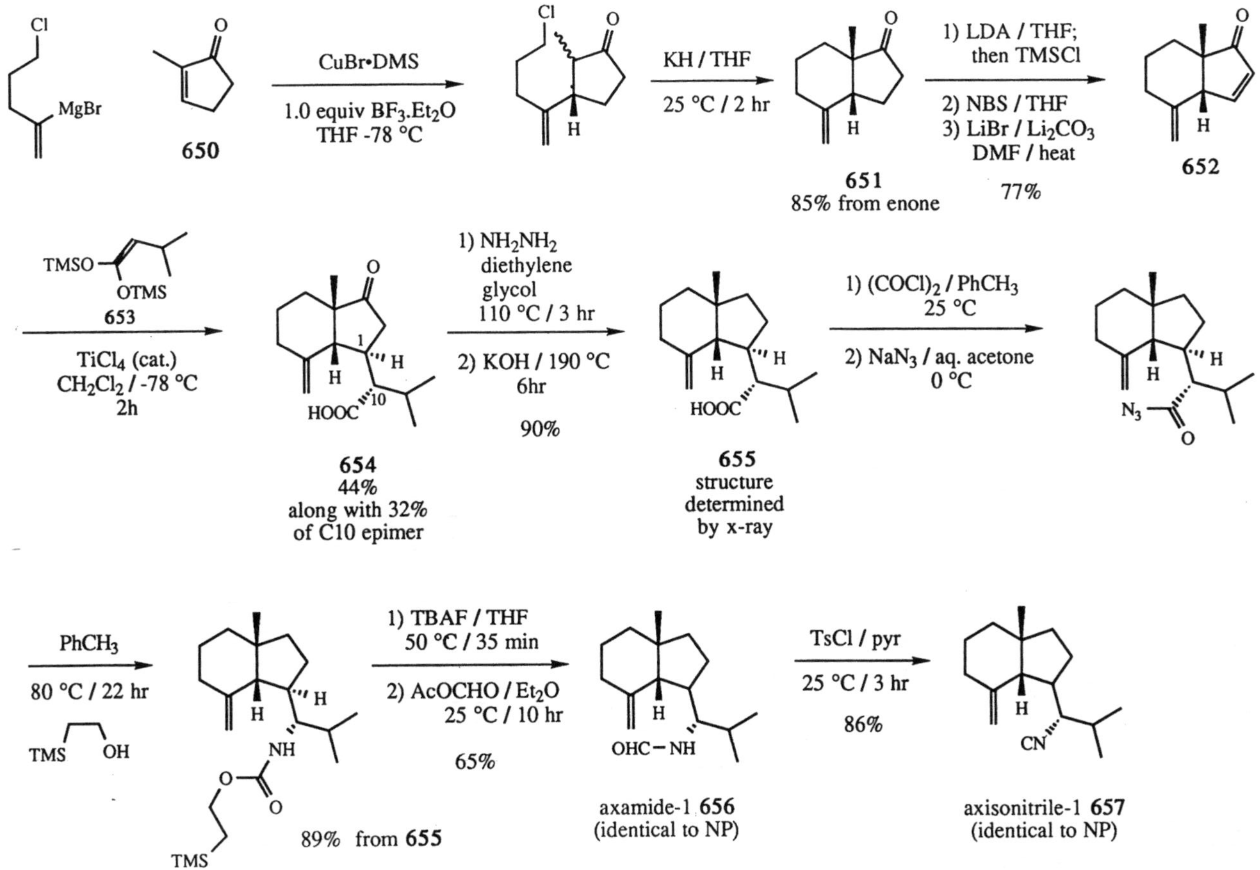

Scheme 108. Piers Synthesis of Racemic Axisonitrile-1 and Axamide-1

chemistry at C1 but as a mixture of isomers at C10. Separation of **654** and Wolff-Kishner reduction gave the acid **655**, the structure of which was confirmed by X-ray crystallography. At this point the carboxyl group had to be excised and converted to the amine with retention of stereochemistry. This was accomplished in a three-step procedure, involving Curtius rearrangement, giving the primary amine (not shown). Treatment with formic-acetic anhydride gave racemic axamide-1 (**656**) which could be dehydrated to axisonitrile-1 (**657**) completing the 13-15 step pathway in about 14% overall yield.

Hart has utilized the same strategy to synthesize three members of the axamide-4 family (Scheme 109) [178]. The basic strategy for constructing the perhydroindane ring system involved intramolecular conjugate addition of the ketone enolate of **658** which gave three isomers (**659**–**661**) in the indicated yields. Isomerization experiments showed that isomerization of the two minor isomers to the major conjugated isomer was not feasible. The C15 methylene group was installed by the Wittig reagent giving **662**. Curtius rearrangement led to the vinyl isocyanate **663**. Reduction with $LiBH(Et)_3$ gave racemic axamide-4 (**664**). Dehydration then provided axisonitrile-4 (**665**) and subsequent treatment with sulfur gave axisothiocyanate-4 (**666**).

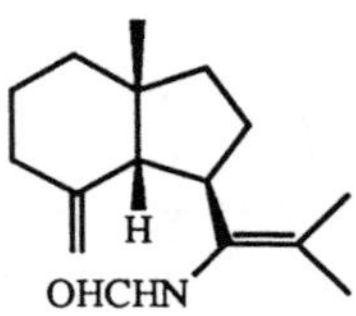

Axamide-4 **664**

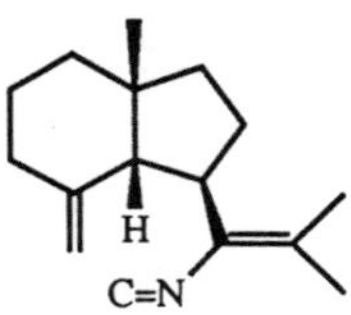

Axisonitrile-4 **665**

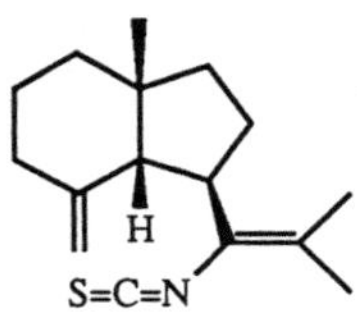

Axisothiocyanate-4 **666**

The spirocyclic axisonitrile-3 has been synthesized by Caine as the (−)-isomer (Scheme 110) [179]. Robinson annulation of **668** derived from (+)-carvone gave the ketol **669**. Conversion to the conjugated dienyl acetate **670** took place with Ac_2O and catalytic acid. Standard chemistry followed, including installation of an α-hydroxyl at C6 which was destined to be the isonitrile-bearing carbon. Irradiation of the cross-conjugated enone followed well-known precedent providing the cyclopropane **671** via σ2a + π2a cycloaddition. Hydrogenation gave the saturated analog **672**. Methylenation of the carbonyl was followed by Birch reductive cleavage of the cyclopropane leading to the tosylate **673** containing the spirocyclic skeleton of the target. Conversion of the tosylate

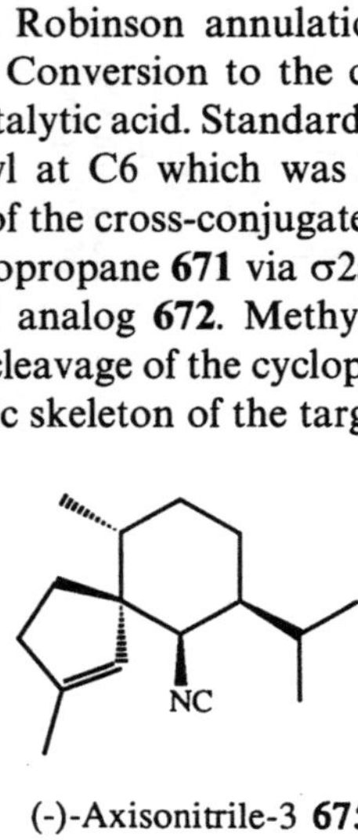

(-)-Axisonitrile-3 **675**

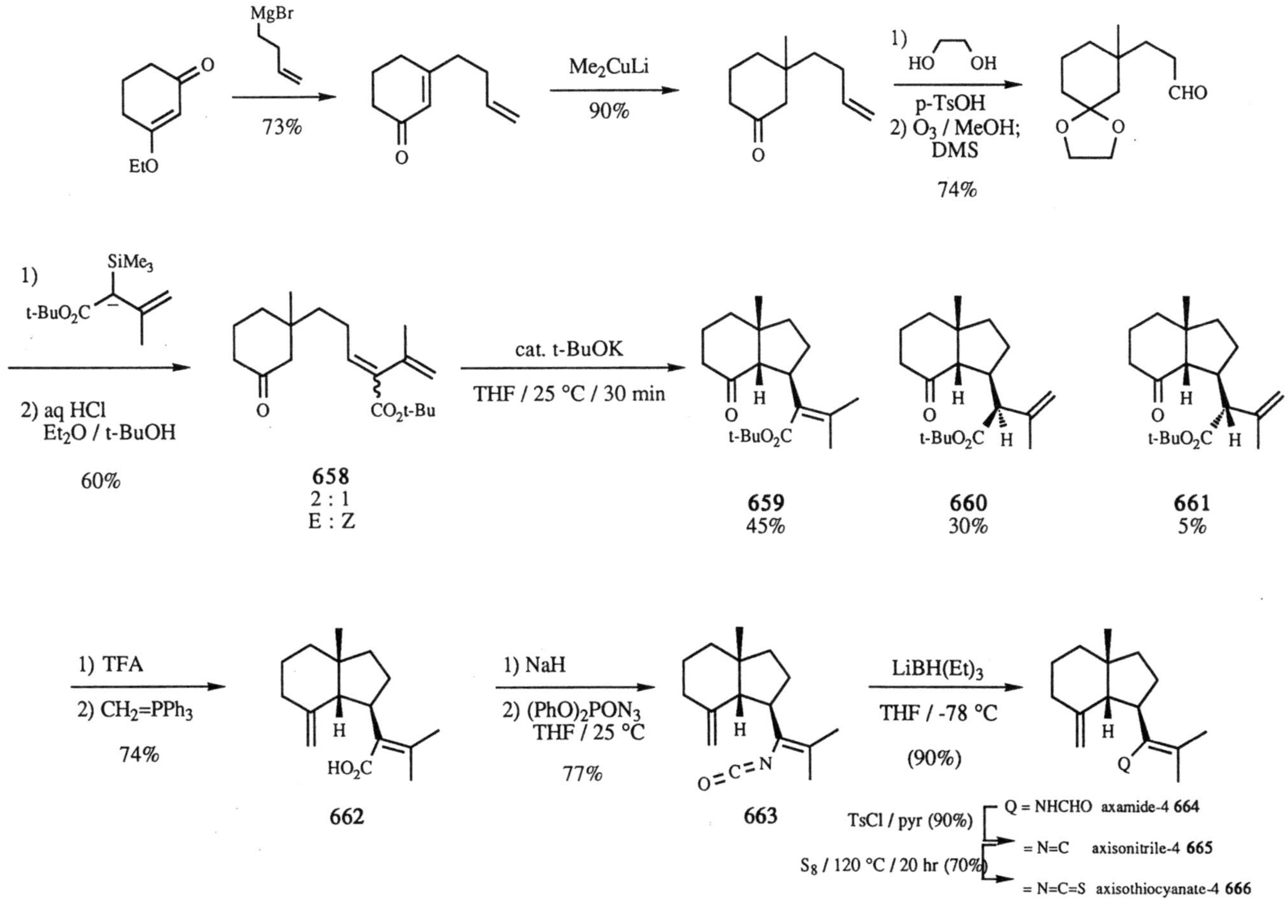

Scheme 109. Hart Synthesis of Axisonitrile-4 and Derivatives

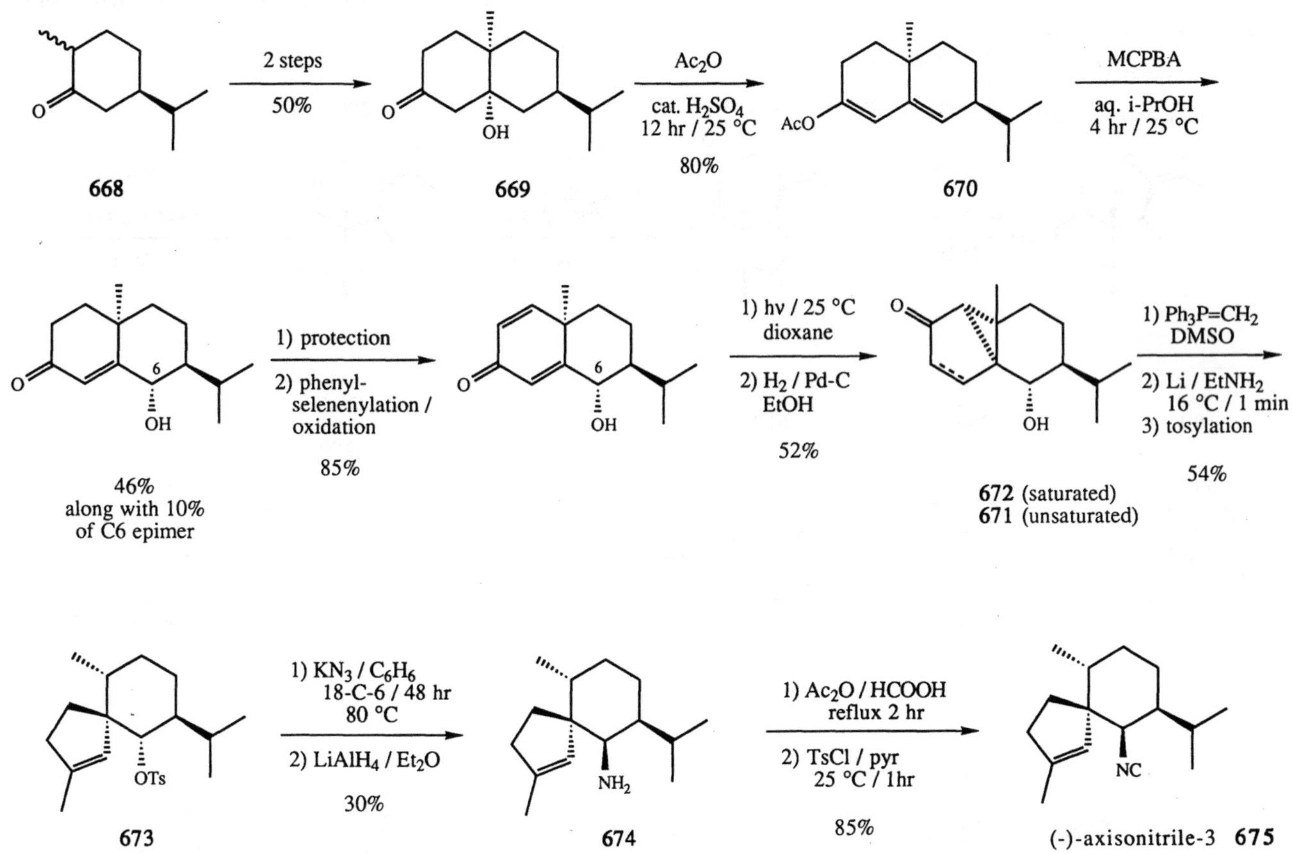

Scheme 110. Caine Synthesis of (−)-Axisonitrile-3

to the inverted amine was accomplished by displacement with azide giving the amine **674**. Formylation of **674** gave (−)-axamide-3 which was dehydrated to (−)-axisonitrile-3 (**675**).

2.2.11.2 9- and 2-Isocyanopupukeananes

The bridged bicyclic isonitriles 2- and 9-isocyanopupukeanane (**695**) and (**683**) have been the targets of many studies. Formal syntheses of **683** have been reported by Chang [180], Piers [181] and White [182], in the form of 9-pupukeanone (**681**), which was an intermediate in the Yamamoto and Corey syntheses of **683**. Yamamoto [183] utilized an intramolecular Diels-Alder strategy to construct the tricyclic ring system (Scheme 111). The silyl enol ether **677** was made from the enone **676** via a standard series of manipulations. Heating of **677** to 160 °C for 30 min and hydrolysis of the silyl ether gave rise to the hydroxy ketone **678** in quantitative yield. Ketalization and oxidation of the alcohol led to ketone **679** to which was added isopropenyllithium, eventually resulting in the diene **680**. Hydrogenation of this substance proceeded with excellent stereoselectivity to the endo isopropyl isomer **681** (9-pupukeanone). Conversion to the oxime and reduction with DIBAL gave the amine **682** with the correct stereochemistry at C9. Conversion to the isonitrile by dehydration of the corresponding formamide followed, giving the racemic target molecule in about 5% overall yield for 19 steps.

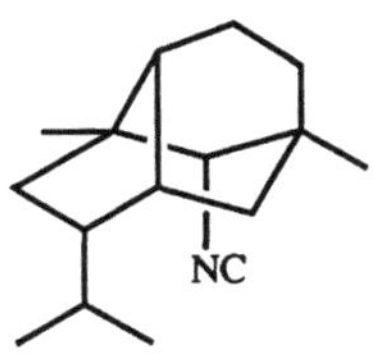

(±)-2-Isocyanopupukeanone **695**

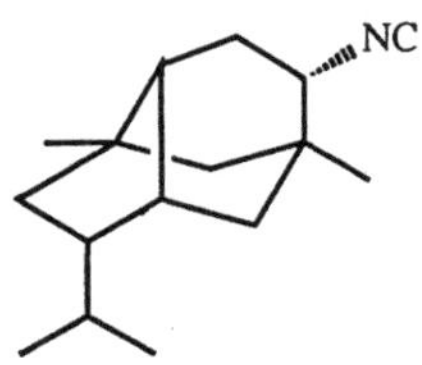

9-Isocyanopupukeanone 683

The Corey synthesis [184] took a much different tack (Scheme 112). The indane **686** was built from the benzaldehyde derivative **685**. Alkylation of the enolate of **686** with MeI gave **687** in which the methyl entered from the side of the enolate away from the isopropyl group. Demethylation gave **688** which was hydrogenated from the β-face to provide the lactone **689**. Preparing for the key step, **689** was manipulated to keto-tosylate **690**. Treatment with base at room temperature provided 9-pupukeanone in good yield. From this point the route is similar to that of Yamamoto, except that reduction of the oxime was accomplished by hydrogenation.

The Corey group also synthesized the 2-isocyano-isomer **695** from the lactone intermediate **689** (Scheme 113) [185]. Intramolecular aldol reaction of the derived keto-aldehyde provided **691** as a mixture of isomers. Removal of the C9 carbonyl group by Raney Nickel desulfurization of the corresponding

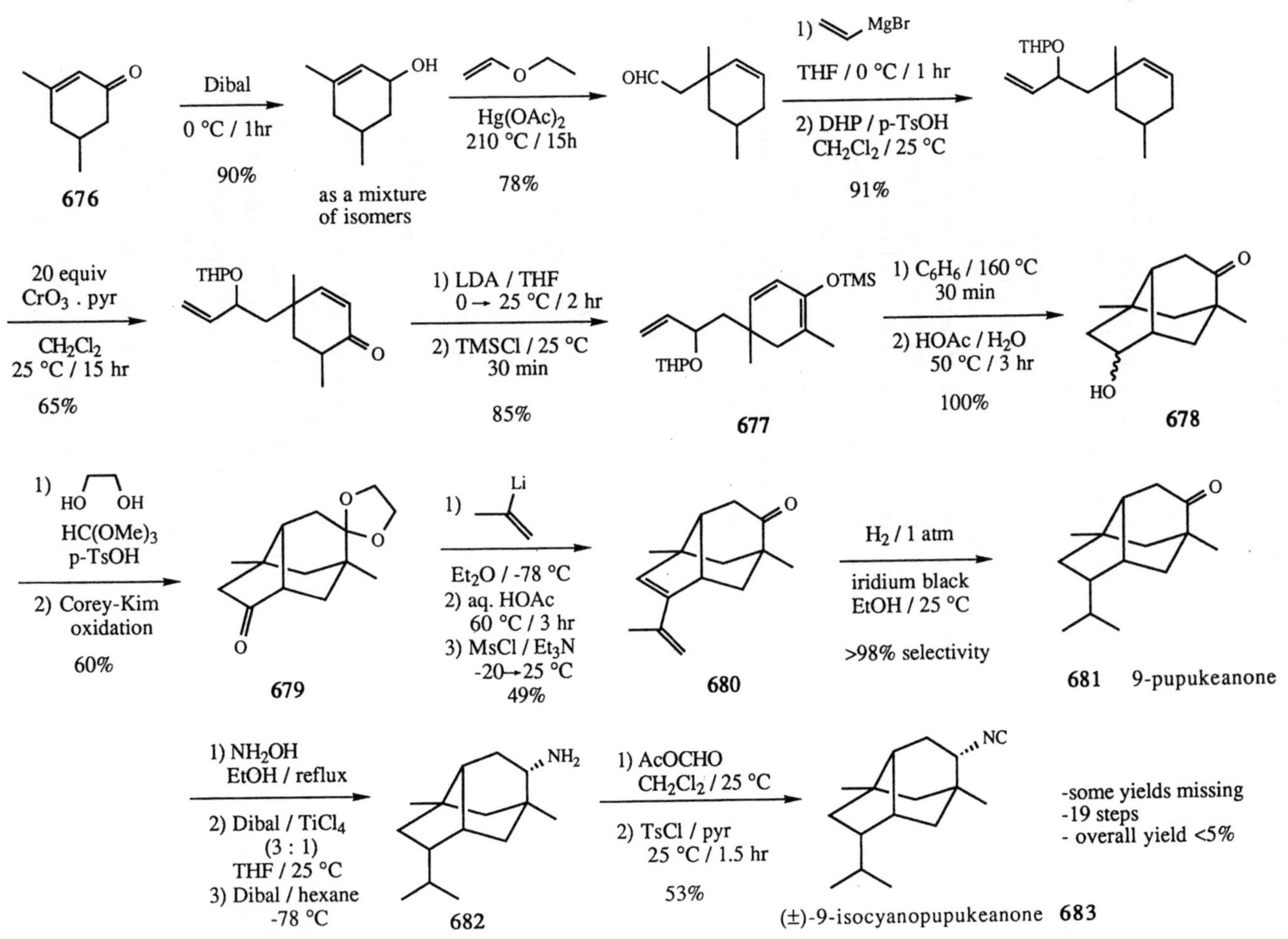

Scheme 111. Yamamoto Synthesis of (±)-9-Isocyanopupukeanane

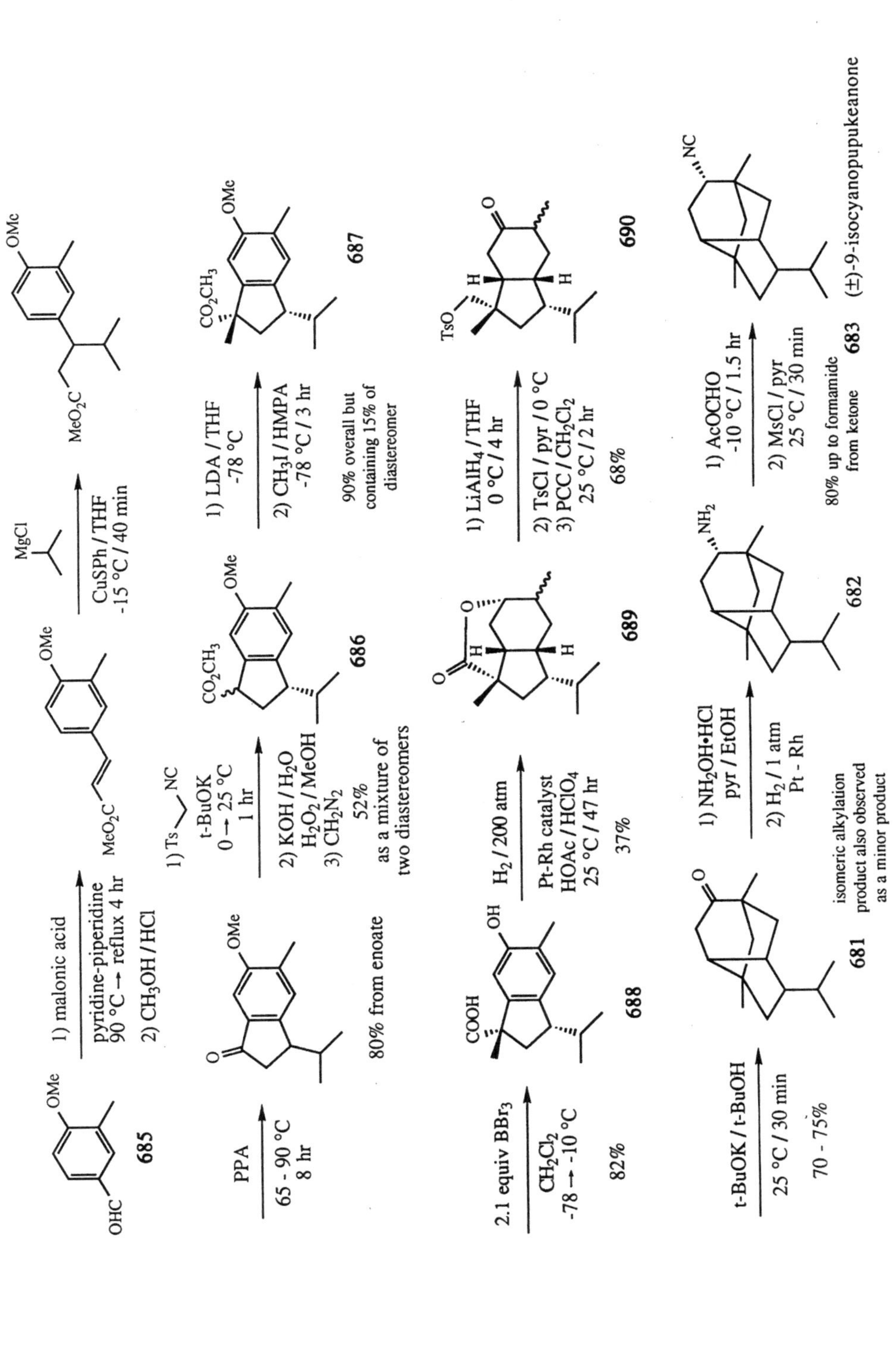

Scheme 112. Corey Synthesis of (±)-Isocyanopupukeanane

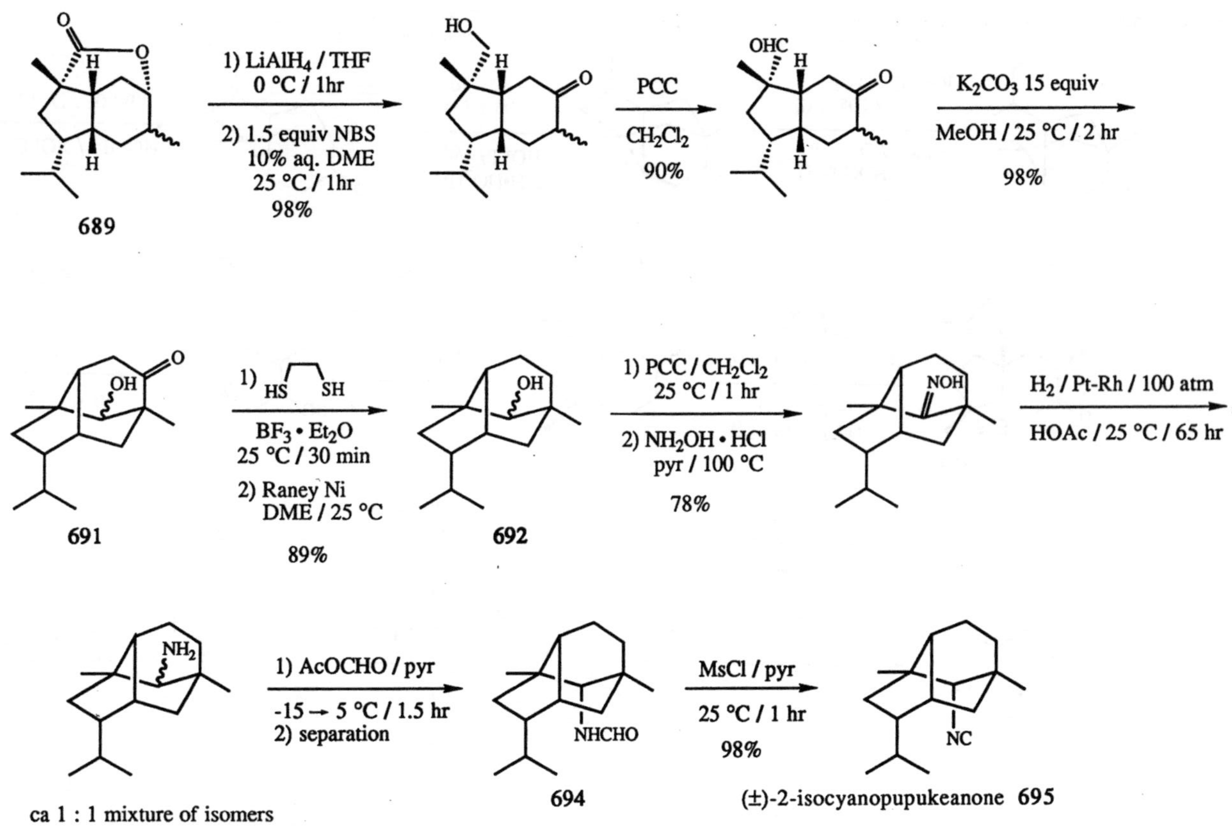

Scheme 113. Corey Synthesis of (±)-2-Isocyanopupukeanane

dithiolane gave alcohol **692** which was oxidized to 2-pupukeanone **693**. Conversion to oxime was again followed by hydrogenation, which lead to a 1:1 mixture of C2 epimeric amines. These were formylated and separated to provide the axial formamide **694** which was dehydrated to racemic 2-isocyanopupukeanane (**695**). A formal synthesis of **695** has been reported by Frater [186] in the preparation of 2-pupukeanone (**693**).

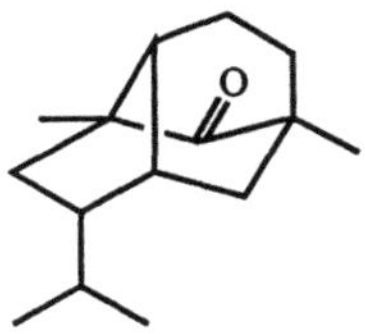

2-Pupukeanone **693**

2.2.12 Prenylated Hydroquinones and Derivatives

2.2.12.1 Avarol

The sesquiterpenoid hydroquinone avarol was isolated [187] from the sponge *Dysidea avara* in 1976. Although doubt has been cast upon the structural assignment [188] of **701**, the original structural assignment was confirmed by the total synthesis of **701** by Sarma [189] (Scheme 114). Reductive alkylation of enone **697** by sequential treatment with lithium in ammonia and 2,5-dimethoxybenzyl bromide produces ketone **698**. Wittig methylenation of **698** followed by hydrogenation over palladium on carbon gives alcohol **699** as an 85:15 mixture of β-Me/α-Me diastereomers. PCC oxidation of the β-Me isomer and addition of methyllithium to the resulting ketone produces a mixture of diastereomeric alcohols **700**. Dehydration of **700** with phosphorus oxychloride gives a mixture of olefin isomers. Isomerization of this mixture to the more stable endocyclic double bond and demethylation with lithium *n*-butylmercaptide affords avarol (**701**) in 64% yield from **700**. The stereochemistry at C-8 is confirmed and avarol obtained in 28% yield over 8 steps.

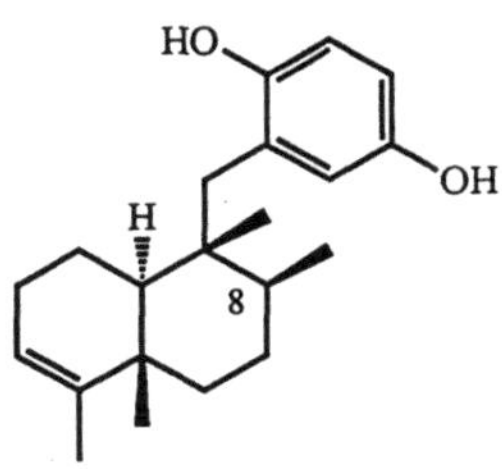

Avarol **701**

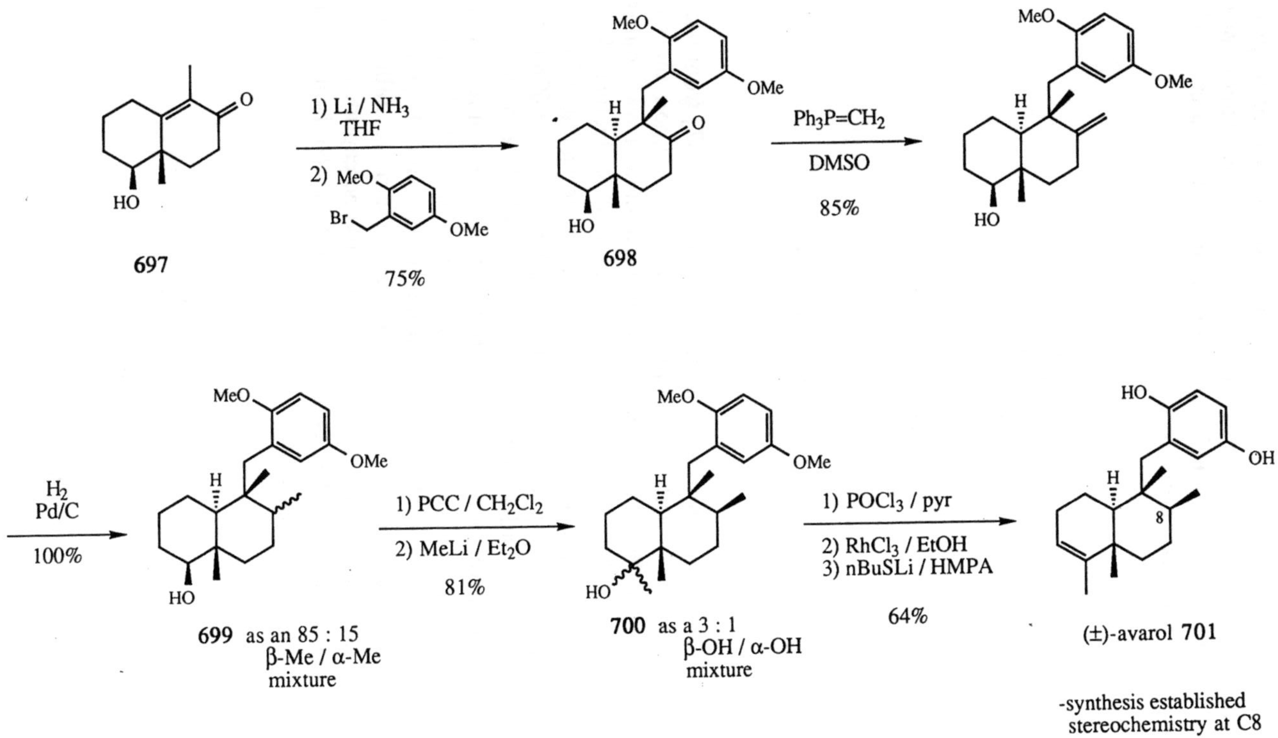

Scheme 114. Sarma Synthesis of (±)-Avarol

2.2.12.2 *Zonarol and Isozonarol*

Zonarol and isozonarol were originally isolated [190] from the Pacific brown seaweed *Dictyopteris undulata* by Fenical in 1973. The structure and absolute configuration of both have been rigorously defined by degradation and spectroscopy. Zonarol and isozonarol have been found to be active against several pathogenic fungi. Welch's racemic synthesis [191] of both **707** and **708** is depicted in Scheme 115. Optically pure ketol **702** is readily available from the Wieland-Miescher ketone. Wolff-Kishner reduction of **702** followed by Jones oxidation and addition of methyllithium to the resulting ketone affords alcohol **703**. Thermal dehydration of **703** in anhydrous DMSO provides the endocyclic alkene with few of the rearrangement products that are obtained using other elimination methods. Epoxidation of the alkene with a buffered solution of MCPBA produces epoxide **704** as a mixture of diastereomers. Eliminative epoxide opening with LDA followed by Collins oxidation gives enone **705**, along with 22% of the endocyclic olefin isomer. Conjugate addition of 2,5-dimethoxyphenylmagnesium bromide to enone **705**, followed by quenching with acetic anhydride produces the enol acetate, which is hydrolyzed to afford ketone **706**. This substance is converted to zonarol (**707**) via Wittig olefination and demethylation, giving a yield of 14% over 10 steps. Isozonarol (**708**) is prepared from ketone **706** by addition of methyllithium, thermal dehydration of the resulting alcohol and demethylation to give **708** along with 10% zonarol **707**. Isozonarol is thus obtained in 11 steps and 6.6% overall yield.

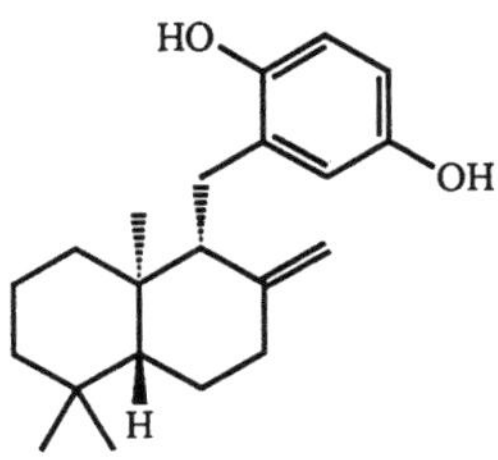

Zonarol **707**

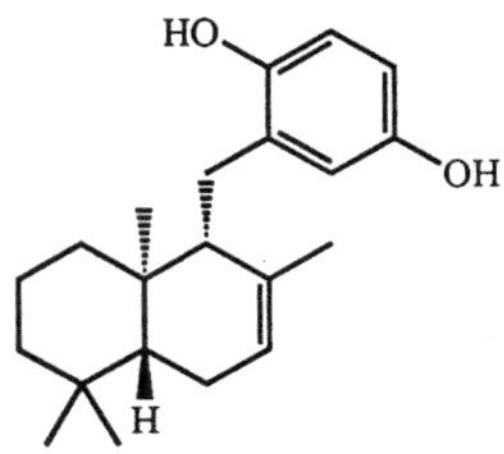

Isozonarol **708**

2.2.12.3 *Puupehenone*

Puupehenone (**714**) was isolated from a sponge of unknown genus, collected near the Blowhole area of Oahu and the Needles of Lanai [192]. Both in vivo and in vitro assays show **714** to be active against a variety of bacteria. Puupehenone has been synthesized by Trammell [193] as shown in Scheme 116. Alkylation of the lithium salt of sesamol with farnesyl bromide proceeds with exclusively C-alkylation to give polyene **710**. This interesting result appears to be due to the disposition of the three oxygen atoms in sesamol, as alkylation of lithium *p*-methoxyphenoxide under identical conditions gives only *O*-alkylation. Acylation of **710**, to prevent participation of the phenolic oxygen in the

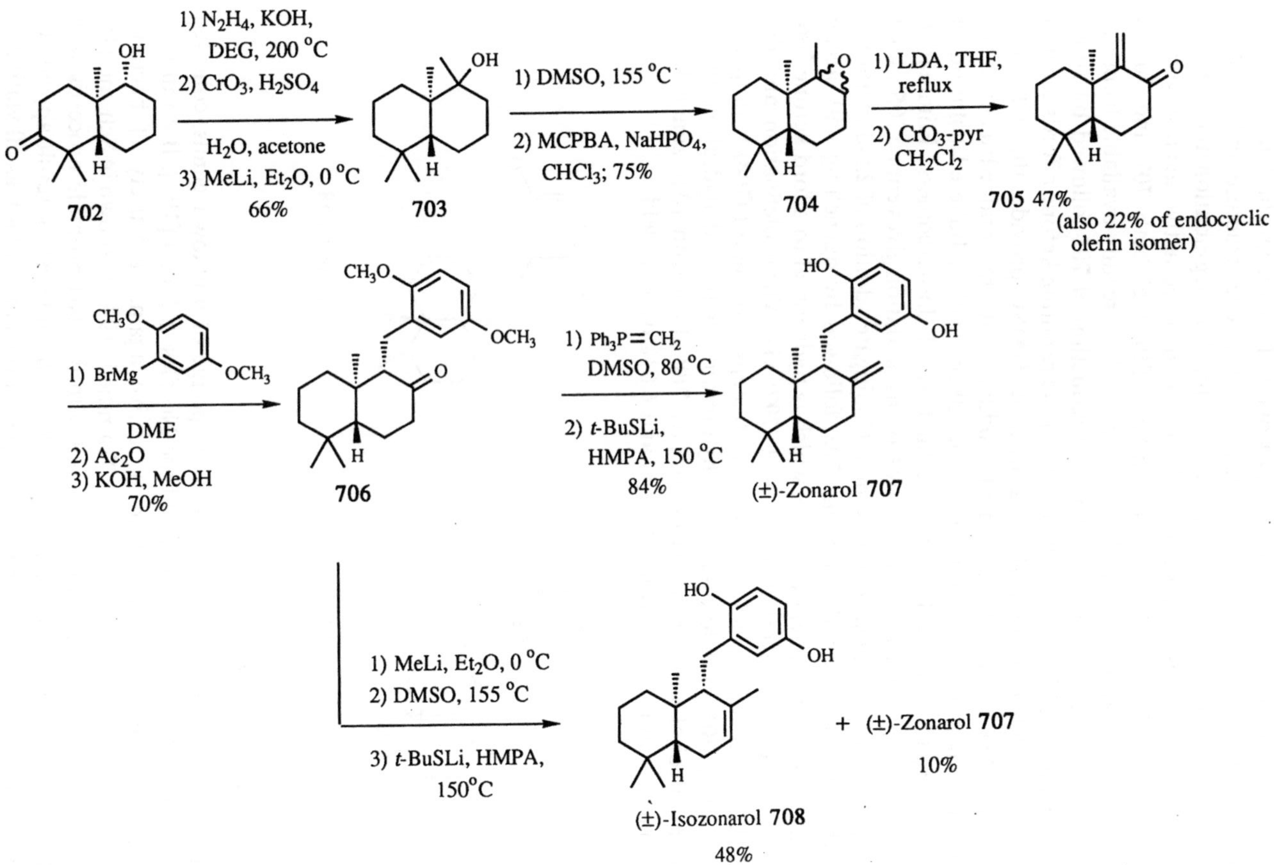

Scheme 115. Welch Synthesis of (±)-Zonarol and (±)-Isozonarol

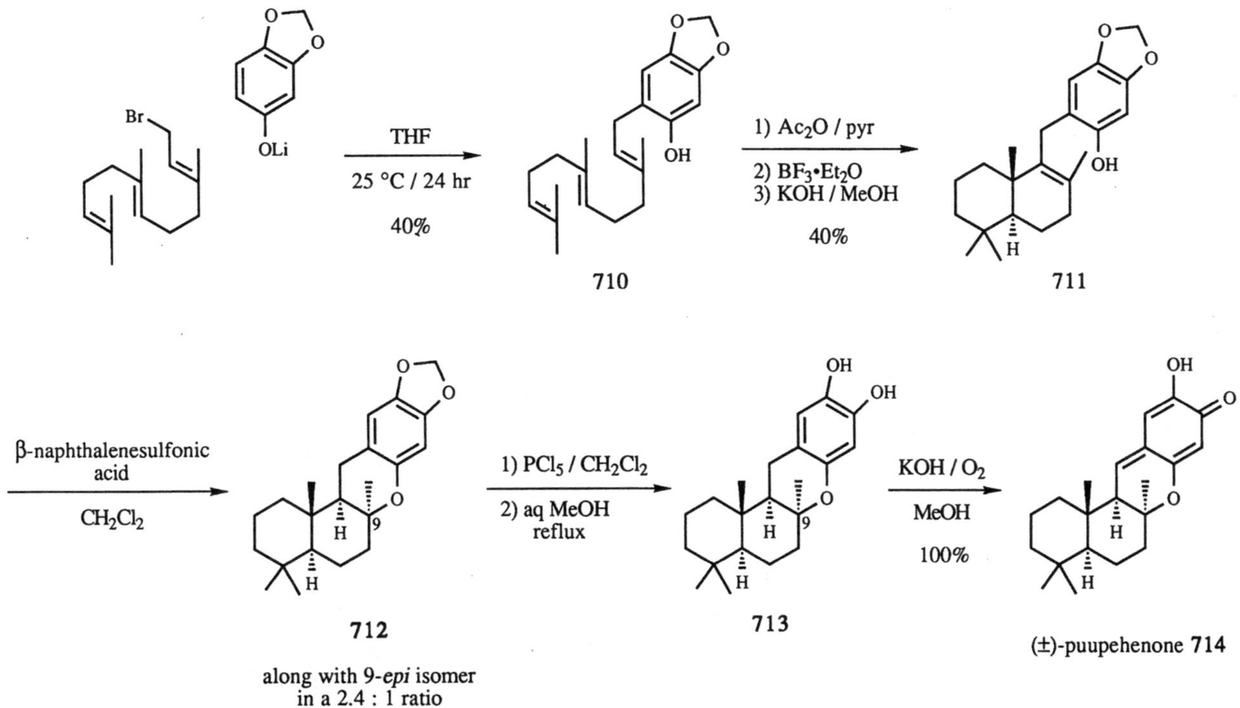

Scheme 116. Trammel Synthesis of (±)-Puupehenone

cyclization process, followed by treatment with $BF_3 \cdot Et_2O$ provides phenol **711** after saponification. Treatment of **711** with catalytic β-naphthalenesulfonic acid affords the puupehenone ring system **712** as a 2.4:1 ratio of epimers at C-9. Cleavage of the acetal was achieved by chlorination and hydrolysis to give catechol **713**. Oxidation of **713** yields puupehenone (**714**) in 8 steps.

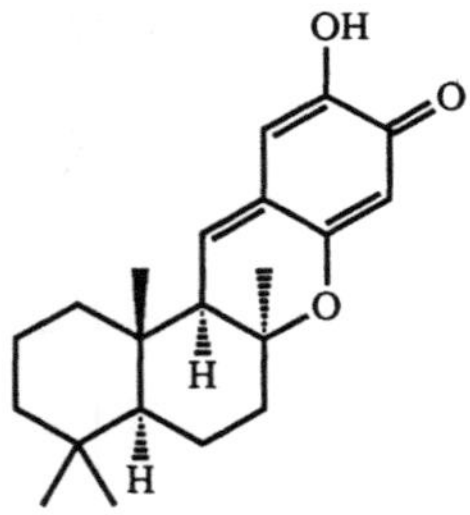

Puupehenone **714**

2.2.12.4 Halenaquinol and Halenaquinone

Halenaquinone (**719a**) is an unusual metabolite, at first appearing to be polyketide in origin. Closer examination reveals the skeleton to be similar to zonarol, puupehenone and others of this class, lacking only one methyl group. Additionally, the oxidation pattern of the quinone ring matches those of phenolic sesquiterpenes. This secondary sponge metabolite possesses in vitro antibiotic activity against *Staphylococcus aureus* and *Bacillus subtilis*. The pentacyclic ring system of **719a** represents an entirely new heterocyclic ring system. It has been isolated [194] from the sponge *Xestospongia exigua*, collected in Palau, Western Caroline Islands. Harada's enantiospecific synthesis [195] of (+)-halenaquinone (**719a**) and (+)-halenaquinol (**719b**) is illustrated in Scheme 117. The optically active Wieland-Miescher ketone (**715**) provides the source of optical activity for the synthesis of **719a–b**. Conversion of **715** to diol **716** occurs in eight steps via standard transformations. Protection of the diol and allylic oxidation with chromium trioxide/3,5-dimethylpyrazole complex provides α,β-unsaturated ketone **717**. Diels-Alder reaction of **717** with 4,5-dimethoxybenzocyclobutene in a sealed tube at 215 °C affords the adduct **718** in 33% yield. The synthesis is completed by four successive oxidations to give

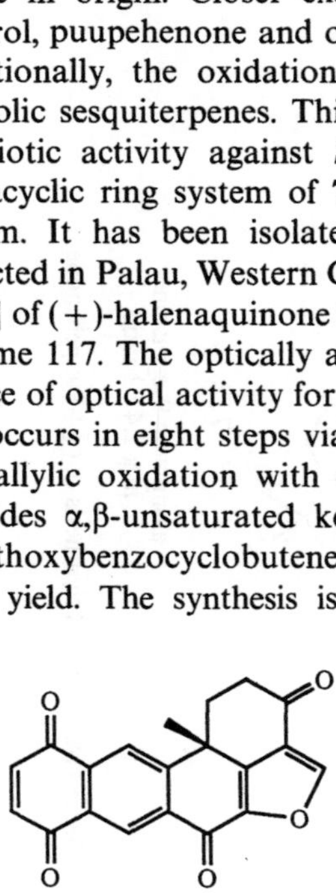

(+)-Halenaquinone **719a**

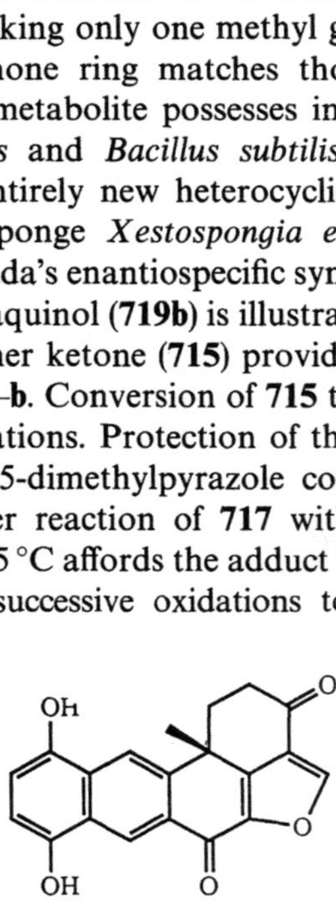

(+)-Halenaquinol **719b**

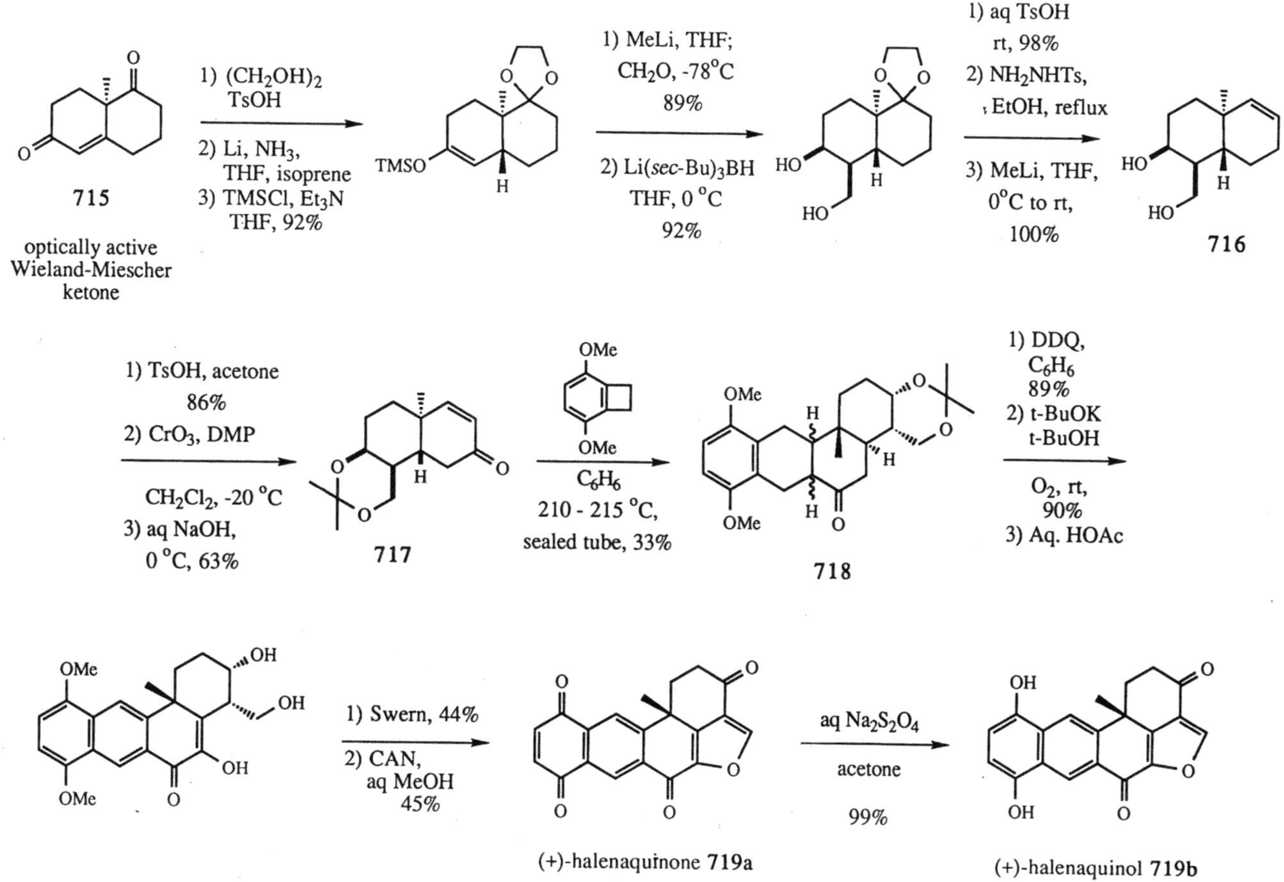

Scheme 117. Harada Synthesis of (+)-Halenaquinol and (+)-Halenaquinone

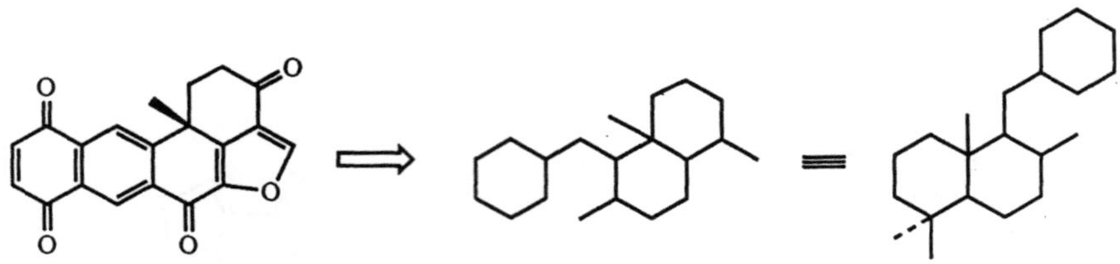

719a halenaquinone

cf. zonarol

(+)-halenaquinone (**719a**) in 17 steps and 2.1% overall yield. Treatment of **719a** with aqueous $Na_2S_2O_4$ produces (+)-halenaquinol (**719b**) in almost quantitative yield.

2.2.13 Miscellaneous Sesquiterpenoids

A number of sesquiterpenes do not easily fit into any of the previous classifications either biogenetically or structurally and thus have been relegated to a miscellaneous category.

2.2.13.1 Upial

Upial is an irregular sesquiterpene isolated from the sponge *Dysidea fragilis* [196]. An enantioselective synthesis of this substance was reported by Taschner [197] (Scheme 118). Key stereochemical control was exerted in the first half of the synthesis. Birch reduction of (−)-carvone and trapping of the resulting enolate with ethyl bromoacetate produces the keto-acid **721** (after hydrolysis). Birch reduction of the ketone and lactonization followed by stereoselective alkylation of the lactone enolate on the convex face provides stereocontrol in the formation of stereocenters at C3, 6 and 7. Conversion of **722** → **723** proceeded in a standard manner. Treatment of vinyl ether **723** with acid effected cyclization to the bicyclo[3.3.1]nonane system providing **724** in 70% yield. Routine chemistry followed, allowing the construction of (−)-upial in 17 steps from (−)-carvone with good stereocontrol. The absolute stereochemistry of the (+)-natural product was thus assigned as the enantiomer of **725**. The C14 epimer of racemic upial has been synthesized by Paquette [198].

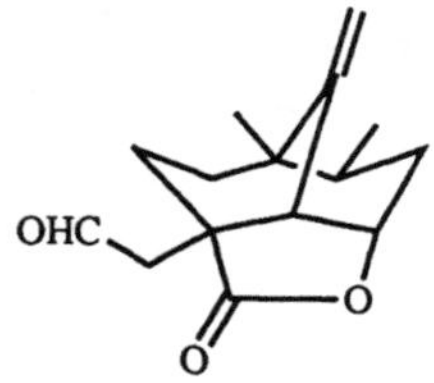

(-)-Upial **725**

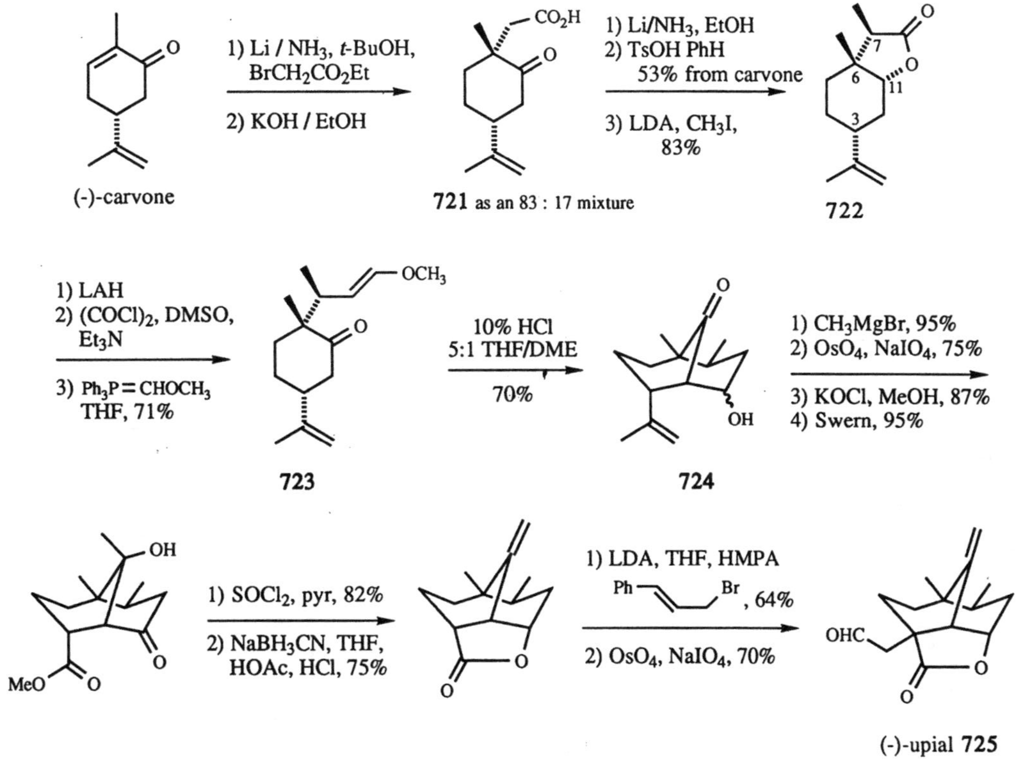

Scheme 118. Taschner Synthesis of (−)-Upial

2.2.13.2 Dactyloxenes-B and -C

The relative configurations of dactyloxenes B and C were determined by Ohloff in an exhaustive synthetic study in which all 8 diastereomeric racemates of the gross structure of spiro compound **732** were produced [199] (Scheme 119). Dactyloxenes B and C were found to arise from the known aldehyde **726**. Addition of ethynylmagnesium bromide to the ketone **727** gave a 1:1 diastereomeric mixture at the newly formed carbinol center which was separated. Acid-promoted cyclization of the indicated isomer resulted in the two isomeric spiro compounds **728** and **729**. Semi-hydrogenation of the individual isomers gave rise to dactyloxenes B (**730**) and C (**731**) which were assigned based on spectroscopic analysis of a large number of isomeric substances.

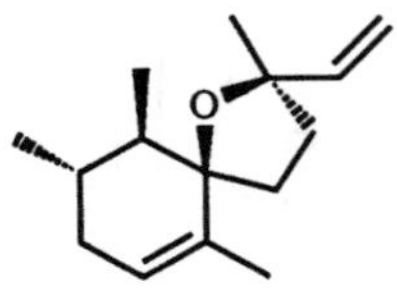

Dactyloxene-B **730**

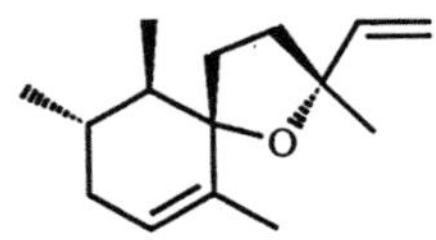

Dactyloxene-C **731**

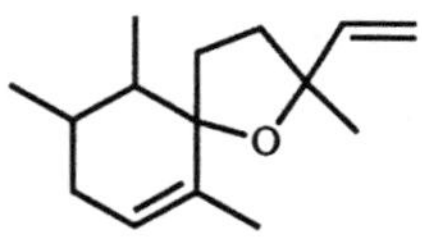

732

2.2.13.3 3-Debromoperforatone, Perforenone and an Unnamed Laurencia Metabolite

Perforenone and 3-debromoperforatone are related metabolites isolated from *Laurencia perforata* by Gonzalez and coworkers [200], who also reported syntheses which confirmed the structures (Scheme 120) [201]. An interesting Claisen rearrangement of **733** led to the cycloheptenone **734** which underwent selective Robinson annulation with **735** to provide a mixture favoring perforenone **736** over its *trans*-dimethyl isomer. Sodium borohydride reduction led

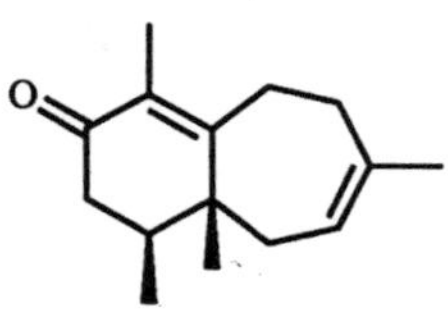

Perforenone **736**

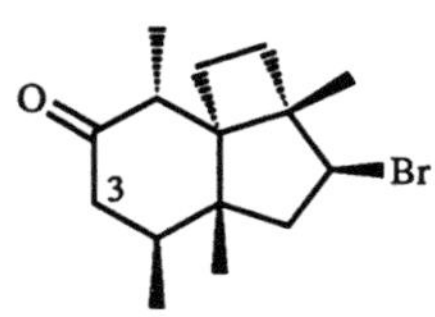

3-Debromoperforatone **739**

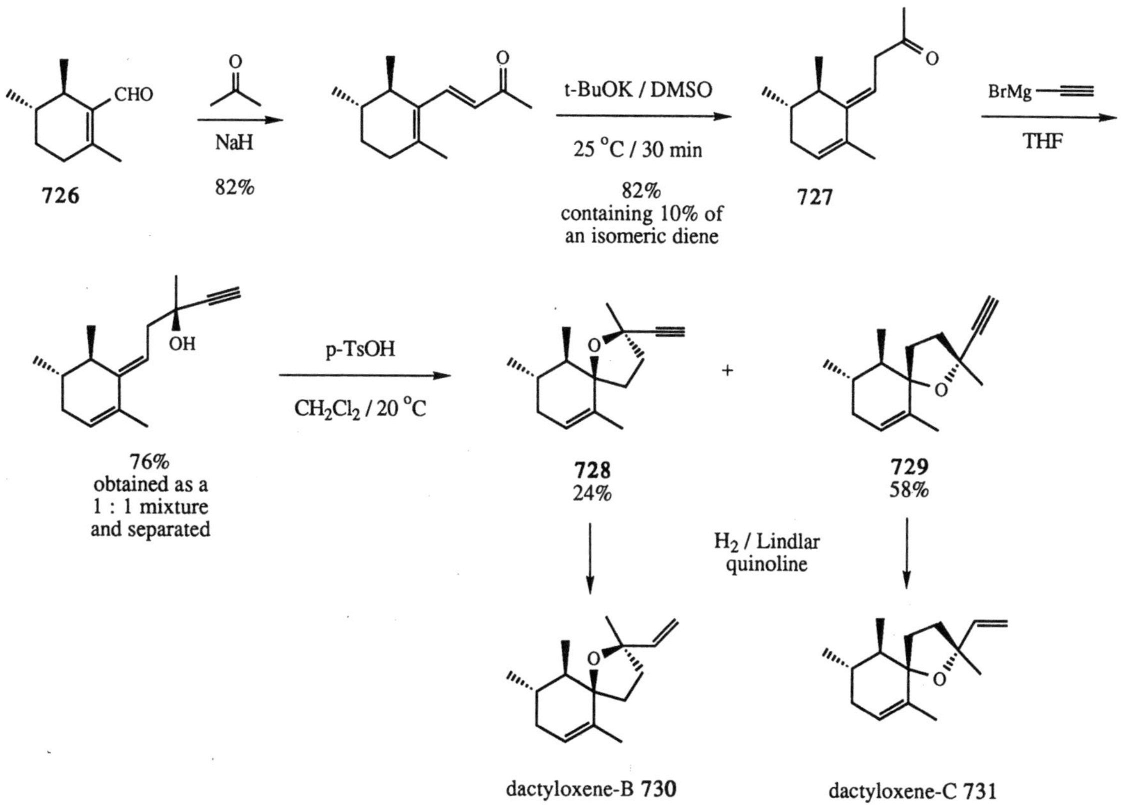

Scheme 119. Ohloff Syntheses of Dactyloxenes B and C

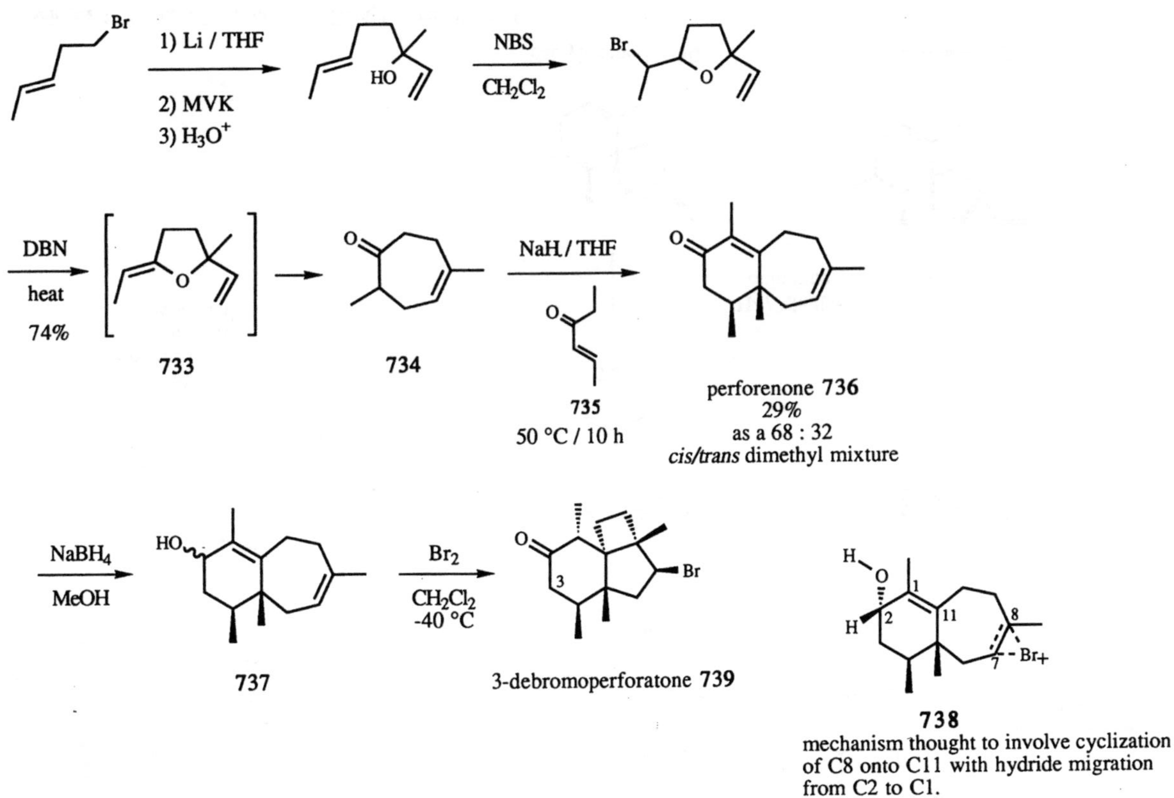

Scheme 120. Gonzalez Syntheses of Perforenone (**736**) and 3-Debromoperforatone (**739**)

to a mixture of allylic alcohols **737**. Treatment of the mixture with bromine at − 40 °C led to a purportedly biomimetic cyclization of the bromonium ion **738** with stereospecific hydride migration from C1 → C2 giving 3-debromo-perforatone (**739**) in unspecified yield. Several biomimetic transformations have been reported in this series.

Majetich [202] has also synthesized (±)-**736** utilizing a Lewis acid-catalyzed conjugate addition of an allylsilane as shown in Scheme 121. Sequential alkylation of the enone **740** allowed stereoselective construction of the two stereogenic centers at C4 and C5 (**741**). Addition of vinyl lithium and rearrangement led to the extended enone **742**. Cyclization to the 7-membered ring occurred on treatment with ethylaluminum dichloride providing an inconsequential mixture of C8 diastereomers. A series of 6 standard maneuvers led to racemic perforenone.

The *Laurencia nidifica* metabolite **747** has been synthesized by Vig [203] and by Oppolzer [204] in racemic form. The tetrasubstituted benzene was constructed by Vig (Scheme 122) beginning with the known tetralone **744**, thus controlling the substitution pattern around the ring. Oxidative cleavage of the olefin **745** led to the dialdehyde **746**, which was transformed via a standard series of steps to the target **747**. Oppolzer used a less conventional pathway to the metabolite (Scheme 123). Selective γ-addition of the pentadienyl anion **750** to **749** gave the blocked phenol **751**. Ionization of the alcohol and rearrangement of the carbocation gave ketone **752** which was converted to **747** by methylmagnesium iodide addition.

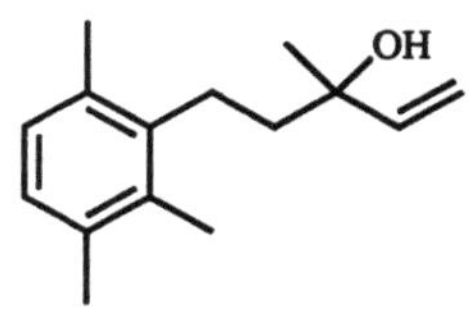

Laurencia Metabolite **747**

2.2.13.4 Aplysistatin

The oxepane-containing sesquiterpene aplysistatin (**761**) was isolated by Pettit from the sea hare *Aplysia angasi* [205]. The compound has been reported to show significant inhibition of murine lymphocytic leukemia [206] and has

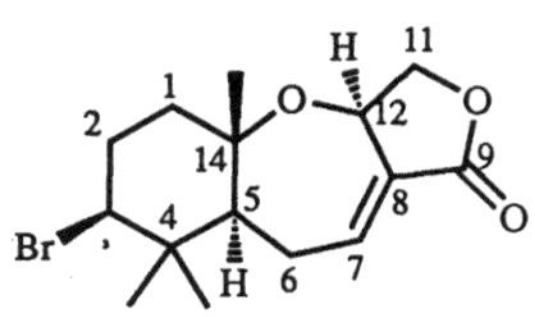

Aplysistatin 761

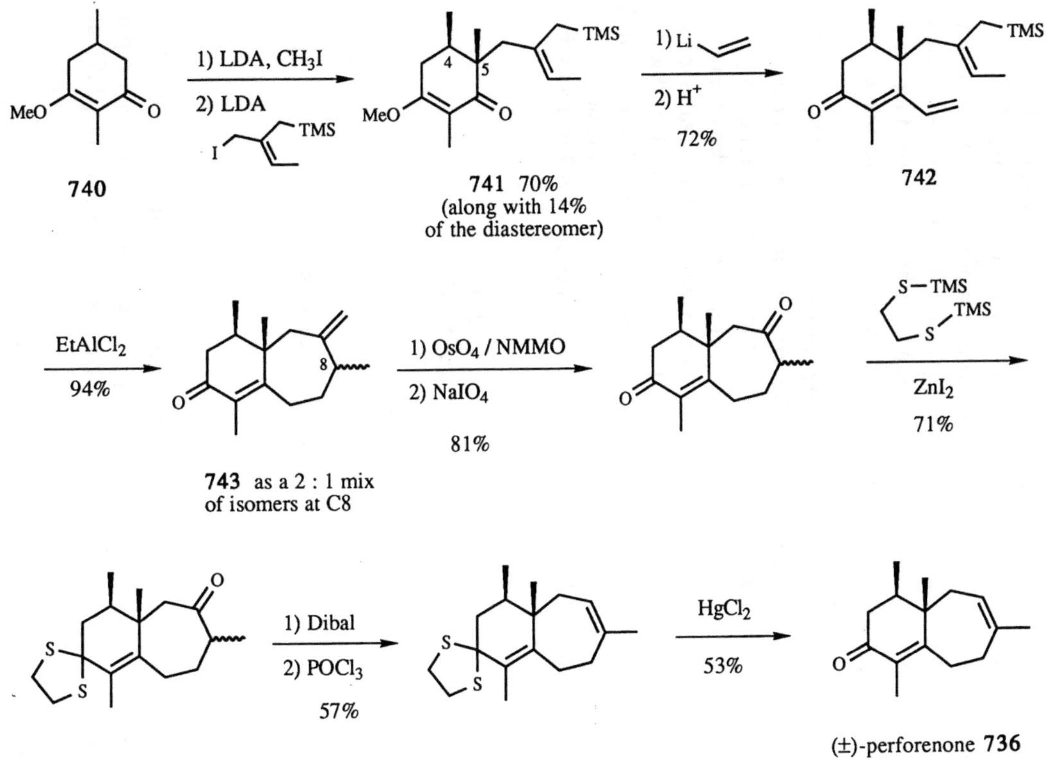

Scheme 121. Majetich Synthesis of (±)-Perforenone

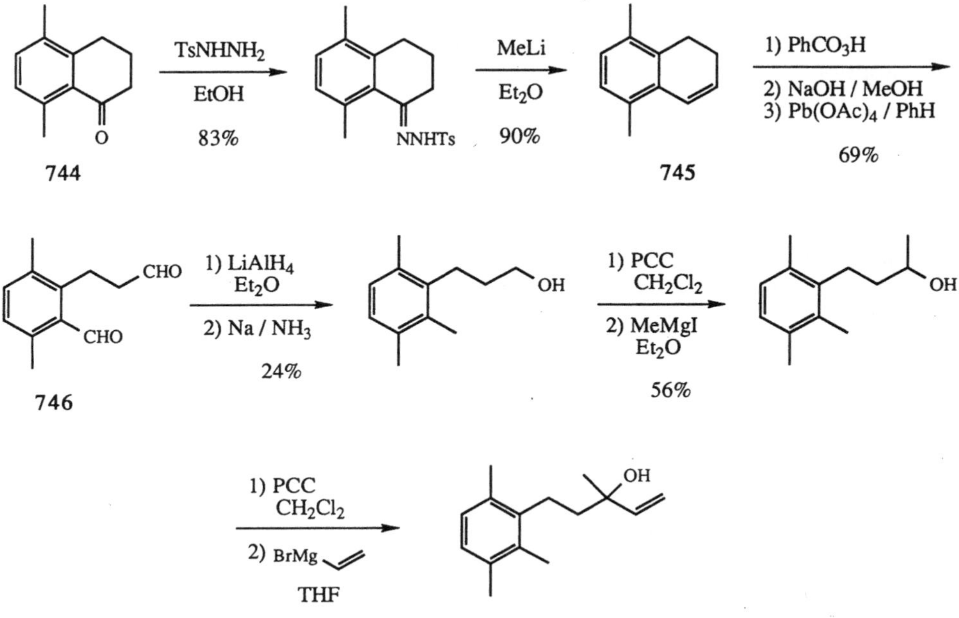

Laurencia metabolite **747**

Scheme 122. Vig Synthesis of *Laurencia* Metabolite **747**

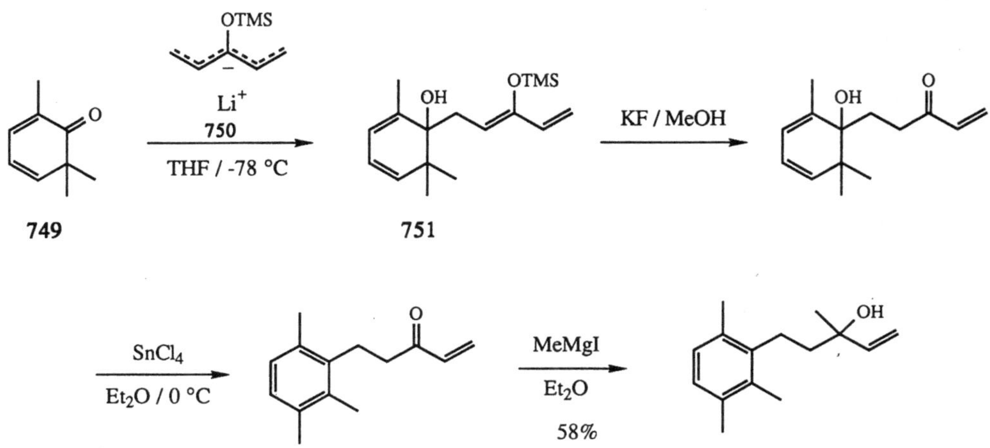

752 47% from **749** *Laurencia* metabolite **747**

Scheme 123. Oppolzer Synthesis of *Laurencia* Metabolite **747**

engendered much synthetic interest. Several syntheses of both racemic and optically active material have been reported, all of which utilize a biomimetic approach involving cation-olefin cyclization.

Hoye [207] reported the first synthesis, which involved cyclization of an extended geraniol derivative. Virtually all succeeding syntheses represent some variation of this process (Scheme 124). Geranyl tosylate was used to alkylate the ester enolate **753** providing **754**. Aldol reaction with α-benzyloxyacetaldehyde provided a 2:1 diastereomeric adduct mixture. When the mixture was treated with mercuric trifluoroacetate and the intermediate cyclized organomercurial was subjected to S_{E2} bromination, a mixture of diastereomeric bicyclic bromides (**755–758**) resulted. The isomer containing the correct relative configuration at C3, C5, C12 and C14 was isolated and carried on in two remaining steps to (±)-aplysistatin. The final lactone forming step was postulated to proceed via benzylic hydride abstraction from **759** leading to the oxenium ion **760** which closed with loss of benzaldehyde eventually resulting in **761**. The lack of stereoselectivity and poor yield of desired product was to be mirrored in the subsequent syntheses and indicates the complexity of the process. In a second approach, Hoye avoided a two-fold cyclization by designing a synthesis utilizing successive single-ring closures (Scheme 125) [208]. The bromolactone **763**, prepared by cyclization of homogeranic acid was reduced and chain extended to the diastereomeric mixture of bromo-alcohols **764**. When the individual *erythro* and *threo* isomers of **764** were subjected to mercurinium ion initiated cyclization the major products possessed the desired configuration at C12 relative to the centers in the starting bromolactone. Conversion of **765** to aplysistatin was accomplished in high yield in three successive steps. This represented a significant improvement over the two-fold cyclization.

The approach of White was quite similar (Scheme 126) [209]. The extended geraniol derivative was converted to sulfide **766**. Sodium borohydride reduction provided the alcohol **767** which underwent cyclization under the same conditions as in the Hoye synthesis to provide the oxepanes **768** and **769** in a 3.6:1 ratio, but in low yield. Conversion of the major isomer to racemic aplysistatin was straightforward. A nearly identical sequence was described by Gosselin and Rouessac [210] (Scheme 127) utilizing the intermediate sulfide **767** which was prepared from geranyl chloride in a slightly different manner. Cyclization of **767** however was accomplished by functionalization of the remote olefin to the bromohydrin mixture **772**. Treatment of **772** with tin(IV) chloride presumably regenerates a bromonium ion which undergoes olefin cyclization and trapping to provide oxepane **768** of the correct configuration in 3% yield. This substance was carried on to aplysistatin by the White procedure.

Two syntheses of optically active **761** have been reported. Prestwich utilized the hydroxylactone **773** (derived from (+)-malic acid) to construct the cyclization substrate **774** (Scheme 128) [211]. Direct bromonium ion initiated cyclization led to an unfavorable mixture of **775** (19%) and **776** (81%) in 12% combined yield. Regiospecific introduction of unsaturation provided (−)-aplysistatin (**761**) and the 12-*epi*-isomer of (+)-aplysistatin (**777**).

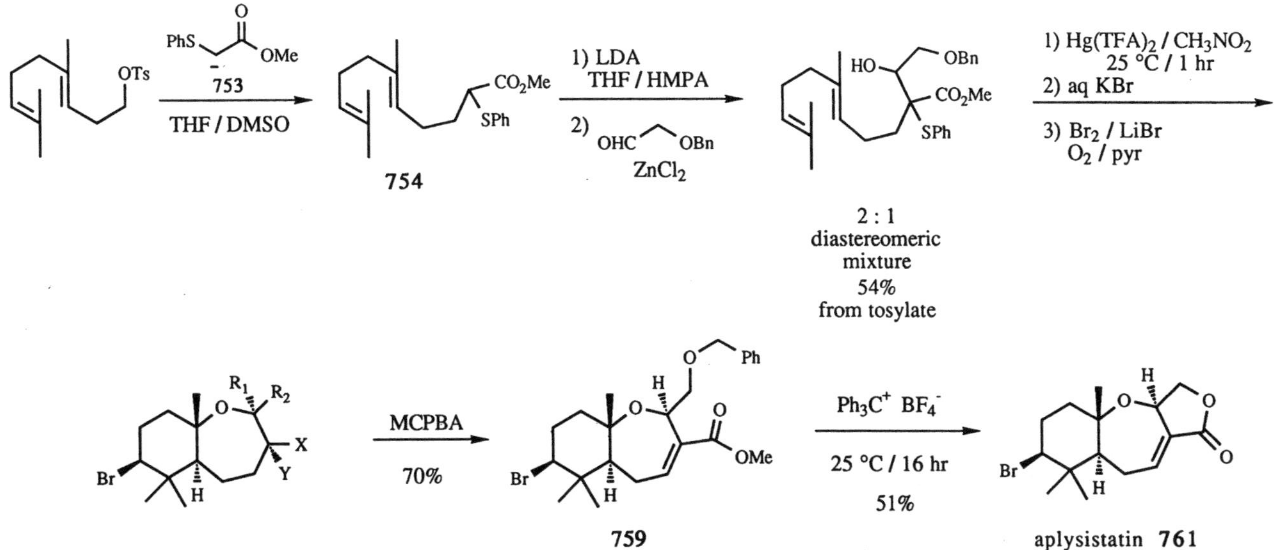

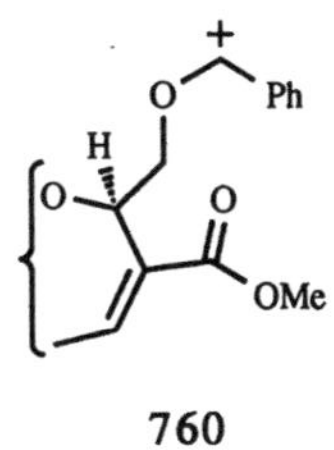

	R_1	R_2	X	Y	
755	H	CH_2OBn	SPh	CO_2Me	11%
756	CH_2OBn	H	CO_2Me	SPh	11%
757	H	CH_2OBn	SPh	CO_2Me	11%
758	CH_2OBn	H	CO_2Me	SPh	11%

Scheme 124. Hoye Aplysistatin Synthesis

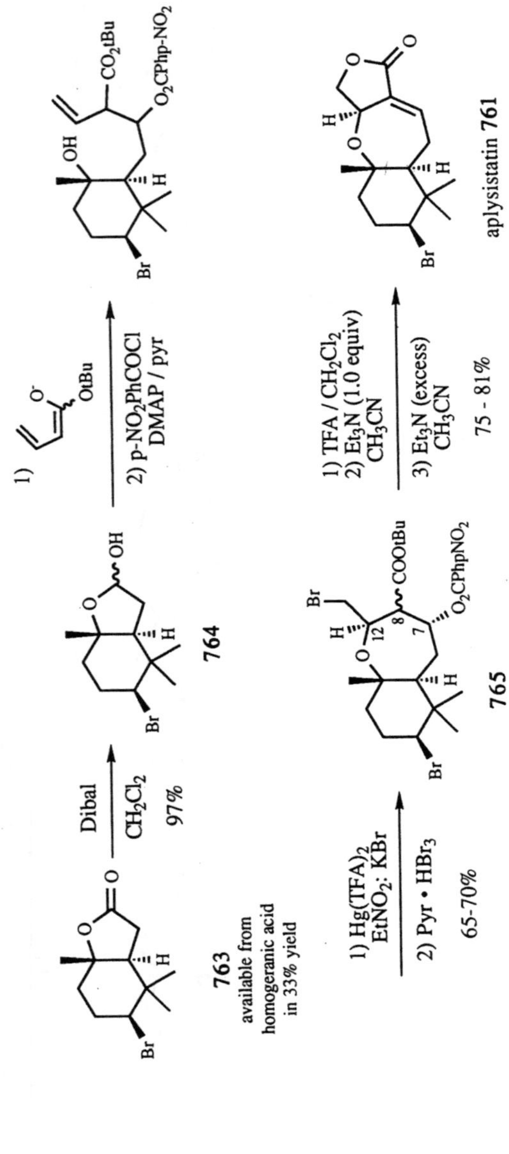

Scheme 125. Hoye Second Aplysistatin Synthesis

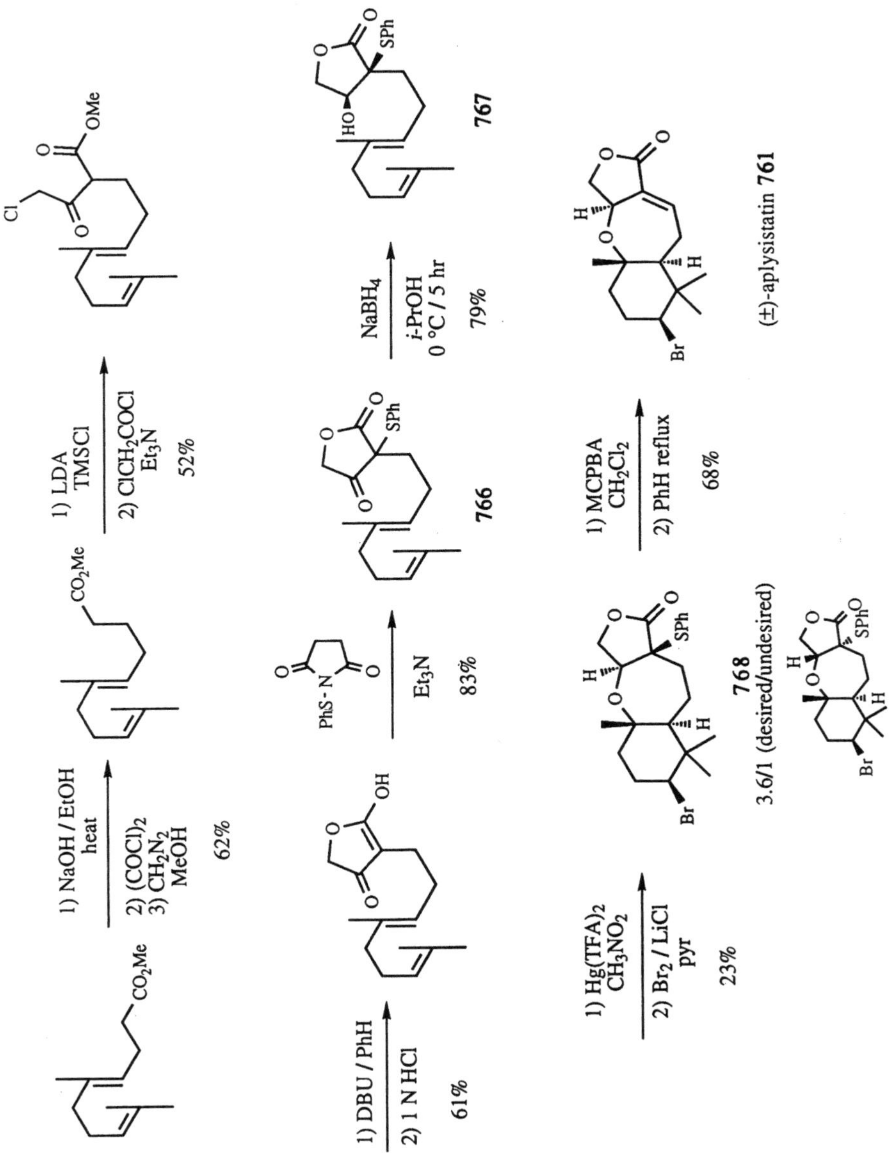

Scheme 126. White (±)-Aplysistatin Synthesis

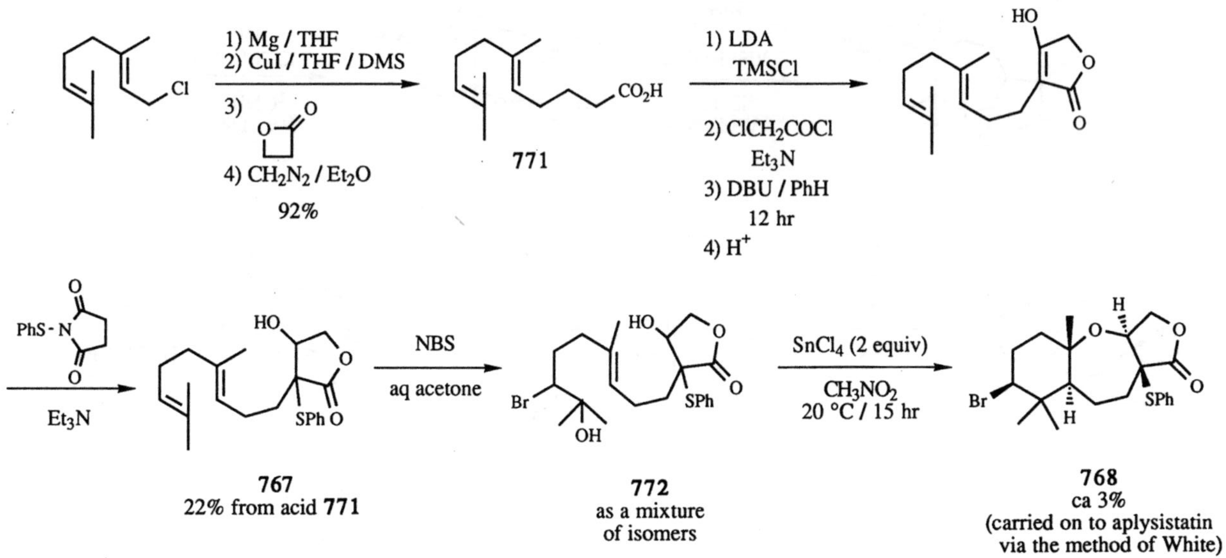

Scheme 127. Gosselin-Rouessac Synthesis of (±)-Aplysistatin

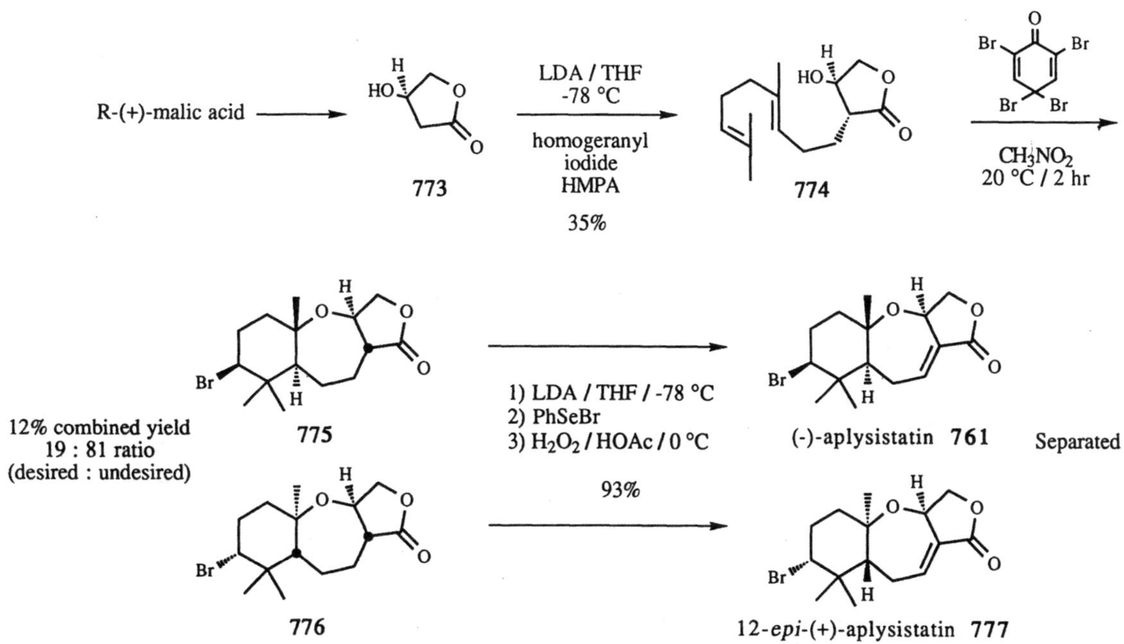

Scheme 128. Prestwich Synthesis of (−)-Aplysistatin

Tanaka also prepared (−)-**761** by cyclization of a hydroxylactone (Scheme 129) [212]. The fully functionalized *bis-seco* cyclization substrate **780** was prepared via standard chemistry from dimethyl *L*-tartrate and geranyl bromide. Hoye-type cyclization of **780** led directly to (−)-aplysistatin in 7% yield along with 45% of 12-*epi*-(+)-aplysistatin, the same two products in the Prestwich cyclization of the saturated substrate.

2.2.13.5 Cavernosine

The icthytoxic C17 terpenoid substance cavernosine (**783**) was isolated by Braekman [213] from the sponge *Fasciospongia cavernosa.* Jefford has efficiently synthesized this substance from dihydro-β-ionone in two steps [214]. Lewis acid-catalyzed alkylation of the furan **781** gave predominantly the erythro isomer **782** corresponding to γ-addition (Scheme 130). Conjugate reduction provided cavernosine.

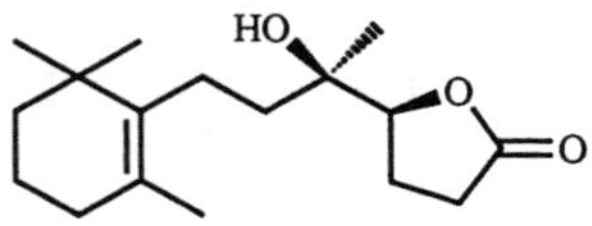

Cavernosine **783**

2.3 Diterpenoids

2.3.1 Dolastanes

The dolastane-type diterpenoids constitute a family of marine natural products possessing a linearly-fused 5,7,6-tricyclic skeleton. They have been isolated from a variety of sources at different geographic locations [215]. The three members of this class that have been synthesized are isoamijiol (**805**), dolasta-1(15), 7,9-trien-14-ol (**791**) and amijitrienol (**799**). Amijitrienol and isoamijiol co-occur as the principal metabolites in the brown seaweed *Dictyota linearis* and have been shown to exhibit moderate antimicrobial activity against *Staphylococcus aureus*

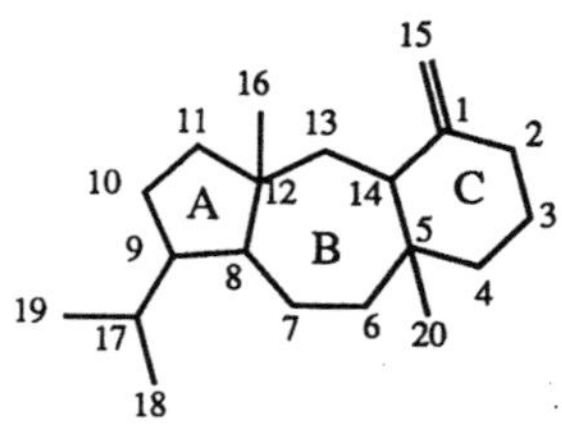

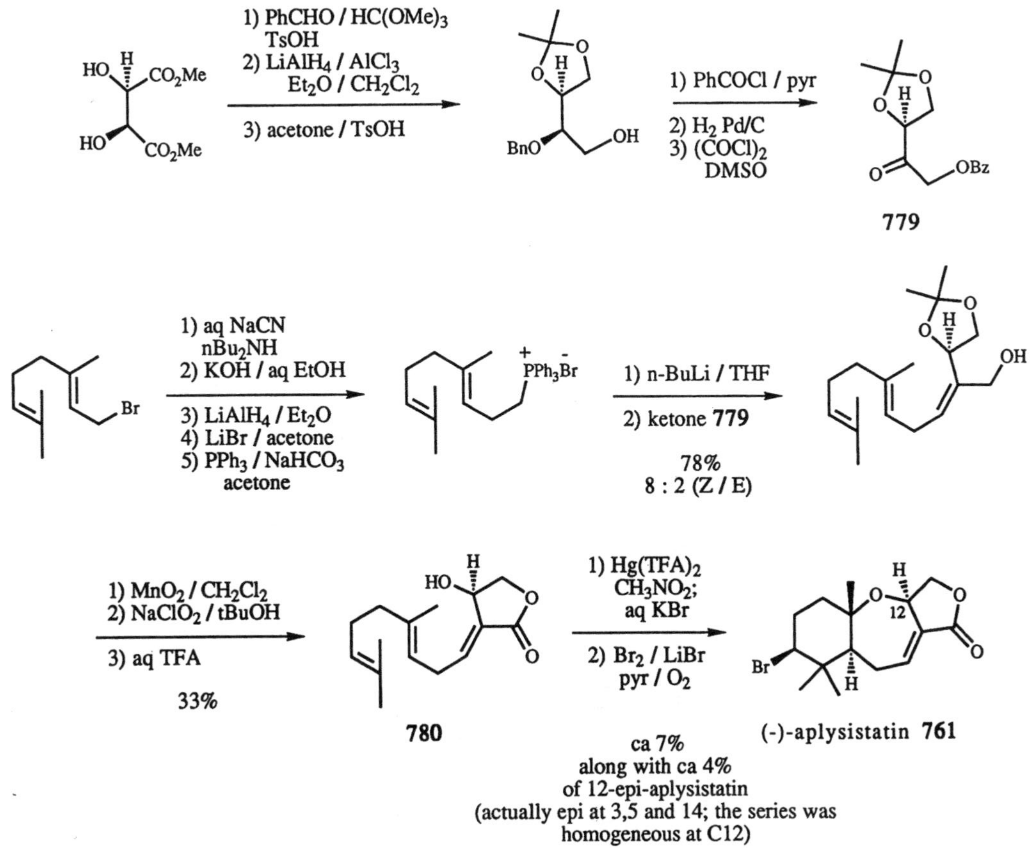

Scheme 129. Tanaka (−)-Aplysistatin Synthesis

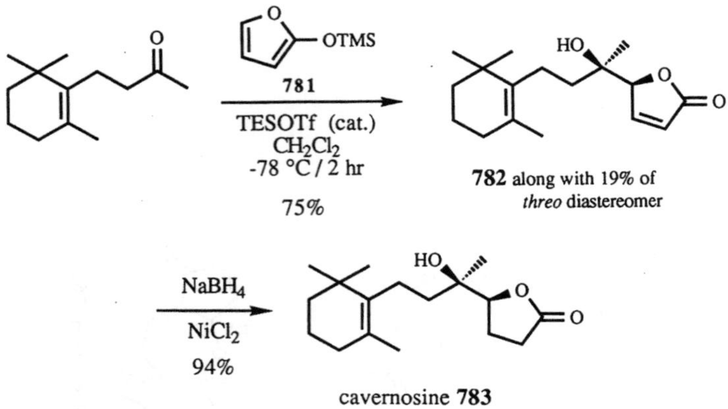

Scheme 130. Jefford Synthesis of Cavernosine

and *Mucor mucedo* [216]. Dolasta-1(15), 7,9-trien-14-ol (**791**) has been isolated [217] from the brown seaweeds *Dictyota linearis* and *D. divaricata.*

2.3.1.1 Dolasta-1(15),7,9-trien-14-ol

The strategy utilized by Piers [218] in the synthesis of (±)-dolasta-1(15), 7,9-trien-14-ol (**791**) (Scheme 131) involves construction of the ring system in the order B → A-B → A-B-C. The B-ring is obtained by ring expansion of a six-membered cyclic ketone. The A-ring is appended by alkylation and palladium-mediated cyclization of an organostannane onto an enol triflate. The C-ring is obtained through an alkylation followed by an intramolecular Grignard reaction. The synthesis is initiated by cyclopropanation of the kinetically formed TMS enol ether of ketone **784** to give **785**. Lewis acid-catalyzed opening of the cyclopropane to give the β-chloroketone followed by dehydrohalogenation and hydrogenation provides the B-ring in the form of ketone **786**. Alkylation of **786** followed by conversion to the enol triflate and palladium-mediated cyclization produces **787** in good yield. The stereochemistry of successive alkylation of **788** with vinylstannane **789** and methyl iodide is controlled by the angular methyl moiety to give **790**, after treatment with iodine. Treatment of **790** with magnesium yields (±)-dolsta-1(15), 7,9-trien-14-ol (**791**) in 15 steps and 3.3% overall yield.

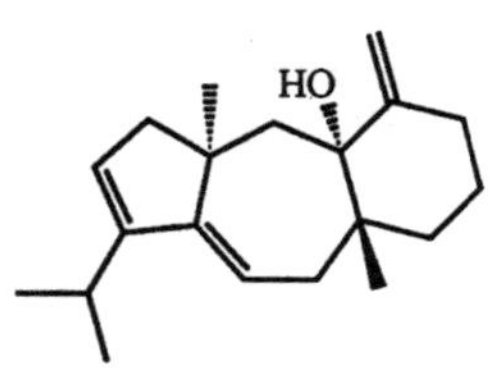

Dolasta-1(15),7,9-trien-14-ol **791**

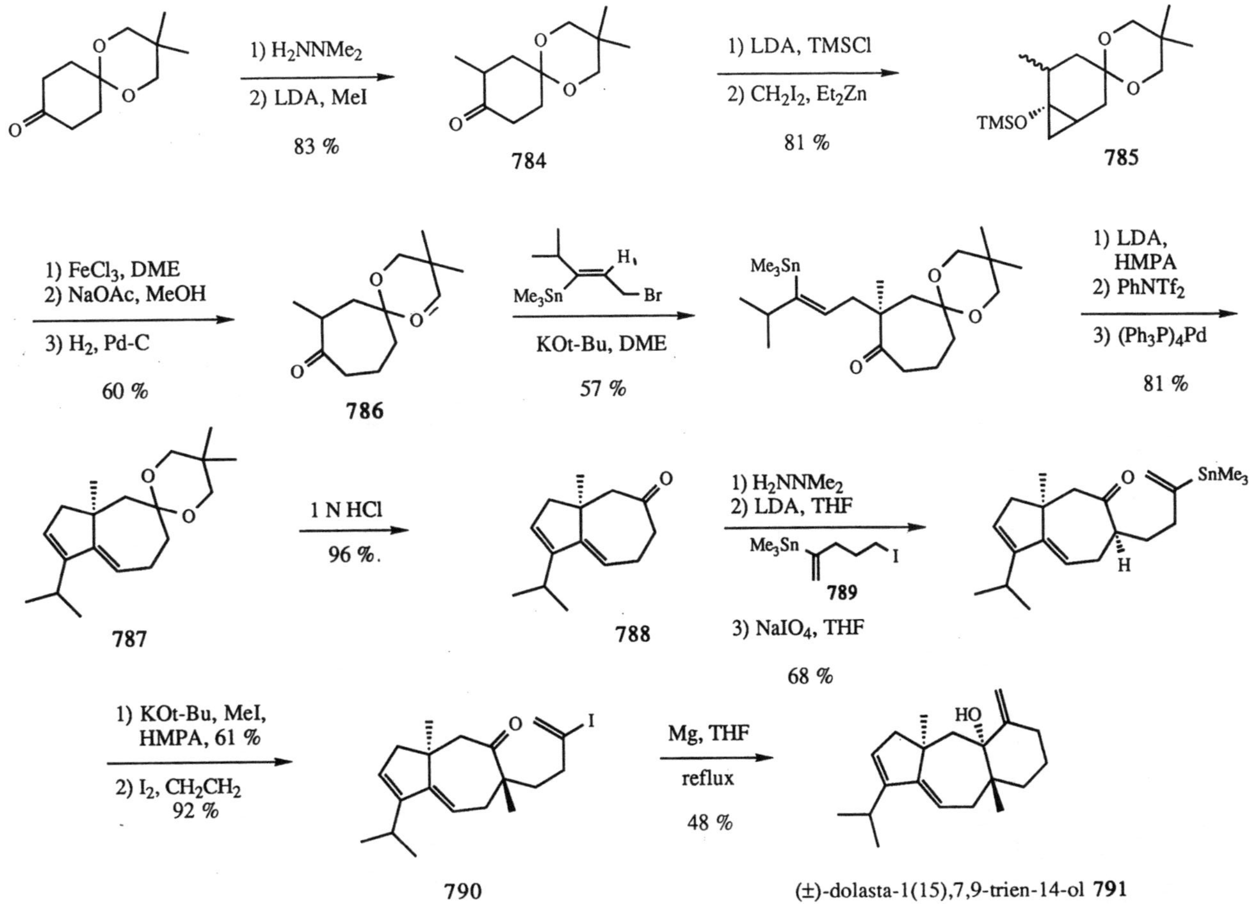

Scheme 131. Piers Synthesis of (±)-Dolasta-1(15),7,9-trien-14-ol

2.3.1.2 *Amijitrienol*

An offshoot of the route used for the preparation of **791** is also utilized by Piers [219] for the preparation of (±)-amijitrienol (**799**) (Scheme 132). The overall strategy is iterative in that the same methodology is utilized for the preparation of both the A- and C-rings. Reduction of intermediate **787** (Scheme 131) under Birch conditions followed by isomerization of the alkene positional isomers gives **793** in 95% yield after hydrolysis. Aldol condensation of **793** with **794** followed by oxidation and alkylation yields the β-diketone **795**. The two carbonyls are differentiated by reduction with DIBAL in the presence of lithium chloride to provide **796** after silylation. Formation of enol triflate **797** and subsequent palladium-induced cyclization produces **798** which is converted to (±)-amijitrienol **799** by treatment with fluoride. The synthesis requires 20 steps from commercially available materials and gives a yield of 1.6%.

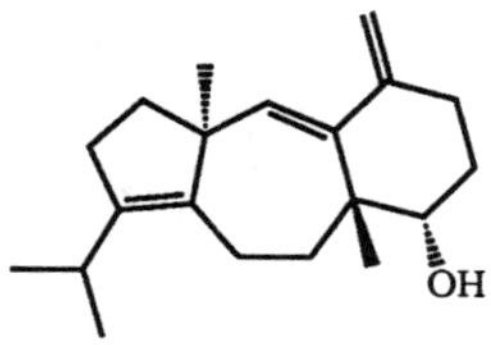

Amijitrienol **799**

2.3.1.3 *Isoamijiol*

Pattenden [220] has prepared (±)-isoamijiol (**805**) by the procedure illustrated in Scheme 133. The general strategy involves formation of the A-B ring system in the form of hydroazulenone **802**, via ring expansion of an angularly fused 5-4-5 tricyclic silyl ether. Addition of the C-ring is completed by a reductive radical cyclization. The key tricyclic intermediate **801** is obtained as a single isomer by irradiation of the mixture of isomers of **800** in hexane. Titanium-induced coupling of **801** with acetone provides the isopropylidene functionality necessary for the acid induced ring expansion to produce **802**. The angular methyl moiety of **802** controls the sequential alkylations to allow formation of **803** stereospecifically, after alkyne deprotection. Reductive radical cyclization secures the penultimate intermediate **804**. Catalytic allylic oxidation with selenium

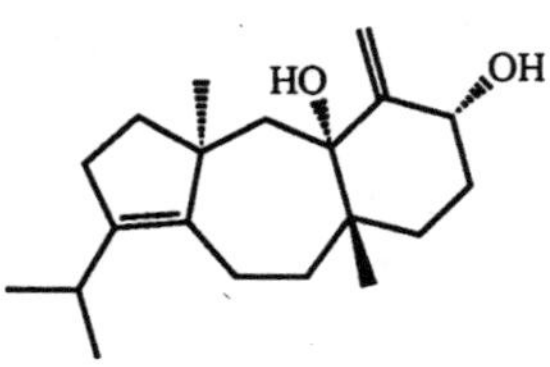

Isoamijiol **805**

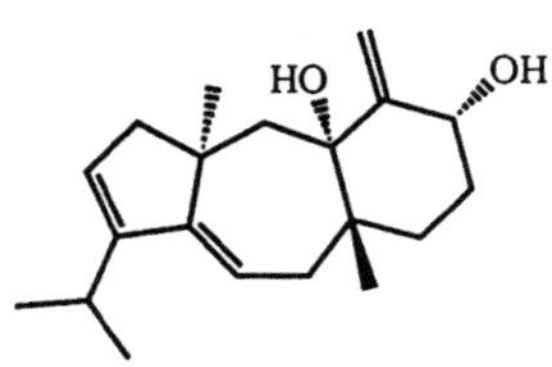

806

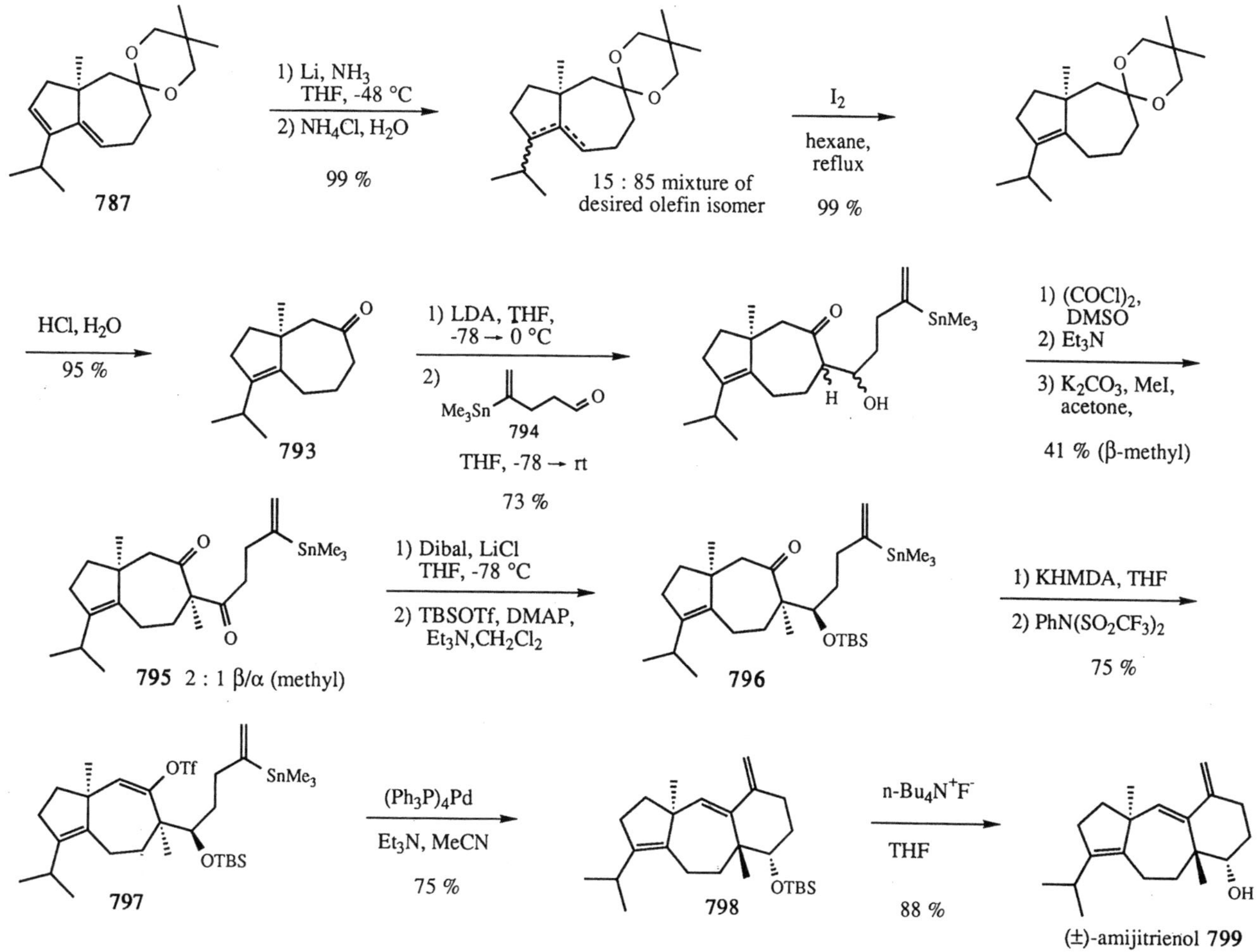

Scheme 132. Piers Synthesis of (±)-Amijitrienol

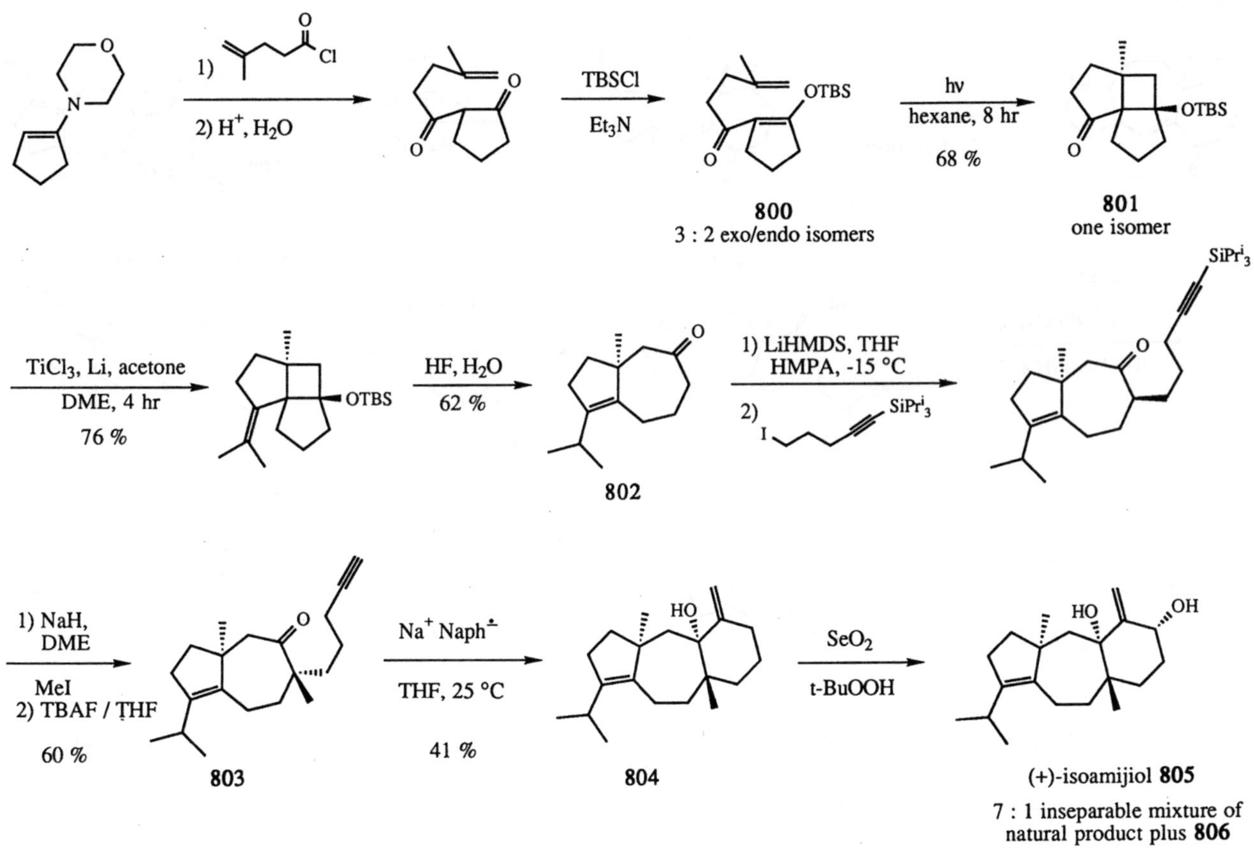

Scheme 133. Pattenden Synthesis of (+)-Isoamijiol

dioxide and *t*-butylhydroperoxide produces (±)-isoamijiol **805** along with an isomeric trienol **806** in a ratio of 7:1 as an inseparable mixture. Isoamijiol is prepared in 10 steps and approximately 5% yield.

Mehta [221] has synthesized both (+)-isoamijiol (**805**) and (±)-dolsta-1(15), 7,9,-trien-14-ol (**791**) along with an isomeric trienol **806** in an enantioselective (but not regioselective) synthesis (Scheme 134). The general strategy involves ring contraction of limonene via known chemistry to give an optically pure A-ring fragment. The B-ring is appended to form the A-B ring system in the guise of hydroazulenone **802**, a common intermediate from the earlier synthesis of Pattenden (Scheme 133). Addition of the C-ring is completed (in identical fashion as in the Pattenden synthesis) by a reductive radical cyclization. Reduction of **808**, vinyl ether formation and stereospecific Claisen rearrangement affords **809** in three steps. Two-carbon homologation and oxidation followed by cation-olefin cyclization of the 1,6-diene formed results in formation of optically pure hydroazulenone **802**. The synthesis is completed in identical fashion as the Pattenden synthesis (Scheme 133) resulting in a 2:3:1 mixture of **805**, **791** and **806** in 60% yield that could be separated by repeated chromatography on silver nitrate-impregnated silica. An overall yield of 2% was obtained over 11 steps.

2.3.2 Adocianes and Amphilectanes

Diterpene isocyanides are an interesting class of metabolites possessing tri- and tetracyclic perhydrophenalene and perhydropyrene ring systems respectively. Two syntheses of metabolites in this class have been reported.

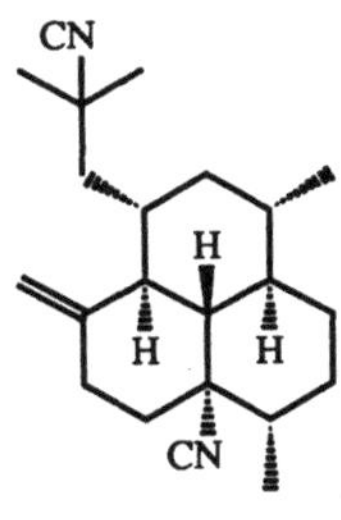

Hymeniacidon amphilecta
Metabolite **822**

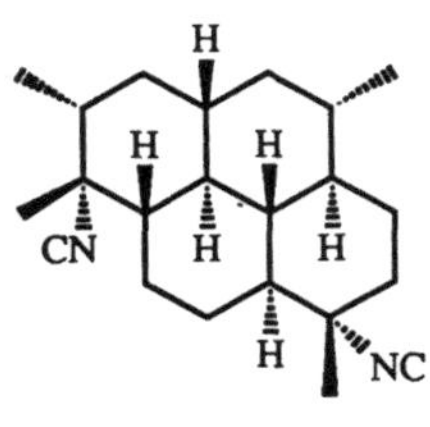

7,20-Diisocyanoadociane **835**

2.3.2.1 *Hymeniacidon amphilecta bis-isonitrile*

Piers [222] has reported the synthesis of the amphilectane (**822**) [223] in only 22 operations (Scheme 135). The β-ketoester **810** (prepared by alkylation) was deprotonated and converted to the vinyl triflate. Intramolecular Stille coupling gave the diene **811** in 86% yield. Diels-Alder cycloaddition of **811** to acrolein

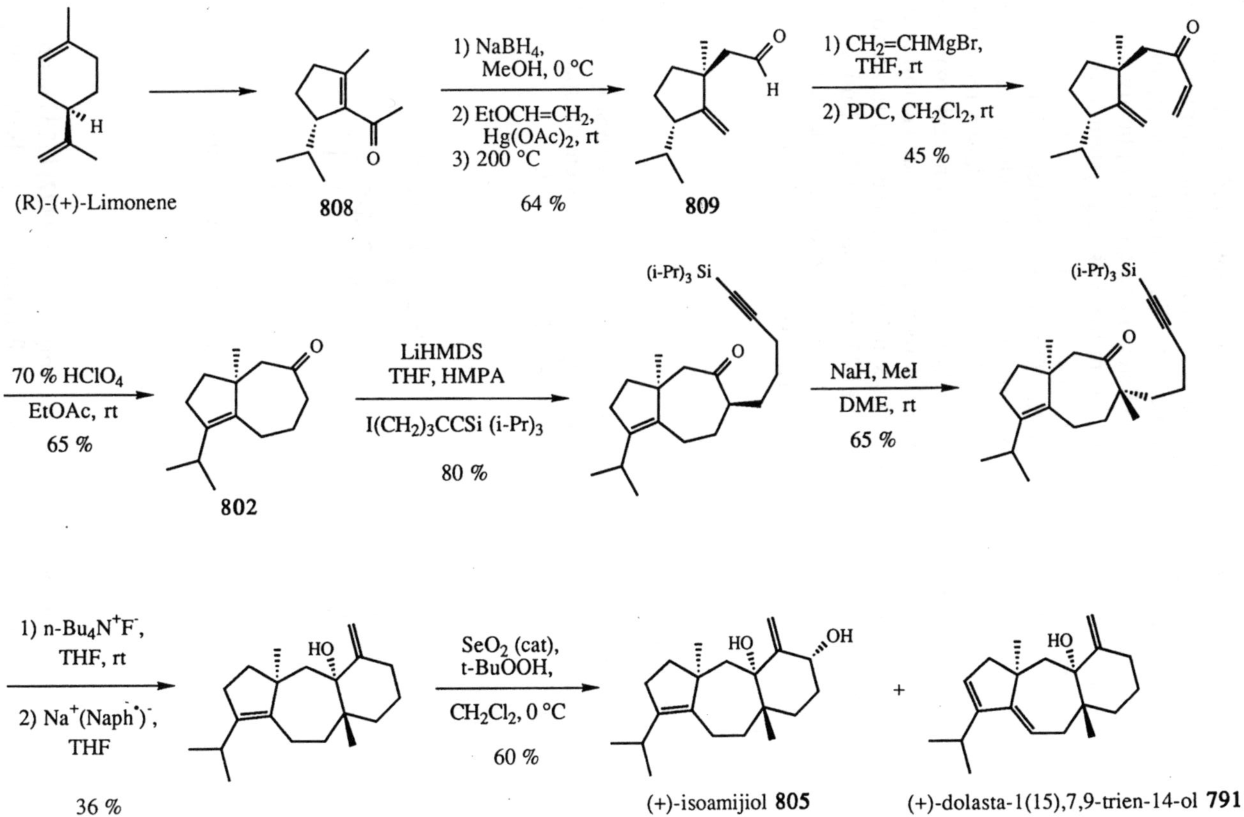

Scheme 134. Mehta Synthesis of (+)-Isoamijiol and (+)-Dolasta-1(15),7,9-trien-14-ol

gave a mixture of four adducts which, after equilibration left only two adducts (**812** and **815**). Adduct **815** was separated and carried on to the intermediate **816** with the C3 methyl group in the proper equatorial orientation. Functionalization of C11 with $CrO_3 \cdot$ pyrazole gave ketone **817**. Routine manipulations led to axial aldehyde **818** which could be epimerized to the equatorial isomer **819** in 95% yield. The isonitriles were introduced simultaneously by Curtius rearrangement of acyl azides prepared from the diacid **820**. Trapping of the resulting diisocyanate with β-trimethylsilylethanol gave intermediate **821**. Deprotection to the amine, formylation to the bis-formamide and dehydration to the bis-isonitrile completed a particularly efficient route in 2% overall yield.

2.3.2.2 7,20-Diisocyanoadociane

Corey [224] has reported the synthesis of the tetracyclic metabolite **835** [225] from *Adocia* sp. (Scheme 136). The menthyl ester **824** was transformed to the ketal **825** with some difficulty. Kinetic conjugate addition of the ester enolate to methyl crotonate gave a favorable mixture of the desired *threo* 1,4-adduct in 60% ee. Reductive de-menthylation along with a standard series of steps led to the triene **827**. Intramolecular Diels-Alder reaction of the α,β-unsaturated ketal gave rise to the *trans*-fused product **828**, whereas cycloaddition of the corresponding α,β-unsaturated ketone gave only a *cis*-fused product. The molecule was next prepared for a second intramolecular [4 + 2] cycloaddition by conversion to triene **829**. Cycloaddition of this substance at 185 °C under dilute conditions led to a 54% yield of the *exo* adduct **830** containing the correct stereochemistry at all ring fusions except C11. This was corrected by oxidatively excising the one-carbon substituent at C10 providing the ketone **831** which could be epimerized at C11 to **832**, possessing what is probably the most stable isomer of this tetracyclic ring system. Methylation of the ketone at C9 gave initially the axial methyl isomer which was epimerized in a subsequent step to equatorial. Treatment of the diketone with MeLi mediated by $CeCl_3$ gave mainly the isomer **833** corresponding to *bis*-equatorial addition of the organometallic reagent. Finally, introduction of the isocyano groups was accomplished by trifluoroacetylation to **834** and treatment with TMSCN and $TiCl_4/CH_2Cl_2$ at room temperature. This latter step gave the natural product (**835**) along with the $C20_{eq}C7_{ax}$ diisonitrile isomer in a combined yield of 55%. Also obtained were the diaxial isonitrile (30%) and the diequatorial diisonitrile (15%). The natural product so produced was obtained in 60% ee based on the early Michael adduct **826**. The absolute configuration of the naturally-derived material was thus as shown in **835**.

2.3.3 Spatanes

The spatane diterpenes originate from species of brown algae in the family *Dictyotaceae* [226]. Possessing the *cis-anti-cis* linear fusion of a (5,4,5)

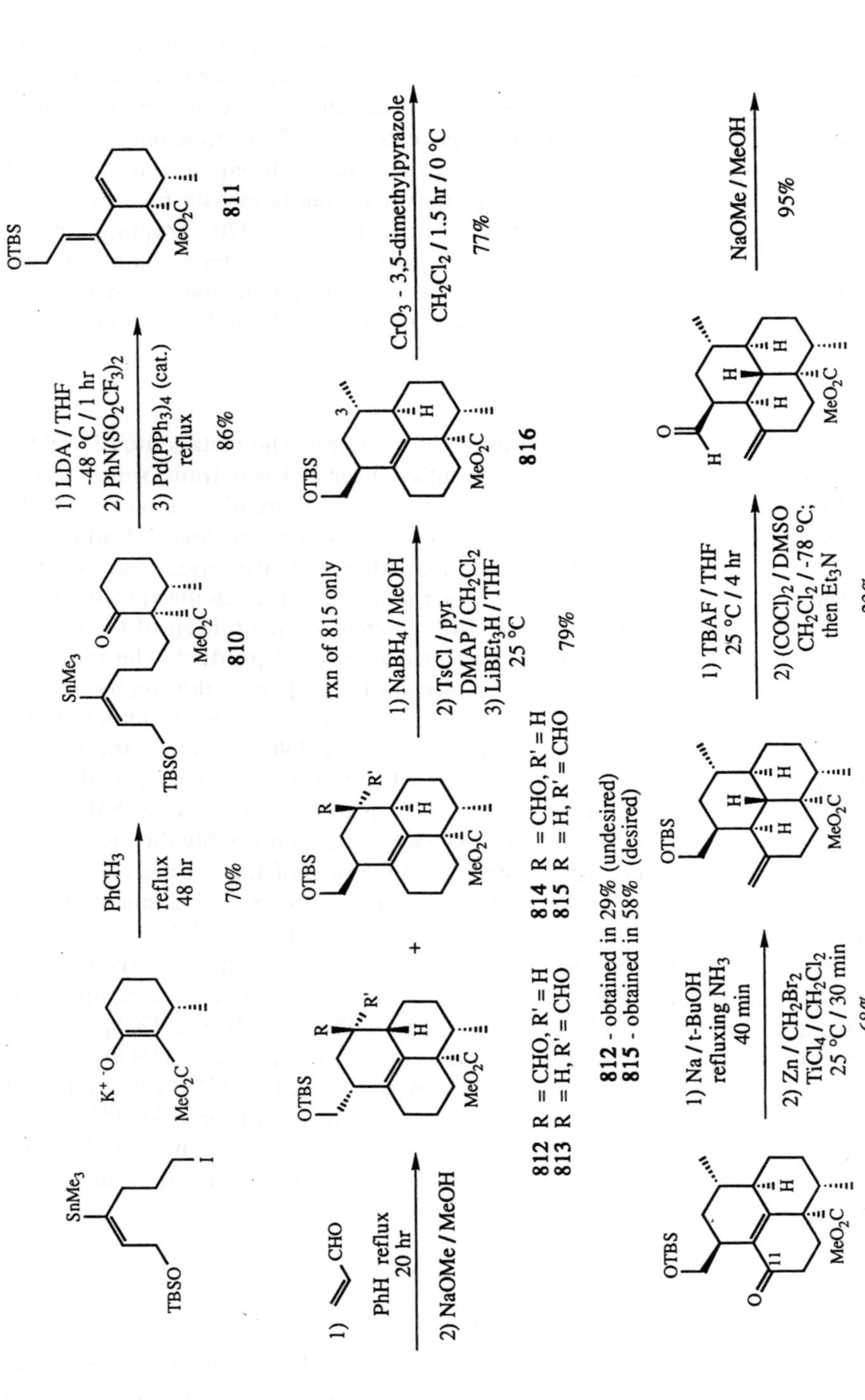
TBSO
SnMe3
I
K+ -O
MeO2C
PhCH3
reflux
48 hr
70%
TBSO
SnMe3
O
MeO2C
810
1) LDA / THF
-48 °C / 1 hr
2) PhN(SO2CF3)2
3) Pd(PPh3)4 (cat.)
reflux
86%
OTBS
MeO2C
811
1) CHO
PhH reflux
20 hr
2) NaOMe / MeOH
OTBS
R
R'
H
MeO2C
+
OTBS
R
R'
H
MeO2C
812 R = CHO, R' = H
813 R = H, R' = CHO
814 R = CHO, R' = H
815 R = H, R' = CHO
812 - obtained in 29% (undesired)
815 - obtained in 58% (desired)
rxn of 815 only
1) NaBH4 / MeOH
2) TsCl / pyr
DMAP / CH2Cl2
3) LiBEt3H / THF
25 °C
79%
OTBS
3
H
MeO2C
816
CrO3 - 3,5-dimethylpyrazole
CH2Cl2 / 1.5 hr / 0 °C
77%
OTBS
O
11
H
MeO2C
817
1) Na / t-BuOH
refluxing NH3
40 min
2) Zn / CH2Br2
TiCl4 / CH2Cl2
25 °C / 30 min
68%
OTBS
H
H
H
MeO2C
1) TBAF / THF
25 °C / 4 hr
2) (COCl)2 / DMSO
CH2Cl2 / -78 °C;
then Et3N
83%
O
H
H
H
H
MeO2C
818
NaOMe / MeOH
95%

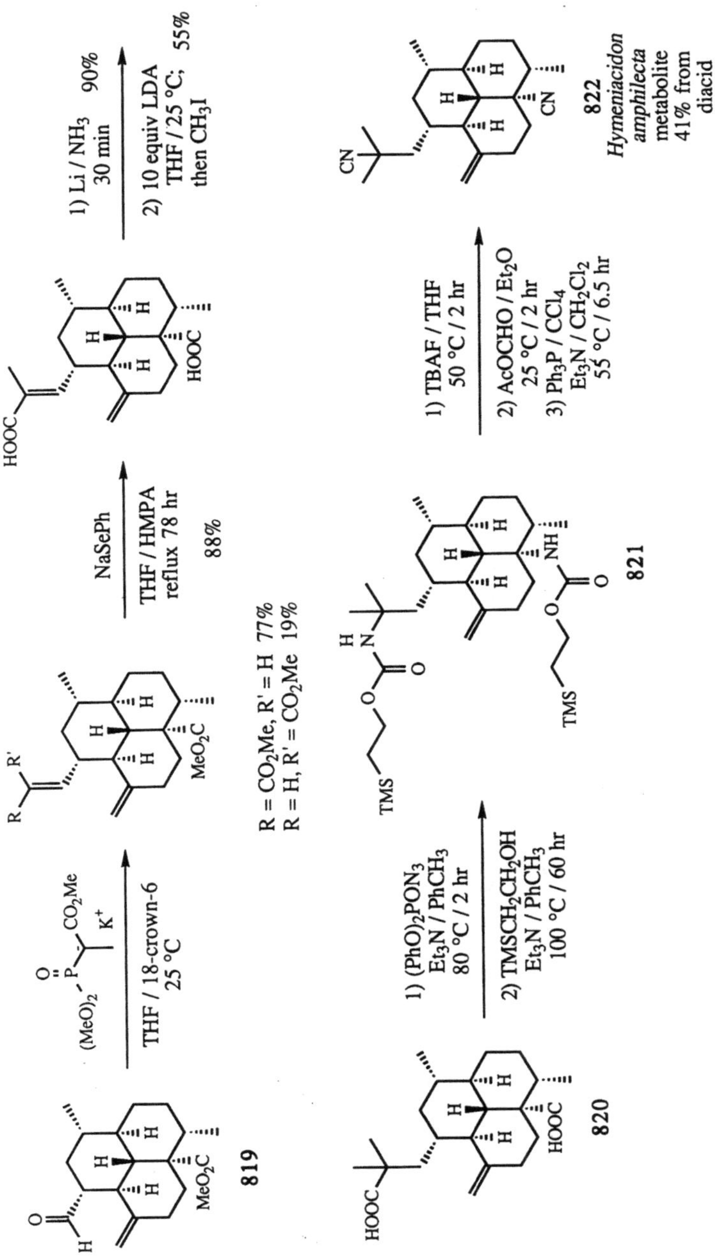

Scheme 135. Piers Synthesis of *Hymeniacidon amphilecta* Metabolite **822**

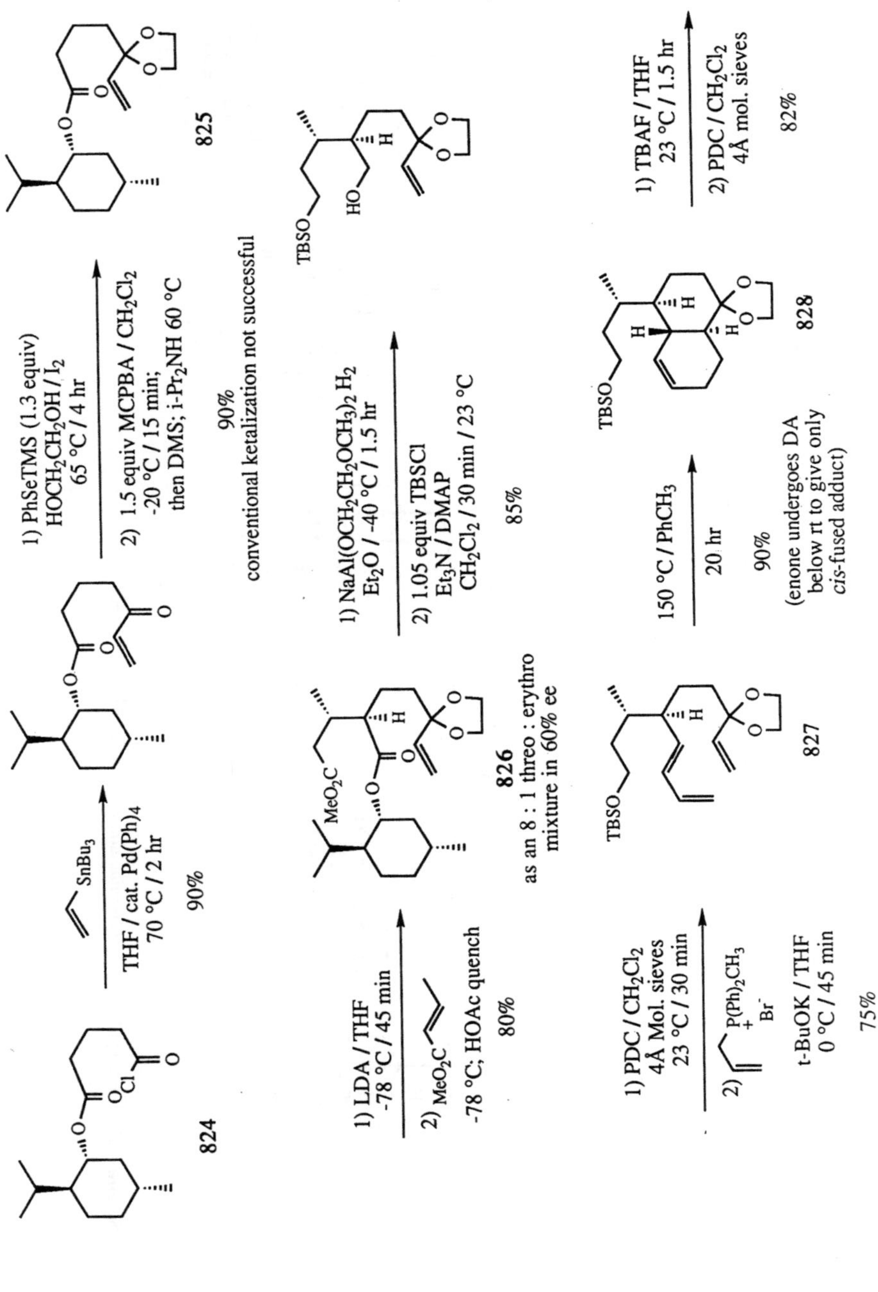
824
SnBu3
THF / cat. Pd(Ph)4
70 °C / 2 hr
90%
1) PhSeTMS (1.3 equiv)
HOCH2CH2OH / I2
65 °C / 4 hr
2) 1.5 equiv MCPBA / CH2Cl2
-20 °C / 15 min;
then DMS; i-Pr2NH 60 °C
90%
conventional ketalization not successful
825
1) LDA / THF
-78 °C / 45 min
2) MeO2C
-78 °C; HOAc quench
80%
MeO2C
826
as an 8 : 1 threo : erythro
mixture in 60% ee
1) NaAl(OCH2CH2OCH3)2 H2
Et2O / -40 °C / 1.5 hr
2) 1.05 equiv TBSCl
Et3N / DMAP
CH2Cl2 / 30 min / 23 °C
85%
TBSO
HO
1) PDC / CH2Cl2
4Å Mol. sieves
23 °C / 30 min
2) P(Ph)2CH3 Br
t-BuOK / THF
0 °C / 45 min
75%
827
150 °C / PhCH3
20 hr
90%
(enone undergoes DA
below rt to give only
cis-fused adduct)
828
1) TBAF / THF
23 °C / 1.5 hr
2) PDC / CH2Cl2
4Å mol. sieves
82%

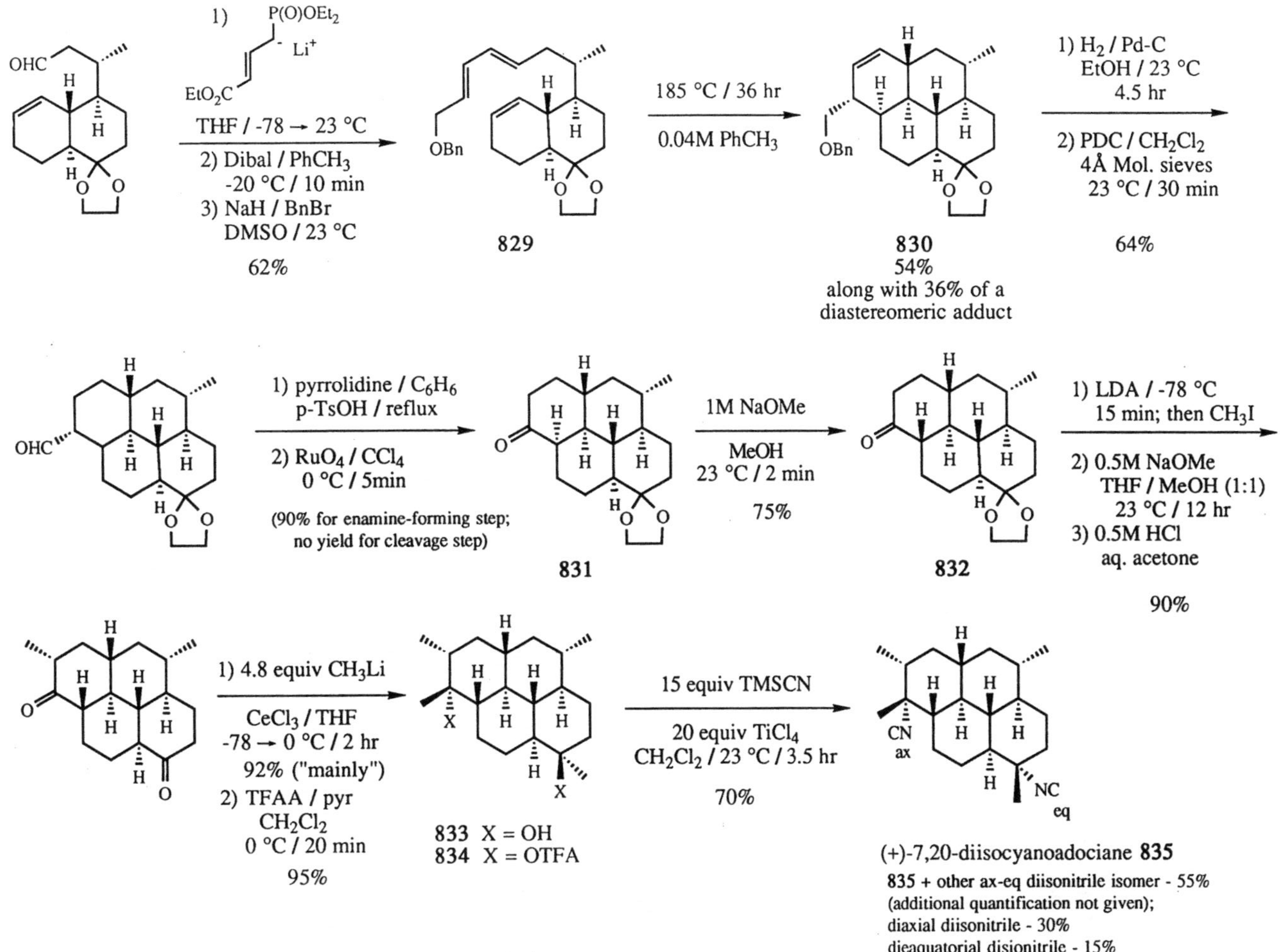

Scheme 136. Corey Synthesis of (+)-7,20-Diisocyanoadociane

carbocyclic ring system, members of this structural family exhibit exceptional toxicity against skin and brain tumor cells [227]. Surprisingly little activity has been reported in this area.

2.3.3.1 Stoechospermol

Not unexpectedly [2 + 2] photocycloaddition has played a major role in syntheses in this series. Syntheses of stoechospermol (**845**), the simplest member of this family, preceded those of spatol (**658**) which possesses a high degree of biological activity.

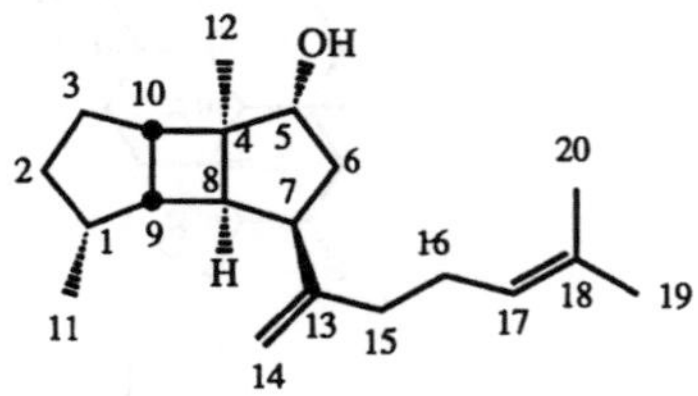

Stoechospermol **845**

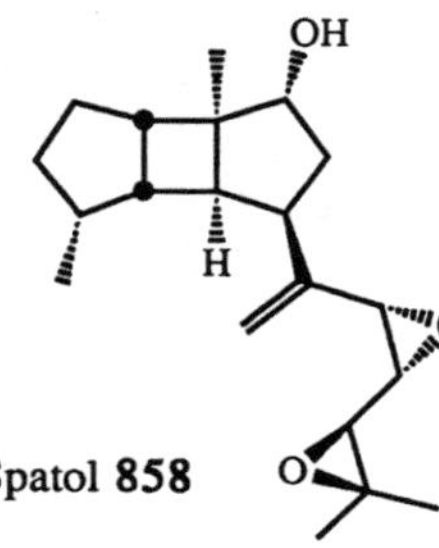

Spatol **858**

The first synthesis in this area was reported by Salomon in 1984 (Scheme 137) [228]. This synthesis of racemic stoechospermol began with the [2 + 2] photocycloaddition of cyclopentenone and bicyclic olefin **837** producing the *cis-anti-cis* tricyclic skeleton of the spatanes in the form of adduct **838**. The regiospecificity of the cycloaddition left a ketone at C1 which was easily transformed into the β-methyl group at this position (**839**) in two steps. Baeyer-Villiger oxidation of **839** introduced the required oxygen functionality at C5 in the correct orientation (**840**). At this point, after only 7 steps (from **836**), a key precursor to (potentially) all of the spatanes was produced, requiring only inversion of configuration at C7 with elaboration of the C8 side chain at this position. Inversion of stereochemistry was accomplished by conversion to the α,β-unsaturated malonate **841**, followed by conjugate reduction by $NaBH_4$ from the α-face producing diester **843**. This reduction was modelled and is believed to occur in an intramolecular fashion via a β-alkoxyborohydride species **842** [229],

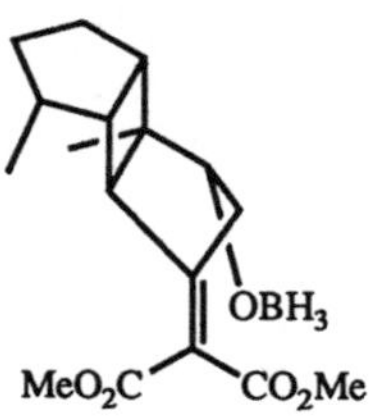

842

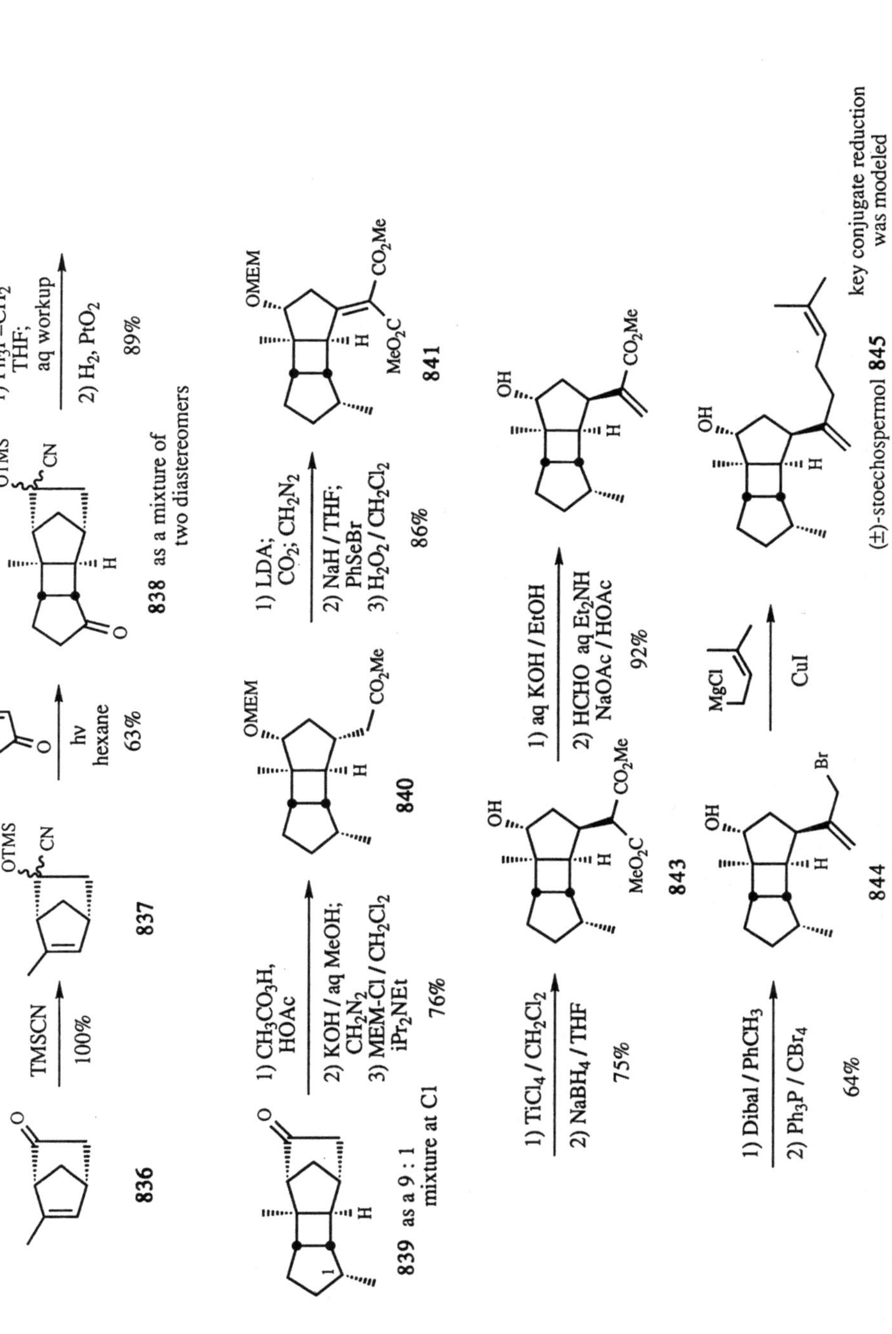

Scheme 137. Salomon Synthesis of (±)-Stoechospermol

thus delivering the hydrogen to the α-face of the conjugated ystem. Straightforward manipulation of the side chain led to the allylic bromide **844**, which led to racemic stoechospermol by Cu^+ catalyzed Grignard coupling.

The Koga approach to stoechospermol [230] also utilizes photocycloaddition, but as an intramolecular variant (Scheme 138). The optically active unsaturated lactone **846** was esterified with the racemic acid **847** producing the diastereomeric mixture **848**. Photocycloaddition was regiospecific leading in 61% yield to **849** as a 1:1 mixture of isomers. From this mixture the isomer with the desired C7 α-OH was separated and carried on. Note the absence of functionality at C5 at this point in the synthesis. Standard chemistry led to the diketone **851** which was cyclized with base providing the complete (5,4,5) ring system **852**. Hydrogenation of the enone followed by Wolff-Kishner reduction provided the intermediate **853** with all stereocenters intact with the undesired configuration at C7, similar to the Salomon synthesis. A series of 9 operations was required to insert the C5 hydroxyl group with the correct relative configuration (**845**). The key β-C–C bond at C7 was constructed by malonate displacement of the tosylate to give **854** whereupon a series of 6 routine transformations were used to convert **854** → (+)-stoechospermol. Overall the synthesis required 33–35 operations from the two cycloadditions partners **846** and **847**.

2.3.3.2 Spatol

Both the Salomon and Koga groups have synthesized (+)-spatol using intermediates from their respective stoechospermol syntheses. Starting from the optically active allylic alcohol intermediate **855**, Koga synthesized the sulfonium salt **856** (Scheme 139) [231]. Deprotonation to the ylid and coupling with the optically active epoxyaldehyde **957** led to 3 diastereomeric diepoxides **858**, **859** and **860** of which (+)-spatol was a minor component. The major product was the sulfide **861**. This allowed definition of the absolute stereochemistry at C17. The Salomon approach to spatol [232] utilized a clever double Payne rearrangement as a method of constructing vicinal diepoxides (Scheme 140). Beginning from the resolved aldehyde **862** addition of the *trans* lithiated alkene **863** produced primarily a single configuration at C15. Epoxidation of the TES ether **864** and desilylation produced the two epoxide diastereomers **865** and **866** in 15 and 55% yield respectively. The major isomer was treated with the Mitsunobu reagent to invert the stereochemistry at C15 and tosylated. Treatment with base effected a Payne rearrangement and subsequent epoxide closure to give the spatol isomer **867** in good overall yield. (+)-Spatol (**658**) could be obtained by mesylation of the minor isomer **865** and double Payne rearrangement. The control inherent in this method holds potential for the synthesis of all possible diepoxide isomers of spatol. Since it is reasonable to assume that the diepoxide functionality is the source of the biological activity of **658**, other isomers of spatol would be desirable for activity evaluation.

2.3.4 Pseudopteranes

The sea whip *Pseudopterogorgia elisabethae* elaborates a series of tricyclic diterpene glycosides reminiscent of the amphilectane diterpenes [233]. Several members of this series possess significant anti-inflammatory properties not related to inhibition of prostaglandin H2 synthase [234].

2.3.4.1 *Pseudopterosin A, E*

Pseudopterosins A (**878**) and E (**888**) have been synthesized by two vastly different strategies.

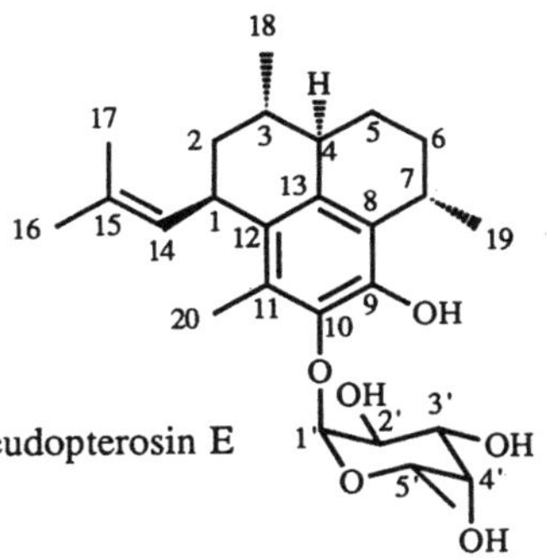

888 Pseudopterosin E

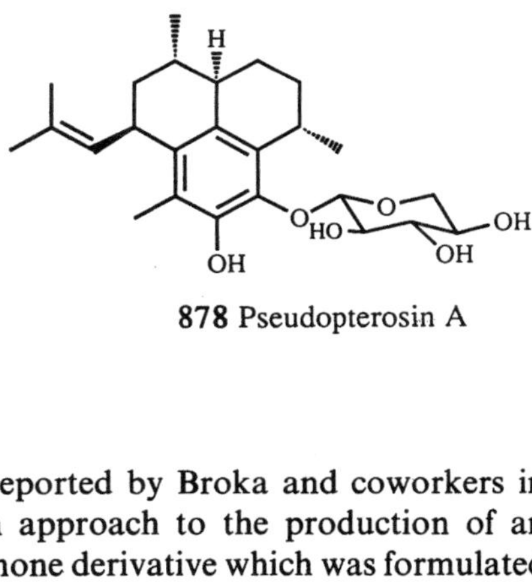

878 Pseudopterosin A

The first synthesis in this series was reported by Broka and coworkers in 1988 [235], who utilized a condensation approach to the production of an appropriately hexasubstituted *o*-hydroquinone derivative which was formulated so as to allow regiospecific introduction of the D-xylose subunit. Beginning with (*S*)-limonene, a routine series of transformations provided the β-trimethylsilyloxy enone **871** in 16 operations (Scheme 141). Cyclocondensation of **871** with the diene **872** (Scheme 142) provided the phenol (**873**) in 66% yield, but with loss of configurational integrity at C7 resulting in a diastereomeric mixture (**873**). Closure of the third ring was achieved via cation-olefin cyclization of the epoxide mixture **874** leading to another disappointing mixture of diastereomers. A second separation was followed by 9 transformations to provide the pseudopterosin aglycone **875** in which the C9 and C10 hydroxyls were differentiated to allow regiospecific C9 glycosylation. This was carried out as shown in Scheme 143 using the bromoxylose derivative **877** providing (−)-pseudopterosin A in 33–35 operations from (*S*)-limonene.

Corey described a second synthesis of **878** as well as a route to pseudopterosin E (**888**) [236]. Both **878** and **888** possess the same aglycone unit. Remote functionalization of the (+)-menthol nitrite ester (**879**) proceeded with good selectivity to a 5:1 mixture of oximes epimeric at C8 (Scheme 144). This was converted to the lactone **880**, possessing 3 of the 4 asymmetric centers present in the aglycone. Standard manipulations led to silyloxy diene **881**, which

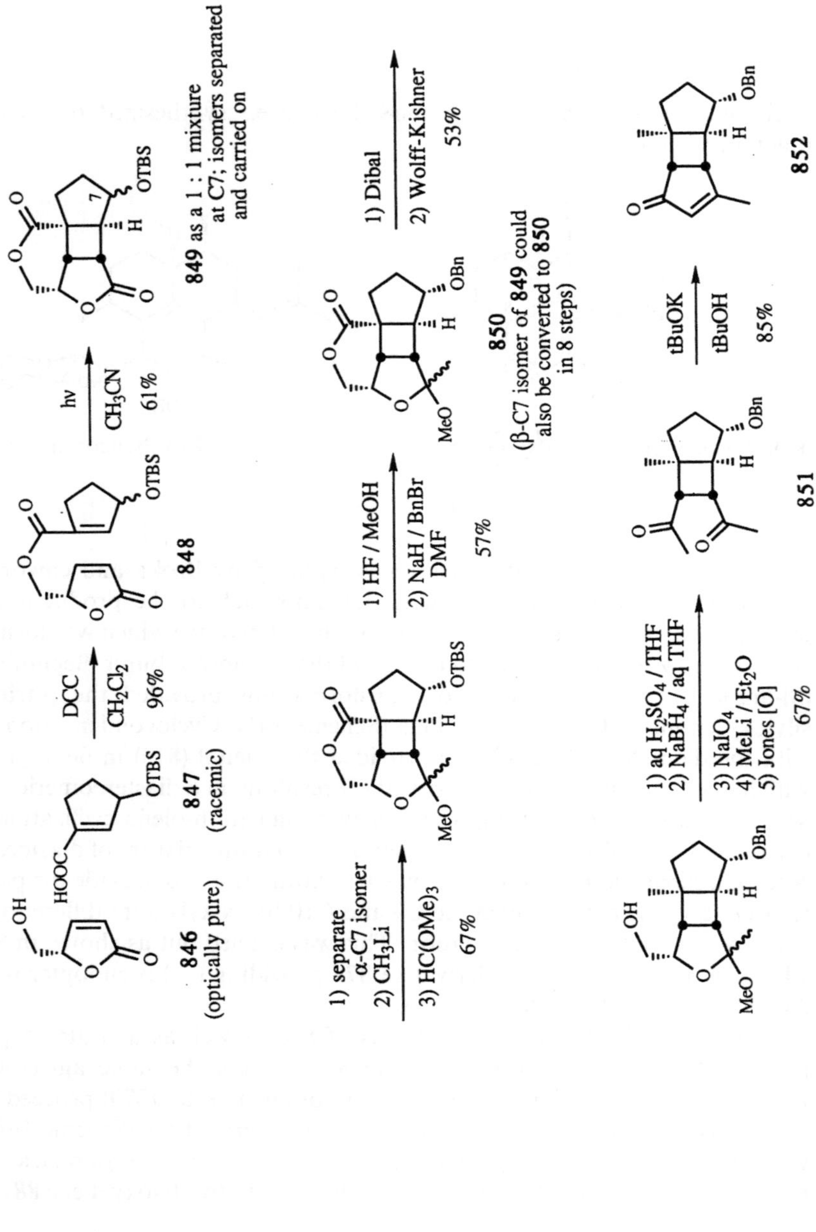
OH
846
(optically pure)
HOOC
OTBS
847
(racemic)
DCC
CH2Cl2
96%
OTBS
848
hv
CH3CN
61%
7
OTBS
H
849 as a 1 : 1 mixture
at C7; isomers separated
and carried on
1) separate
α-C7 isomer
2) CH3Li
3) HC(OMe)3
67%
MeO
OTBS
1) HF / MeOH
2) NaH / BnBr
DMF
57%
OBn
850
(β-C7 isomer of 849 could
also be converted to 850
in 8 steps)
1) Dibal
2) Wolff-Kishner
53%
1) aq H2SO4 / THF
2) NaBH4 / aq THF
3) NaIO4
4) MeLi / Et2O
5) Jones [O]
67%
851
tBuOK
tBuOH
85%
852

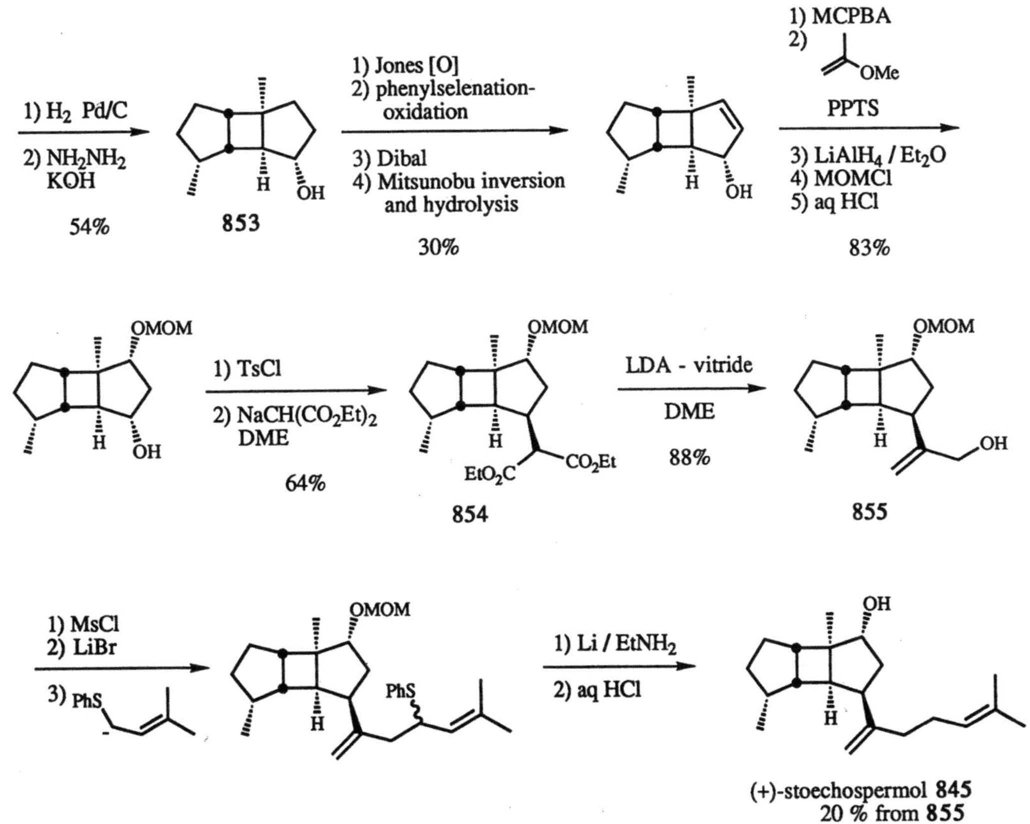

Scheme 138. Koga Synthesis of (+)-Stoechospermol

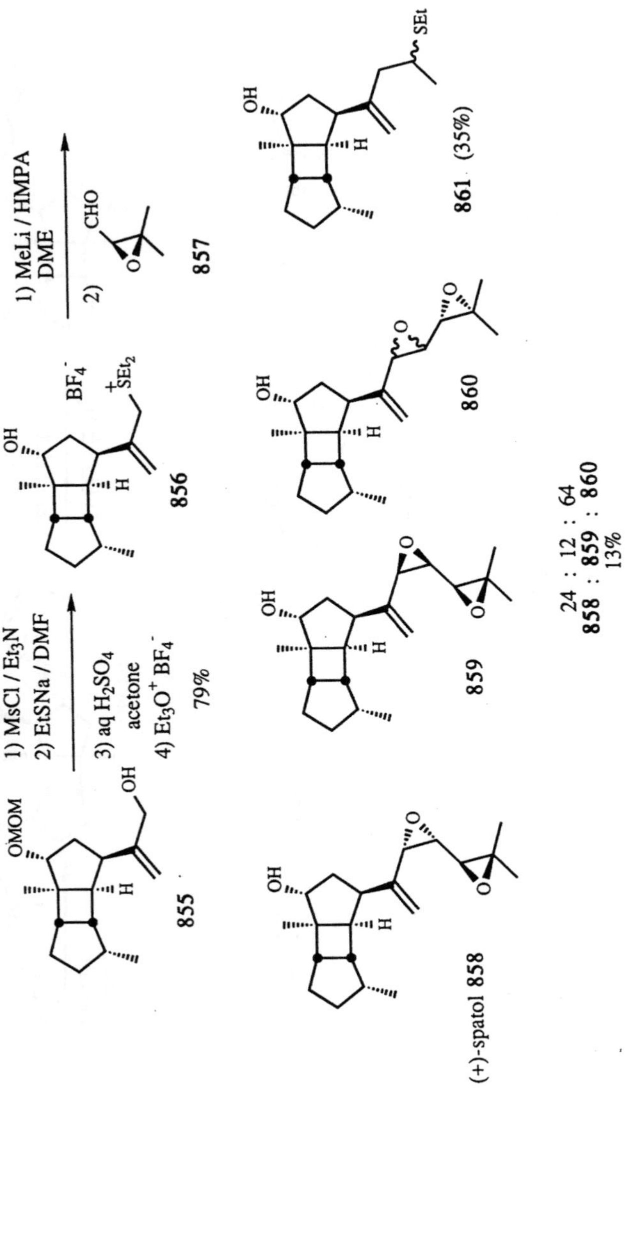

Scheme 139. Koga Synthesis of (+)-Spatol from Intermediate **855**

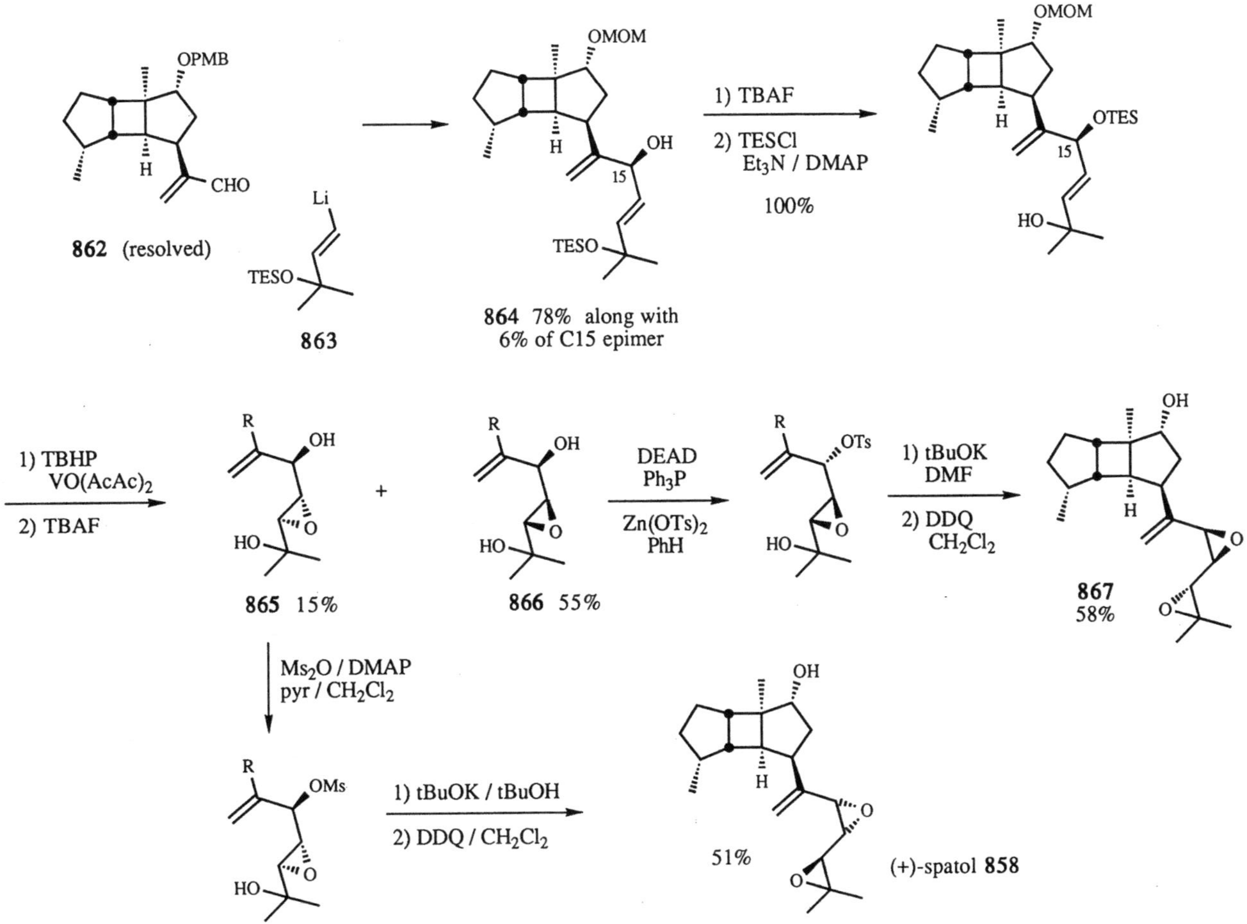

Scheme 140. Salomon Synthesis of (+)-Spatol from **862** by Double Payne Rearrangement

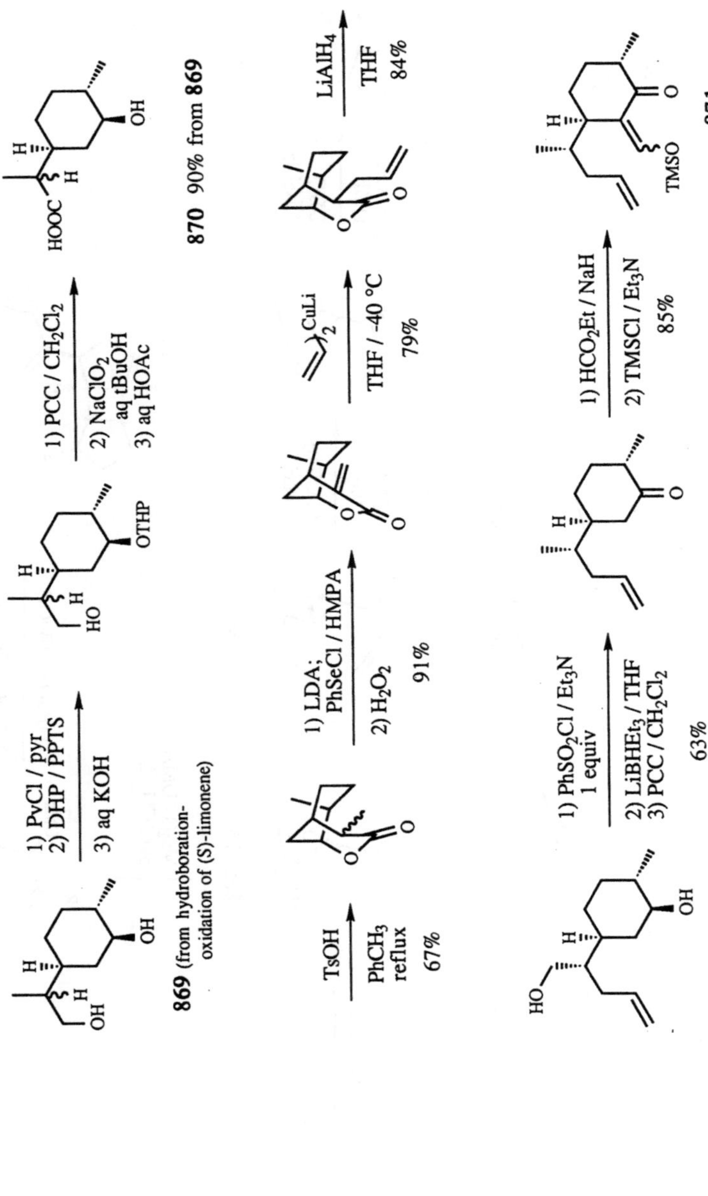

Scheme 141. Broka Synthesis of (−)-Pseudopterosin A

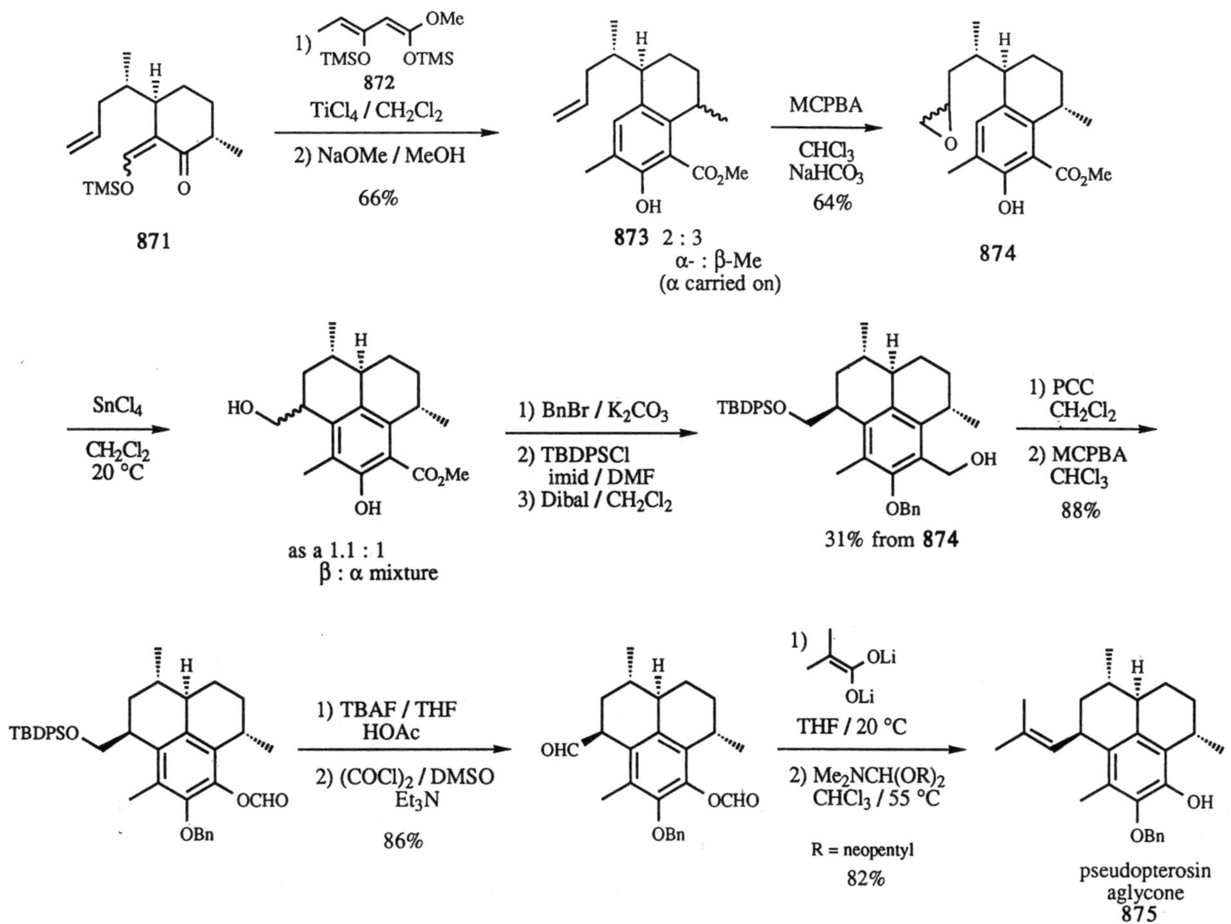

Scheme 142. Conversion of **871** into the Pseudopterosin Aglycone **875**

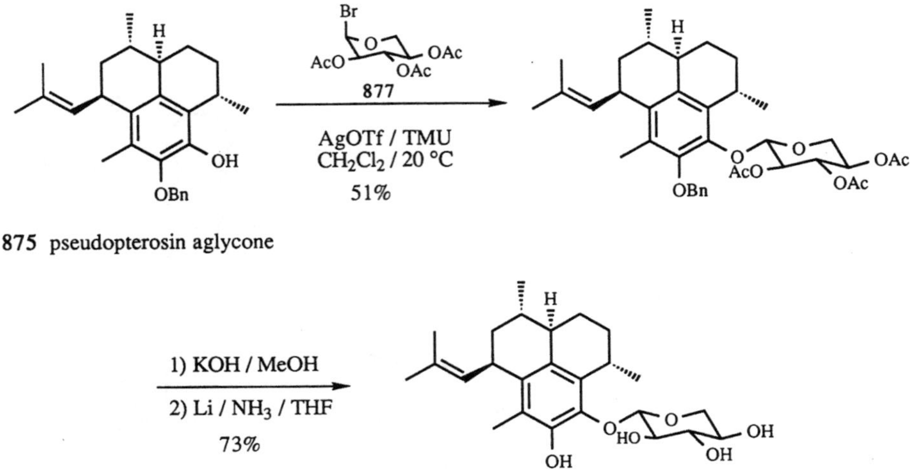

875 pseudopterosin aglycone

(-)-pseudopterosin A 878

Scheme 143. Conversion of Aglycone **875** to (−)-Pseudopterosin A (Broka)

underwent γ-selective Mukaiyama-type aldol reaction with 2-butynal to give the dione **882** after oxidation of the intermediate alcohol. Treatment of this substance with KH in THF at room termperature led to the phenol **883** in 70% yield. It was suggested from earlier work that this transformation might proceed via electron transfer within an extended π-system to a biradical species (at C11 and C12) which undergoes closure and aromatization. Oxidation to the *o*-hydroquinone was accomplished via the method of Barton to give **884** which led after five steps to the aglycone **885**. Note that **885** contains free hydroxyls at C9 and C10. It was found that the C10 hydroxyl could be selectively derivatized, thus obviating the need for a regiospecifically protected intermediate (Scheme 145). Direct glycosylation of **885** with bromide **886** provided **887** selectively which was deprotected to give pseudopterosin E in 53% yield from the aglycone **885**. Preparation of pseudopterosin A required initial protection of C10 which was accomplished by tosylation under standard conditions. Glycosylation of **889** with bromide **890** led eventually to **878** in 46% yield from the aglycone. Both pseudopteranes were produced in 25–27 steps overall by this route.

2.3.5 Aplysin-20 and Derivatives

Aplysin-20 (**894**), isoaplysin-20 (**904**) and concinndiol (**900**) were some of the first examples of brominated diterpenes obtained from *Laurencia*. Not surprisingly, all approaches to these metabolites so far have used cation-olefin cyclization of a geranylgeraniol derivative or some homolog thereof.

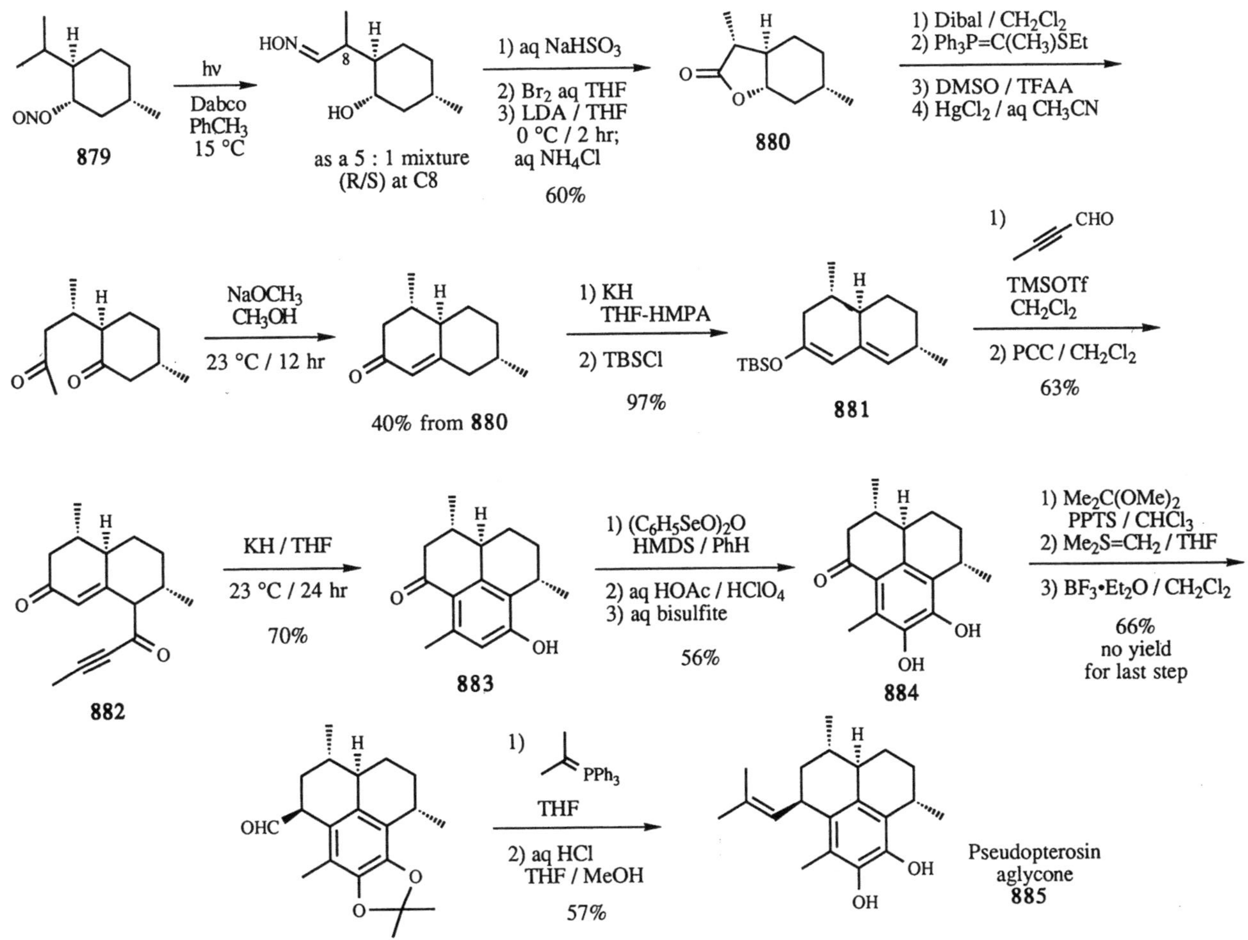

Scheme 144. Corey Synthesis of Pseudopterosin Aglycone

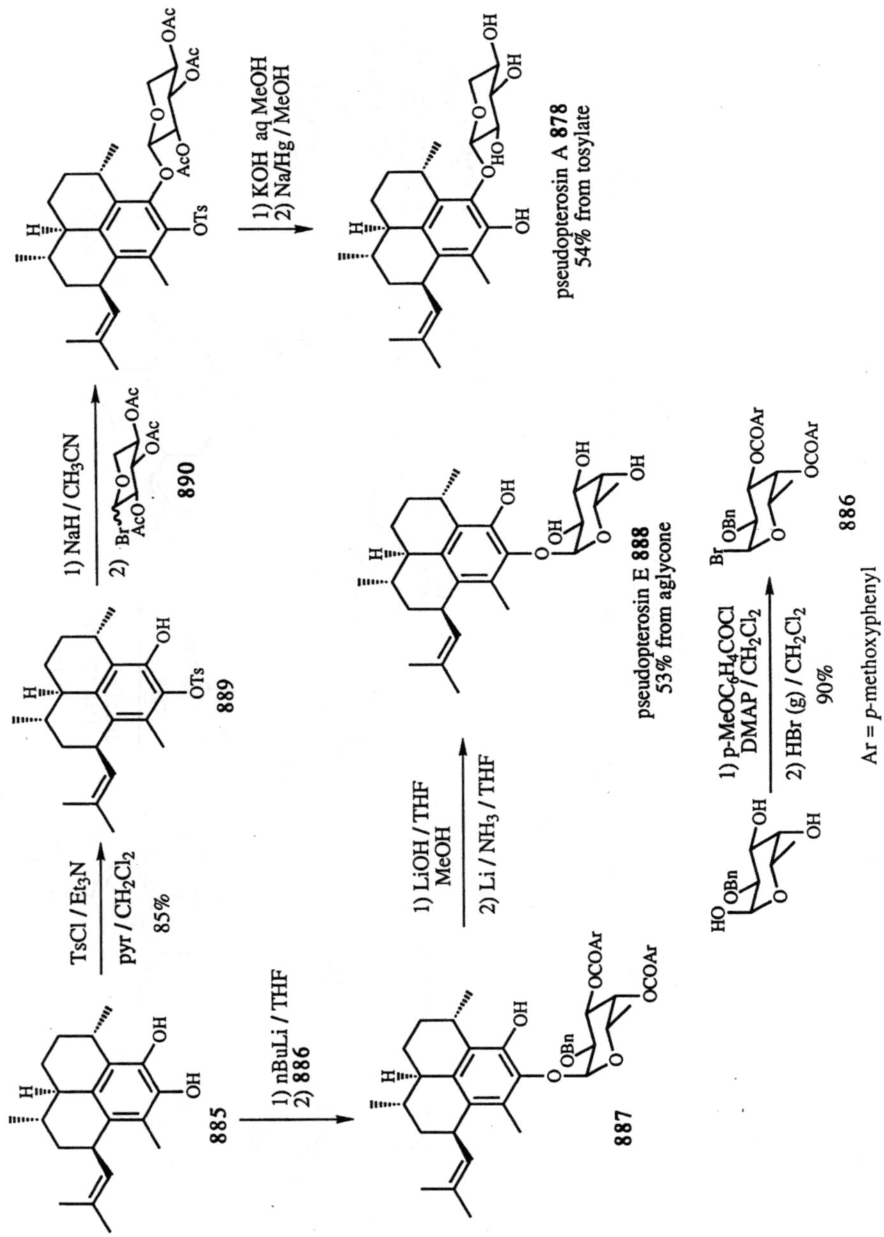

Scheme 145. Conversion of the Pseudopterosin Aglycone to Pseudopterosins A and E

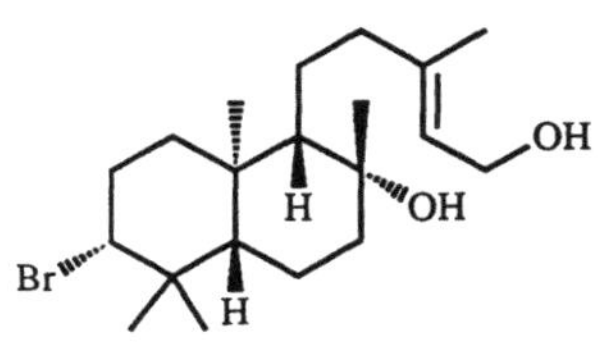

Aplysin-20 **894**

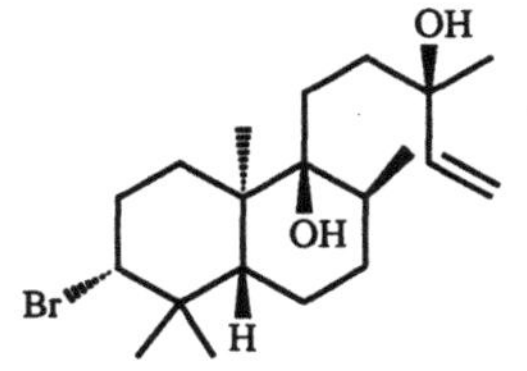

Concinndiol **900**

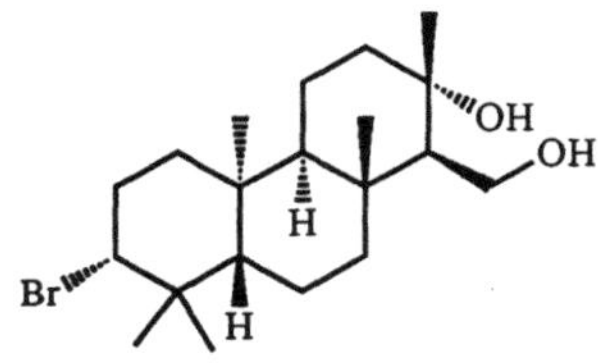

Isoaplysin-20 **904**

To synthesize aplysin-20 (**894**) Murai [237] treated the acyclic bromo-trifluoroacetate (Scheme 146) with $BF_3 \cdot Et_2O$ to effect cyclization affording **891** in 22% yield in a non-stereoselective process. From this point the remainder of the synthesis involved extension of the C9 side chain and installation of the C8 alcohol with its attendant stereochemistry. This latter goal was accomplished by acid-promoted cyclization of the unsaturated acid **892** to the lactone **893**. The synthesis was approximately 16 steps (from nerolidol) and was the first synthesis in the series.

Cylization of a larger substrate (**895**, Scheme 147) provided Kato [238] with a 20% yield of bromoterpene **896**. This substance was a branchpoint intermediate for the syntheses of both aplysin-20 and concinndiol (**900**). Epoxidation of **896** gave the epoxide isomers **897** and **898**, the major of which was carried on to **894** aplysin-20 by reduction with AlH_3. The minor isomer **898** could be converted to **900** by a route which involved an allylic rearrangement and the stereospecific axial opening of the epoxide **899**. In a study of the mechanistic nature of biomimetic polyolefin cyclization [239]. Nishizawa cyclized geranylgeraniol acetate (Scheme 148) and obtained the bicyclic organomercurial **902** and the tricyclic product **903**, both in rather low yield. These compounds could be converted to aplysin-20 and isoaplysin-20 in two additional steps each.

A series of related diterpenes isolated from Dictyoceratid and Dendroceratid sponges have been produced by partial synthesis and will not be discussed. [240]

2.3.6 Agelas Diterpenoids

2.3.6.1 Ageline A

Ageline A has been constructed by Tokoroyama [241], in a non-convergent synthesis, as shown in Scheme 149. The cyclic terpene portion **909** is prepared from acyclic precursors and homologated in a stepwise fashion to provide the diterpene skeleton **917**. In the final step the 9-methyladenine ring is appended to **918**. The cyclic terpene **909**, is prepared as shown in Scheme 150. Alkyne **905** is stereospecifically converted to ester **906** by lithiation, methoxycarbonylation with methyl chloroformate and addition of lithium dimethylcuprate. Ester reduction and reaction with phosphorus tribromide yields allylic bromide **907**.

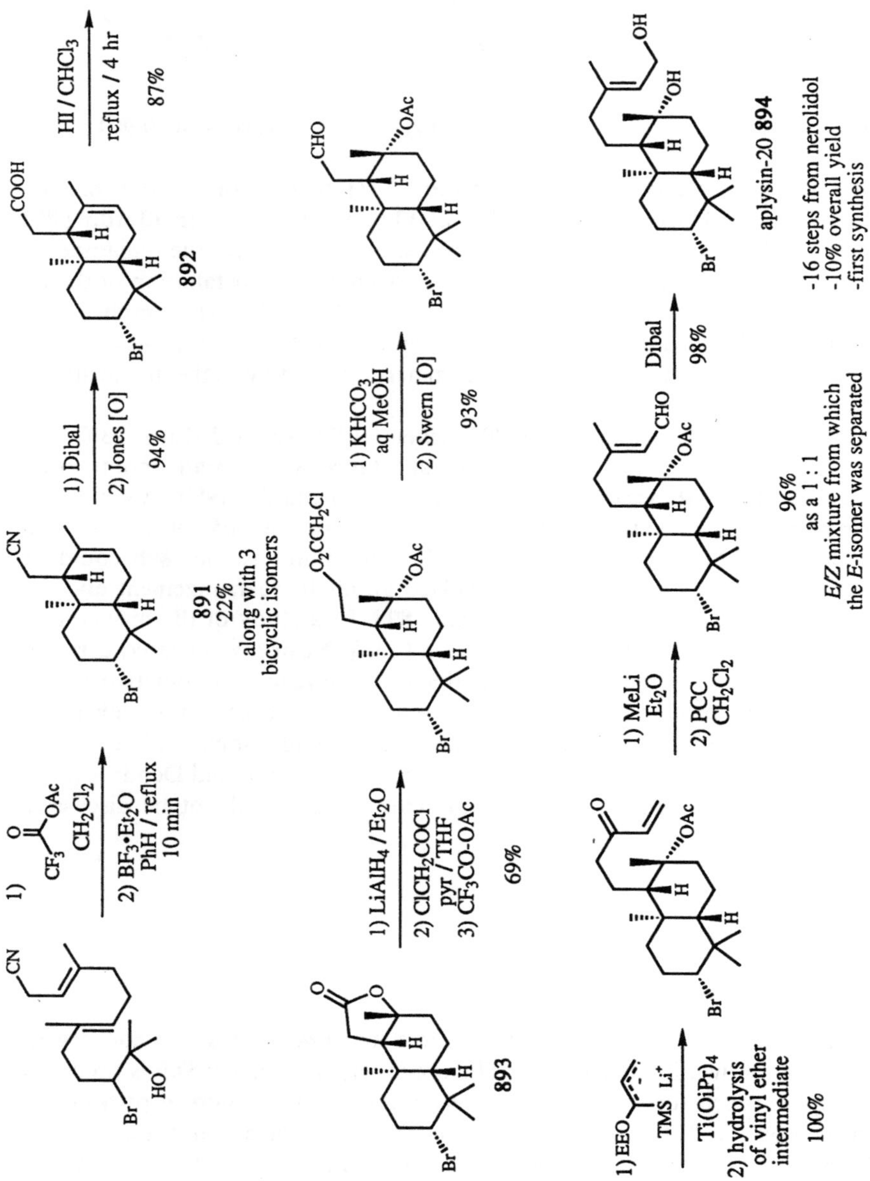

Scheme 146. Murai Synthesis of Aplysin-20

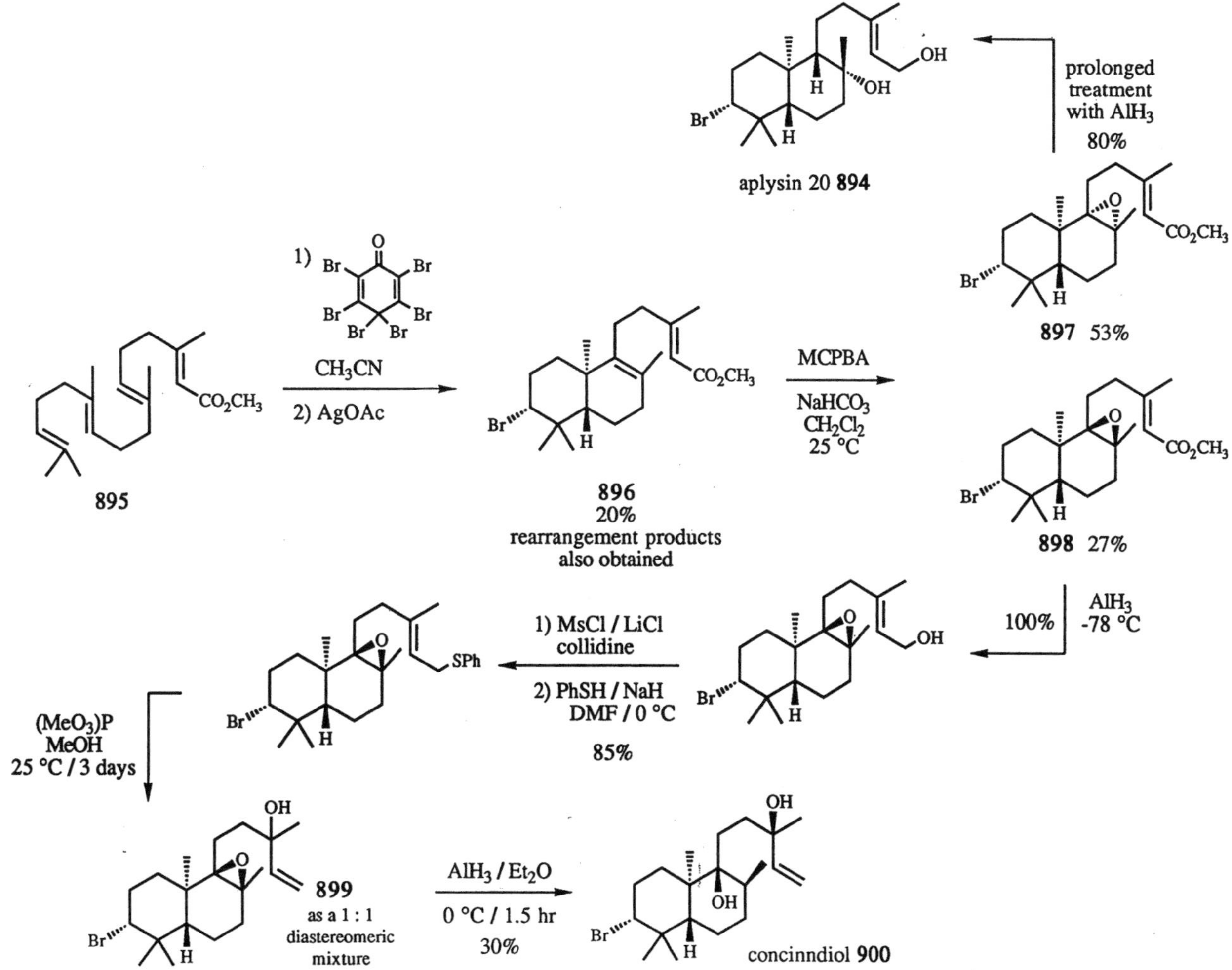

Scheme 147. Kato Synthesis of Racemic Aplysin 20 and Concinndiol

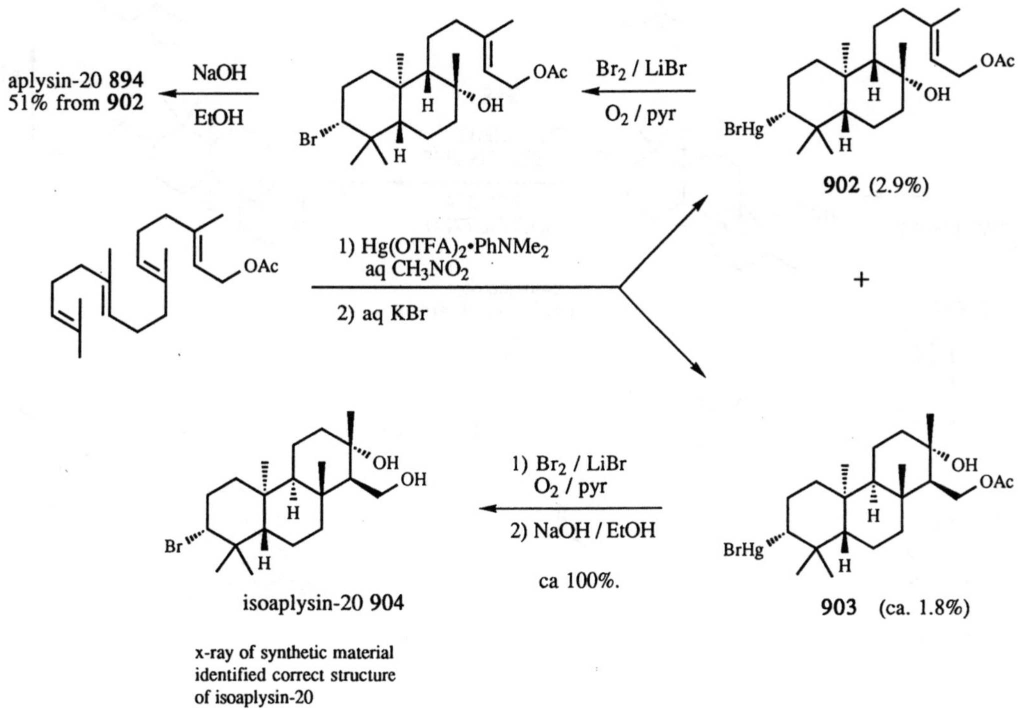

Scheme 148. Nishizawa Synthesis of Aplysin-20 and Isoaplysin-20

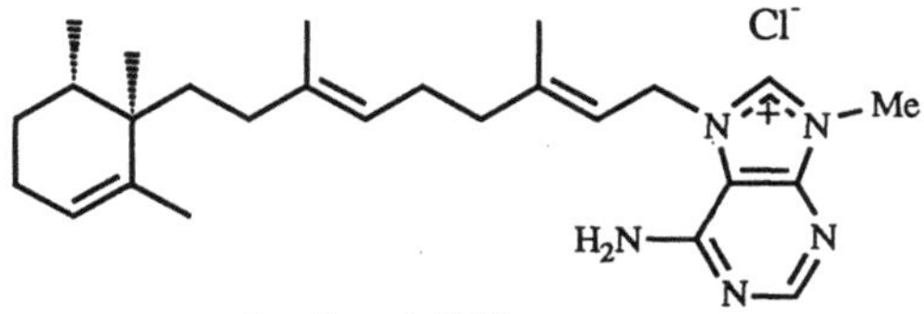

Ageline A **919**

Treatment with trimethylsilyllithium, removal of the benzyl protecting group, and oxidation affords aldehyde **908** in 25% yield. The cyclized product **909** with the correct *cis* methyl stereochemistry is obtained by treatment of **908** with either boron trifluoride etherate or tin(IV) chloride. Before the homologation process can take place, the vinyl methyl functionality must be added to the cyclic terpene. This is done in eight steps, beginning with Jones oxidation of **909**, acetalization of the resulting ketone, and hydroboration-oxidation of the vinyl appendage to provide alcohol **910**. Protection of the alcohol as its benzyl ether and hydrolysis of the acetal yields ketone **911**. Addition of methylmagnesium iodide to **911** and successive treatment with thionyl chloride in pyridine and TsOH produces substituted cyclohexene **912**. The building up of the terpenoid sidechain begins with deprotection of the alcohol followed by mesylation and treatment with sodium iodide in acetone to give iodide **913** which, upon displacement with lithium acetylide, provides alkyne **914**. The (*E*)-alkenyl iodide **915** is obtained via zirconium-catalyzed carboalumination followed by treatment with iodine. Palladium-catalyzed coupling of iodide **915** with pentenylzinc bromide **916** provides triene **917**. Deprotection of the alcohol and conversion to the bromide **918**, followed by coupling with the 9-methyladenine moiety provides ageline A in 26 steps. As shown in Scheme 150, an alternative procedure for the preparation of key intermediate **917** from iodide **913** was developed. Lithiation of **913**, followed by transmetallation to afford the alkylzinc species allows palladium-catalyzed coupling with alkenyl bromide **920** to provide **917** in 66% yield.

2.3.6.2 Agelasidine C

Tokoroyama [242] has applied the methodology developed for the preparation of ageline A (**919**) (Scheme 149) to the synthesis of agelasidine C (**923**) (Scheme 151). Displacement of allylic bromide **918** (a common intermediate prepared in the synthesis of ageline A) with phthalimide-substituted thiol **921** affords a sulfide which is subsequently oxidized to the sulfone and converted to the free

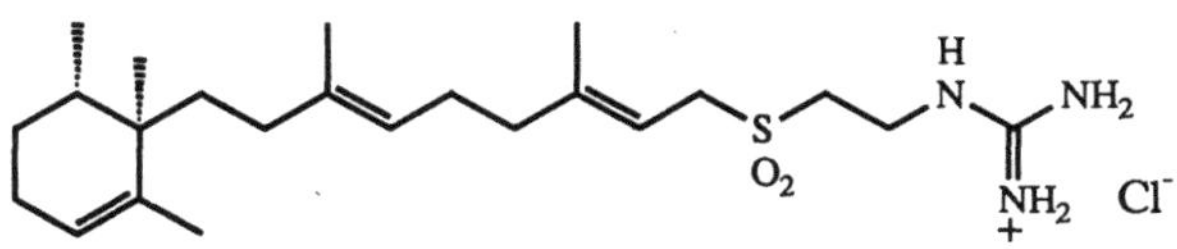

Agelasidine C **923**

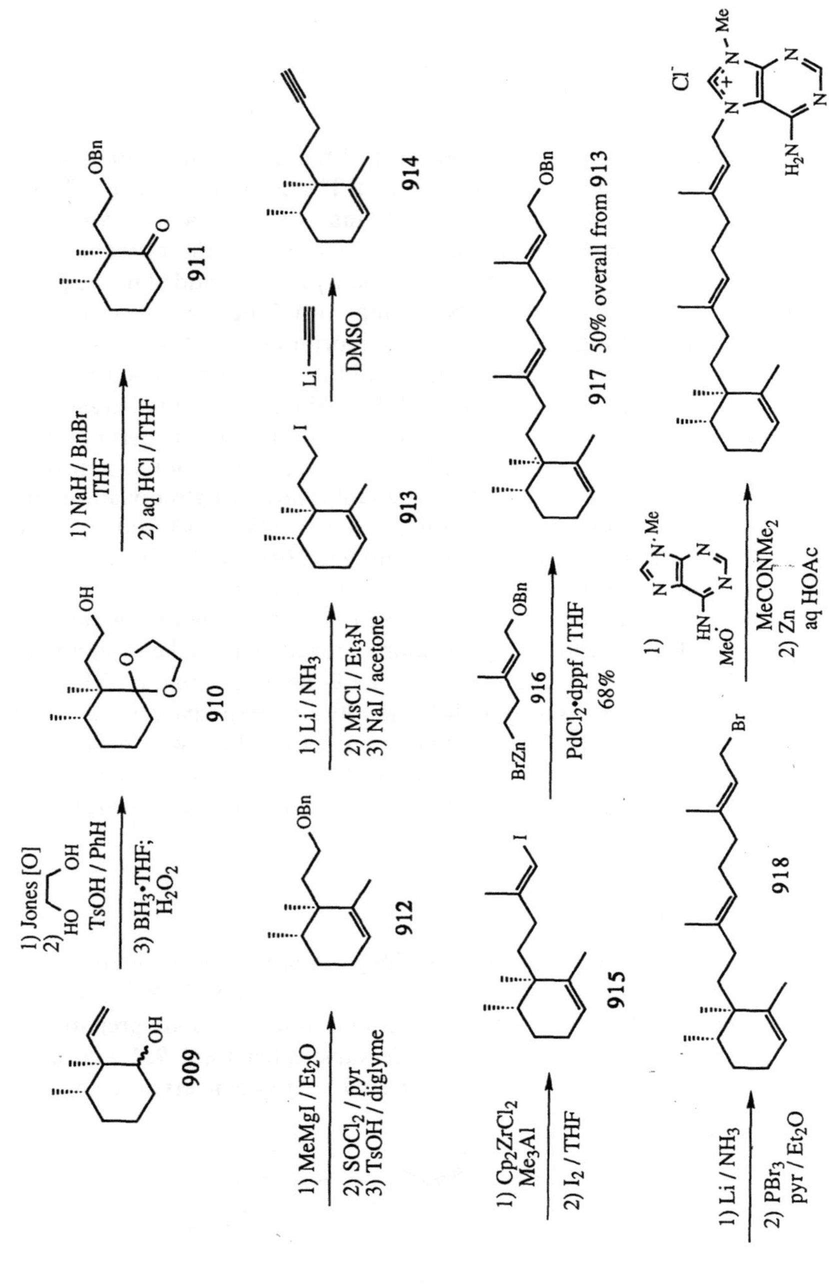

Scheme 149. Tokoroyama Synthesis of (±)-Ageline A

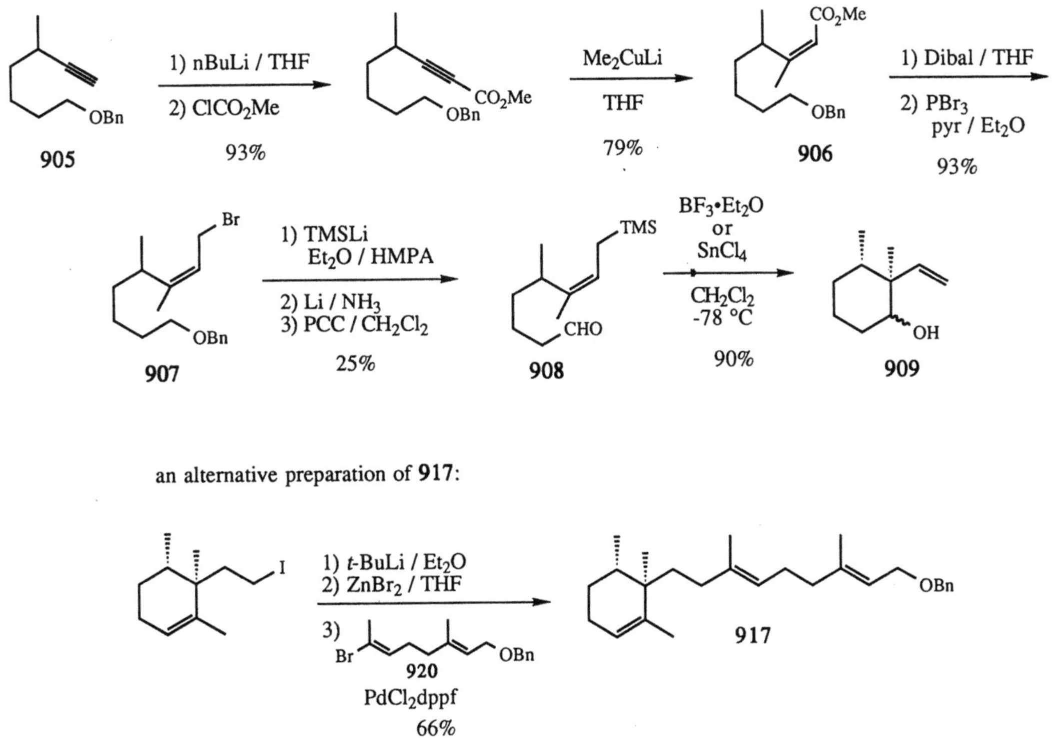

Scheme 150. Synthesis of **909**

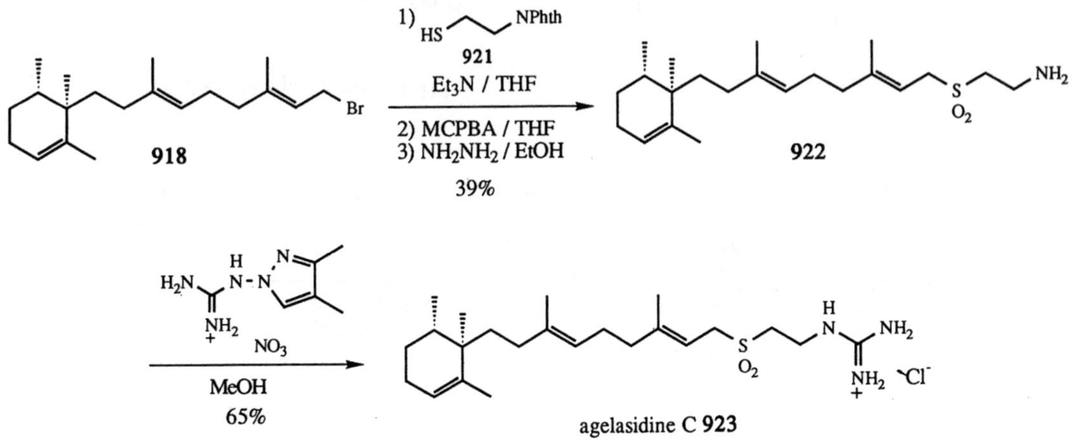

Scheme 151. Tokoroyama Synthesis of Agelasidine C

amine **922** by treatment with hydrazine. Guanylation of **922** is achieved via a transamination reaction with 3,5-dimethylpyrazole-1-carboxamidine nitrate to afford agelasidine C in 4 steps from intermediate **918**.

2.3.6.3 Agelasidine A

Agelasidine A has been prepared by Ichikawa [243], as shown in Scheme 152. The key transformation involves the construction of the tertiary thiol moiety by hetero-Claisen rearrangement of an allyl xanthate to a dithiocarbonate. Farnesol (**924**) is converted into the corresponding xanthate by reaction with potassium *t*-butoxide, carbon disulfide and methyl iodide. Heating the allyl xanthate in refluxing benzene provides the tertiary dithiocarbonate **925** in 90% yield. Reduction of **925** with lithium aluminum hydride and alkylation with 2-bromoethanol provides sulfide **926**. Oxidation of the sulfide to the sulfone and tosylation of the alcohol with elimination gives vinyl sulfone **927**. Conjugate addition of NH_3 to **927** followed by treatment with 1-guanyl-3,5-dimethylpyrazole nitrate affords agelasidine A in 8 steps and 23% overall yield.

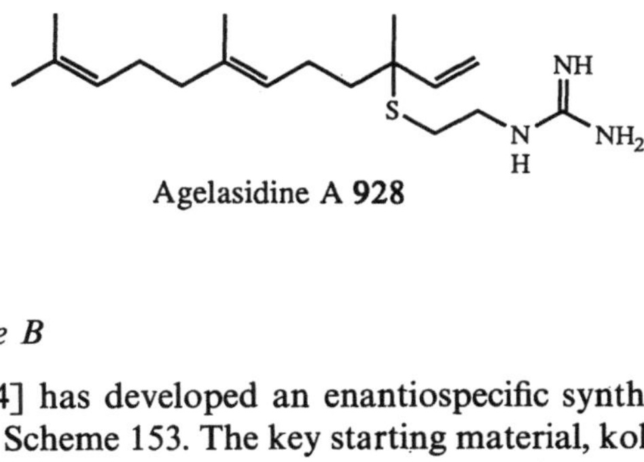

Agelasidine A **928**

2.3.6.4 Agelasine B

Tokoroyama [244] has developed an enantiospecific synthesis of agelasine B (**931**), as shown in Scheme 153. The key starting material, kolavenic acid (**929**) is readily available from the roots of *Solidago* species. Esterification of **929** followed by reduction and treatment with phosphorus tribromide provides allylic bromide **930**. Coupling of **930** with N^6-methoxy-9-methyladenine followed by removal of the methoxy group with zinc in acetic acid yields (−)-agelasine B (**931**) in 5 steps.

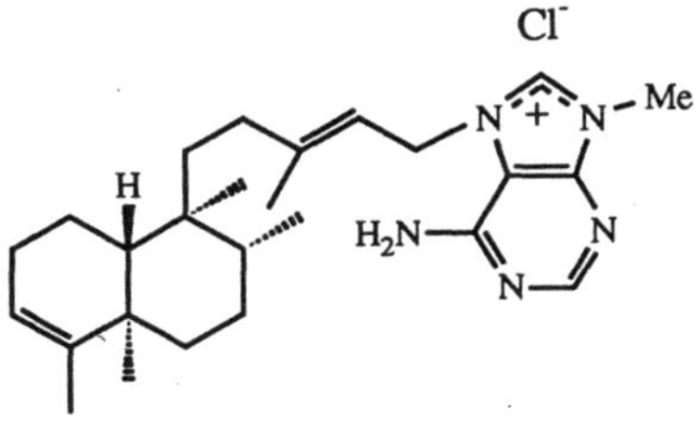

Agelasine B 931

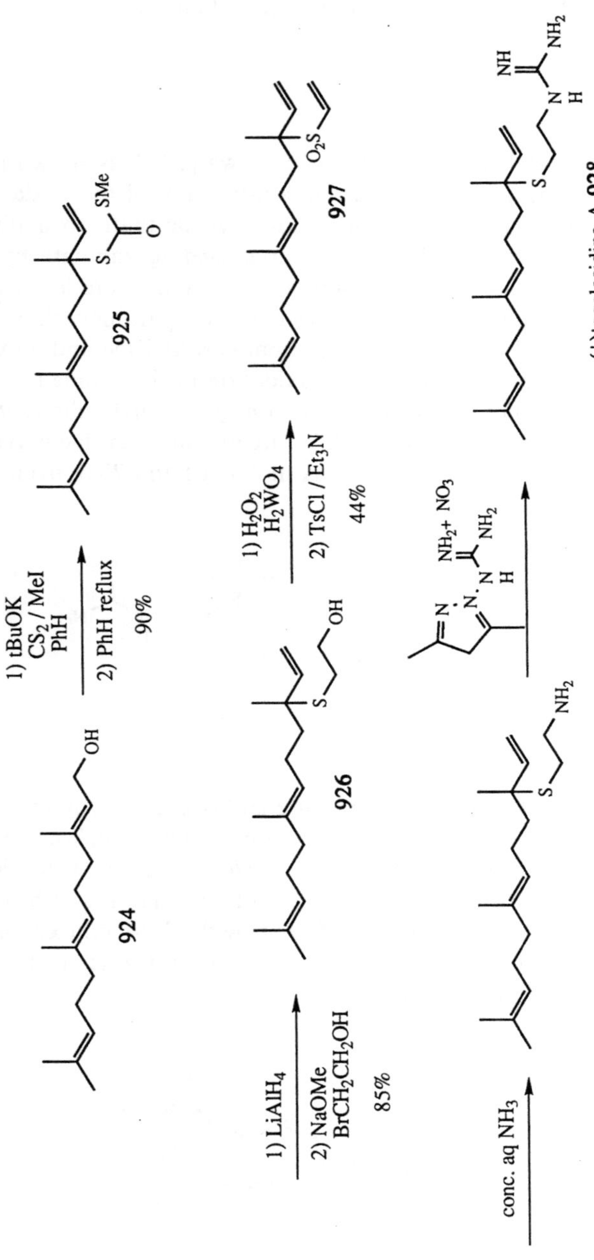

Scheme 152. Ichikawa Synthesis of (±)-Agelasidine A

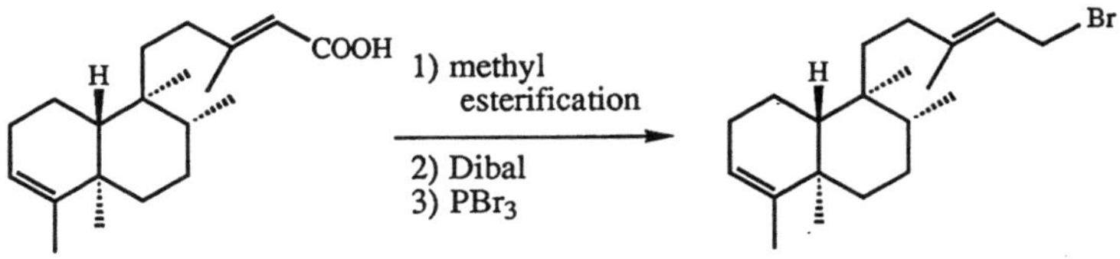

929 readily available from the roots of *Solidago sp.*

930

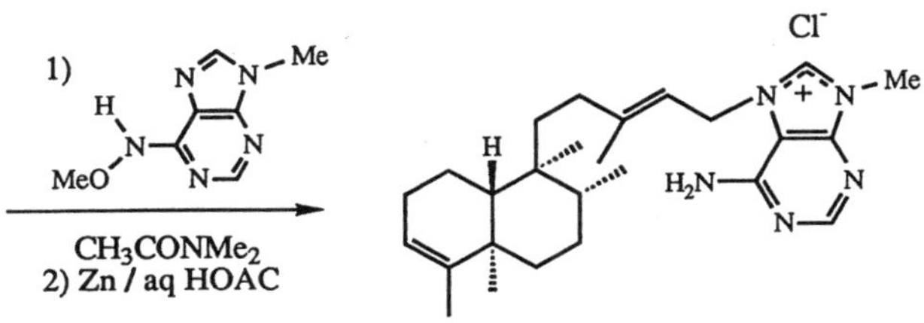

(-)-agelasine B **931**

Scheme 153. Tokoroyama Synthesis of (−)-Agelasine B

2.3.7 Prenylated Phenols and Derivatives

2.3.7.1 Taondiol Methyl Ether

Prenylated phenols are common marine metabolites. The syntheses of two compounds in this class have been reported. (±)-Taondiol methyl ether has been synthesized by Gonzalez (Scheme 154) using a biomimetic approach involving cation-olefin cyclization [245]. *trans,trans*-Farnesol (**933**) was converted to the higher homolog **934** and alkylated with the phenol **935**. This intermediate (not shown) was hydrobrominated at the remote olefin according to well-known precedent, and the epoxide formed by base treatment. Cyclization initiated by $SnCl_4$-catalyzed epoxide opening proceeded in remarkably good yield for a four-fold cyclization to give (±)-taondiol methyl ether.

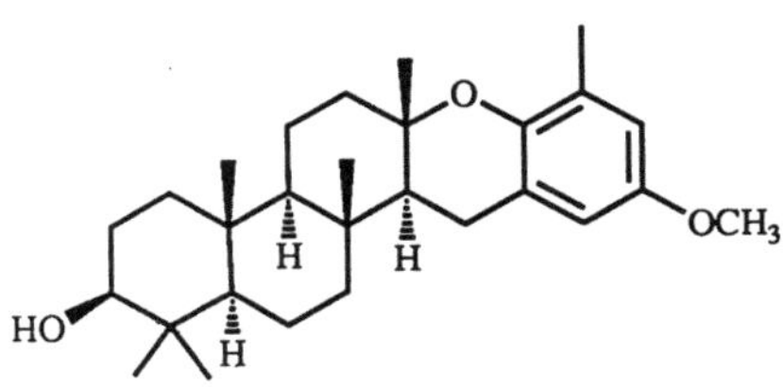

Taondiol methyl ether **936**

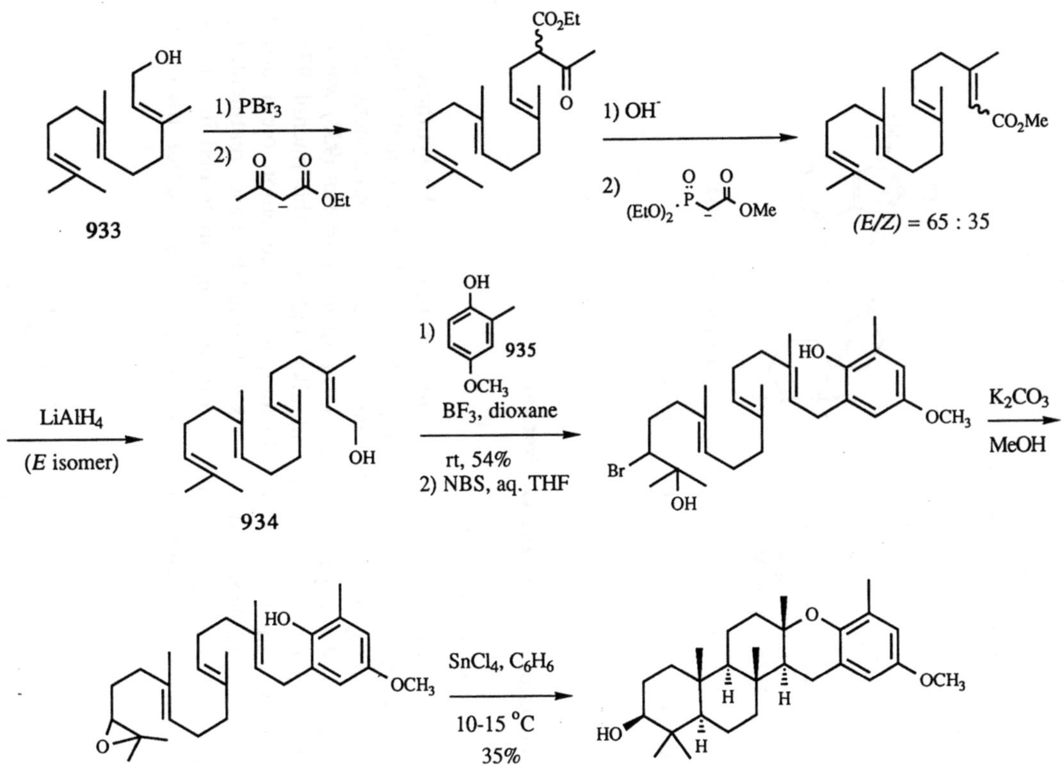

(±)-taondiol methyl ether **936**

Scheme 154. Gonzalez Synthesis of (±)-Taondiol Methyl Ether

2.3.7.2 Bifurcarenone

The prenylated hydroquinone (±)-bifurcarenone acts as an inhibitor of cell division during mitosis. This brown algal metabolite [246] (*Bifurcaria galapagensis*) has been synthesized in racemic form by Mori (Scheme 155) [247]. The cyclopentane ring of **943** was derived from the symmetrical ketone **937**. After Baeyer-Villiger oxidation of **937** and reduction to the alcohol **938**, it was a matter of sequentially adding the two side chains. This was accomplished by coupling the allylic chloride **940** (Scheme 156) to the anion obtained by deprotonation of **938**. The lower side chain was constructed by the addition of the dianion **942** to aldehyde **941**. Refunctionalization led to (±)-**943**. The synthesis served to establish the configuration of the trisubstituted olefin as *E* in the natural product.

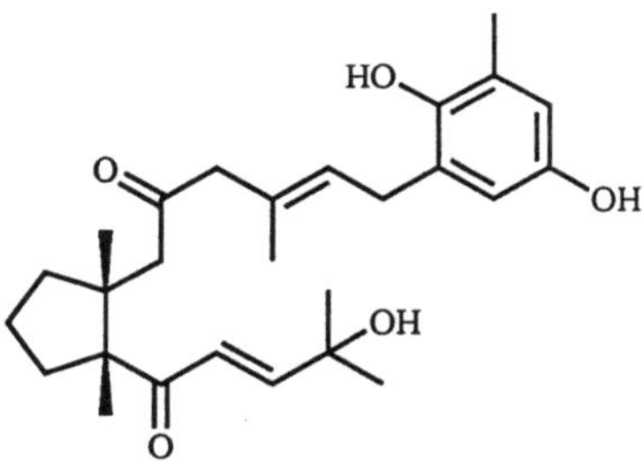

Bifurcarenone **943**

2.3.8 Miscellaneous Diterpenoids

2.3.8.1 Dictyolene, Pachydictyol A and Obscuronatin

The brown algal metabolite dictyolene (**951**) [248] has been synthesized twice. The first synthesis, by Marshall [249] employed the known cyclohexadiene **945** obtained via Birch reduction of *m*-toluic acid (Scheme 157). Alkylation of the enamine of **945** with ethyl bromoacetate gave **946**. Acid-promoted cyclization established the 4 stereocenters of the bicyclo[4.4.0] ring system resulting in **947** in 45% yield along with the epimeric hydroxy ester **948**. It was found that alkylating agents approached the lactone enolate of **947** from the α-face

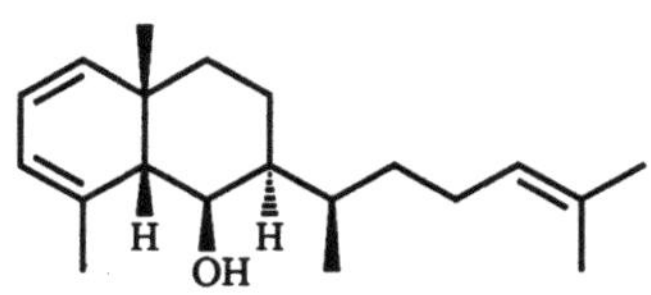

Dictyolene **951**

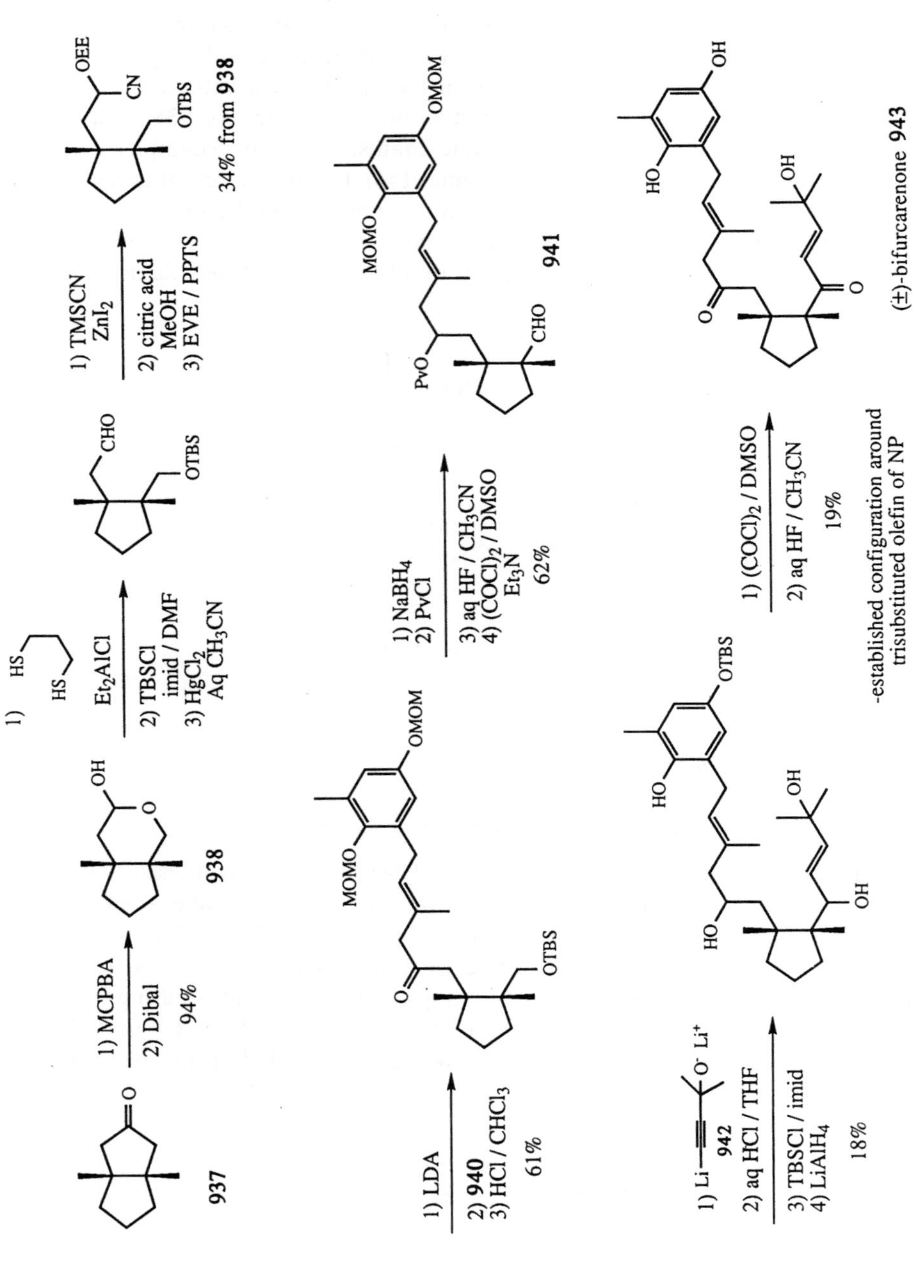

Scheme 155. Mori Synthesis of (±)-Bifurcarenone

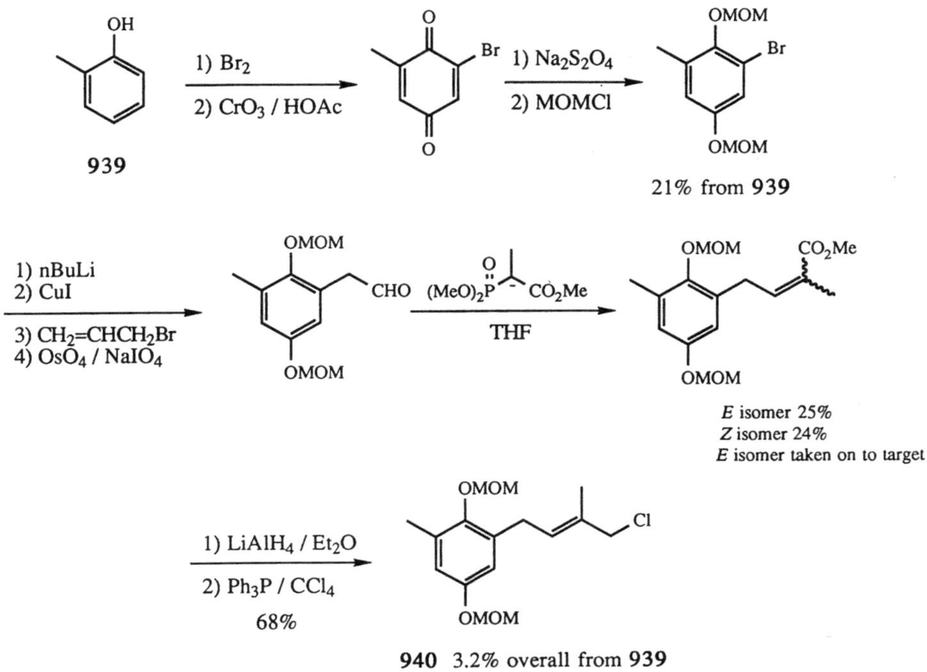

Scheme 156. Synthesis of Bifurcarenone Aromatic Fragment (Mori)

necessitating introduction of a methyl group prior to refunctionalization leading to iodide **950**. Coupling with an organonickel reagent and removal of the acetate gave racemic dictyolene.

An entirely different approach was later described by Greene [250] making astute use of readily available (−)-α-santonin (Scheme 158). Epimerization of C6 with acid provided the more stable *cis*-fused lactone **953**. Treatment of **953** with tosylhydrazine provides the corresponding tosylhydrazone which, on treatment with catecholborane gives solely the olefin **956** without any dienic product. By carefully varying the reaction conditions the allylic tosylhydrazone could be isolated. This is probably the product of conjugate reduction followed by reduction of the imine. Smooth decomposition of **954** → **956** was observed on

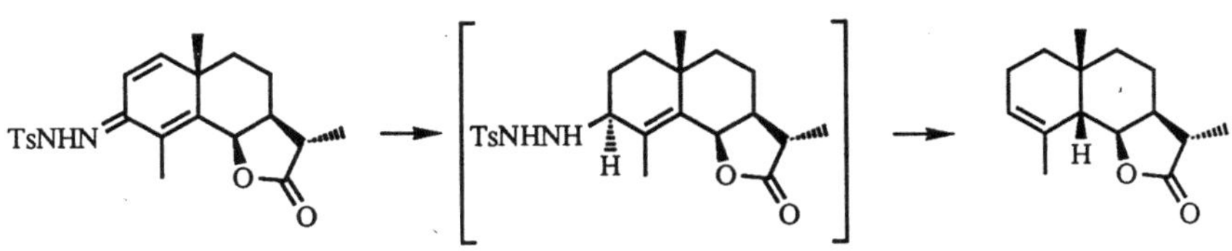

954 **955** **956**

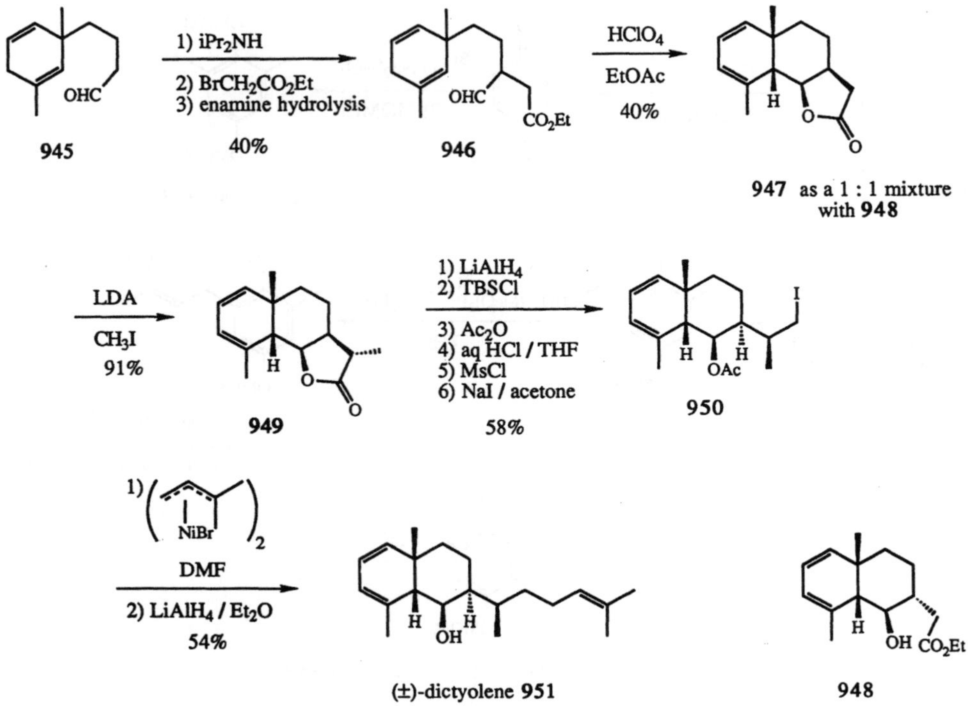

Scheme 157. Marshall Synthesis of (±)-Dictyolene

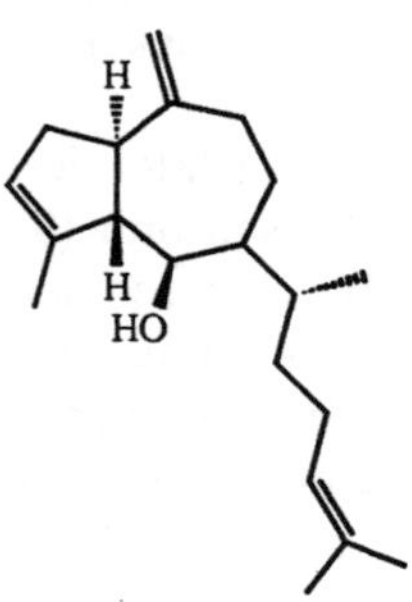

Pachydictyol A **962**

treatment with NaOAc. A three-step oxidation produced the diene-lactone **949** identical to an intermediate in the Marshall synthesis. Conversion of **949** to (−)-dictyolene followed an analogous pathway.

Greene also made use of **952** in a synthesis of (+)-pachydictyol A (Scheme 159). Irradiation of **952** and basic workup provided the well-known photoproduct **958**, which bears a striking structural relationship to **962**. Dehydration

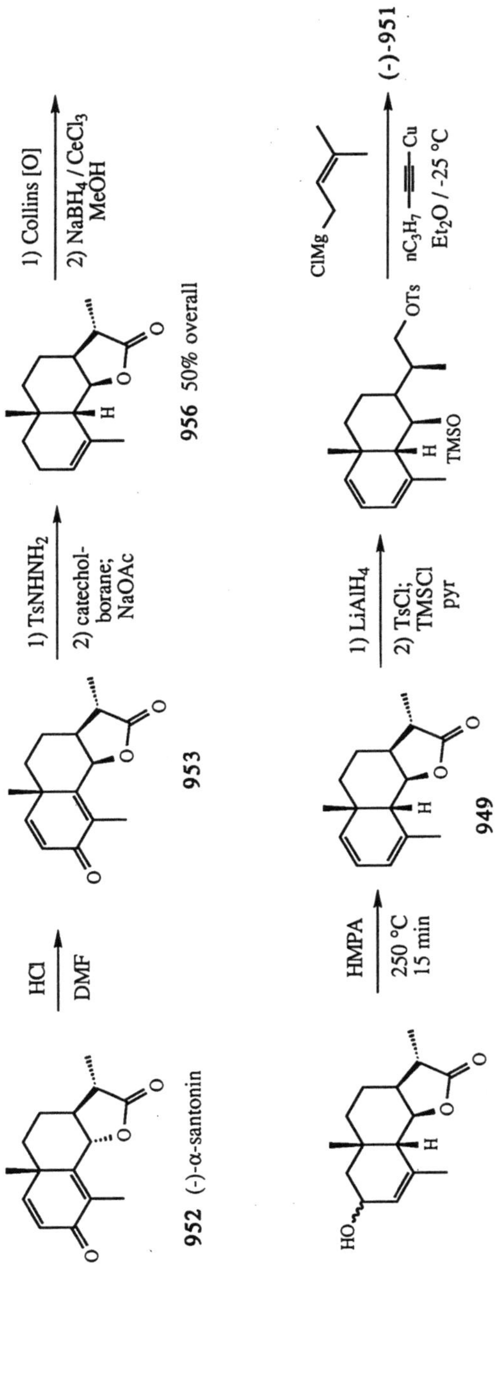

Scheme 158. Greene Synthesis of (−)-Dictyolene

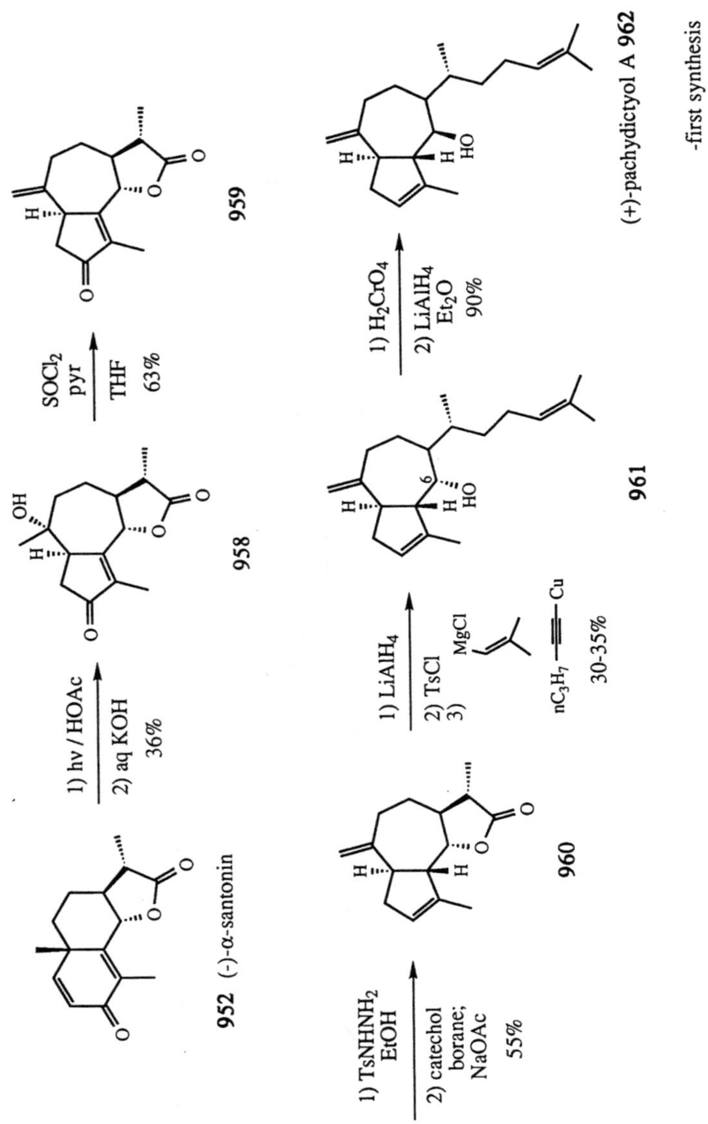

Scheme 159. Greene Synthesis of (+)-Pachydictyol A

with $SOCl_2$ led to exocyclic olefin **959**. Formation of the tosylhydrazine and reduction with catecholborane followed a more predictable pathway this time resulting in **960**. The lactone was then manipulated in a manner analogous to the dictyolene syntheses providing 6-*epi*-pachydictyol A (**961**). A simple oxidation-reduction sequence then provided (+)-**962**.

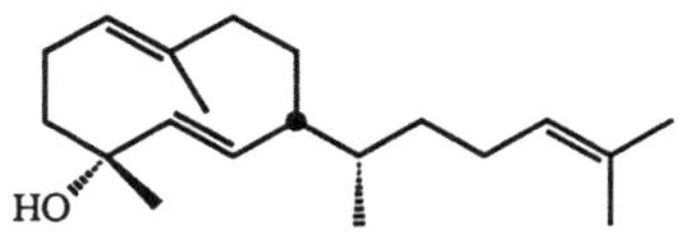

Obscuronatin 972

Obscuronatin (**972**), a structurally related diterpene from the soft coral *Xenia obscuronata* [251] has been synthesized by Kodama [252]. The key steps in this process (Scheme 160) involved the Claisen rearrangement of **965**, which was built from the allylic alcohol **963** and the acetal **964**. The resulting product **966** was transformed to the epoxide mixture **967** and the 10-membered ring was closed by intramolecular opening of the epoxide by an α-phenylsulfide-stabilized allylic carbanion leading to **968**. Dehydration of the alcohol led to a mixture of olefinic isomers which were subjected to hydrogenation. Isomer **969** was selectively reduced to the desired **971** in 19% yield, while the isomer **970** was not reduced and could be isolated from the hydrogenation mixture in 80% yield. 2,3-Sigmatropic rearrangement of the sulfoxide derived from **978** stereospecifically led to (±)-obscuronatin.

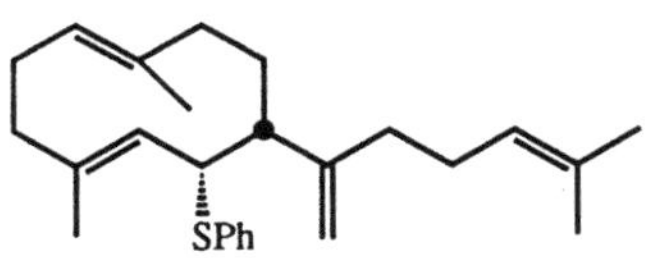

969

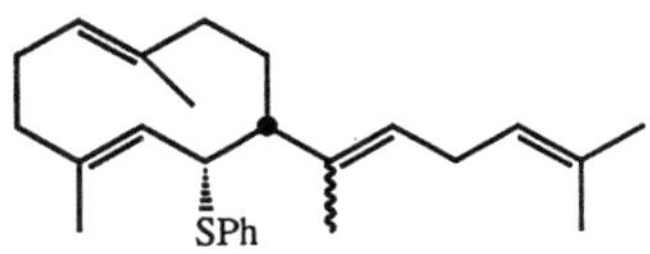

970

2.3.8.2 *Petiodial and Udoteatrial*

The green algae, Family Udoteaceae, are sources of a variety of terpenic cyclopentene aldehydes. Petiodial (**983**) [253] and udoteatrial (**979**) [254] share the same diterpene carbon skeleton and both have been synthesized. The structure of udoteatrial was corrected by Whitesell [255] using a synthetic approach based on the known bicyclic ketone **974** and classical iridoid chemistry (Scheme 161). The two stereocenters in **974** were parlayed into four stereocenters in **975**. Ireland-Claisen rearrangement of the acetate **976** and re-functionalization led to **977**, containing all of the non-ketal stereocenters of udoteatrial. Ozonolysis of **977** led to the tricyclic tosylate. Completion of the side chain was accomplished in a single step providing **978**. Treatment of **978**

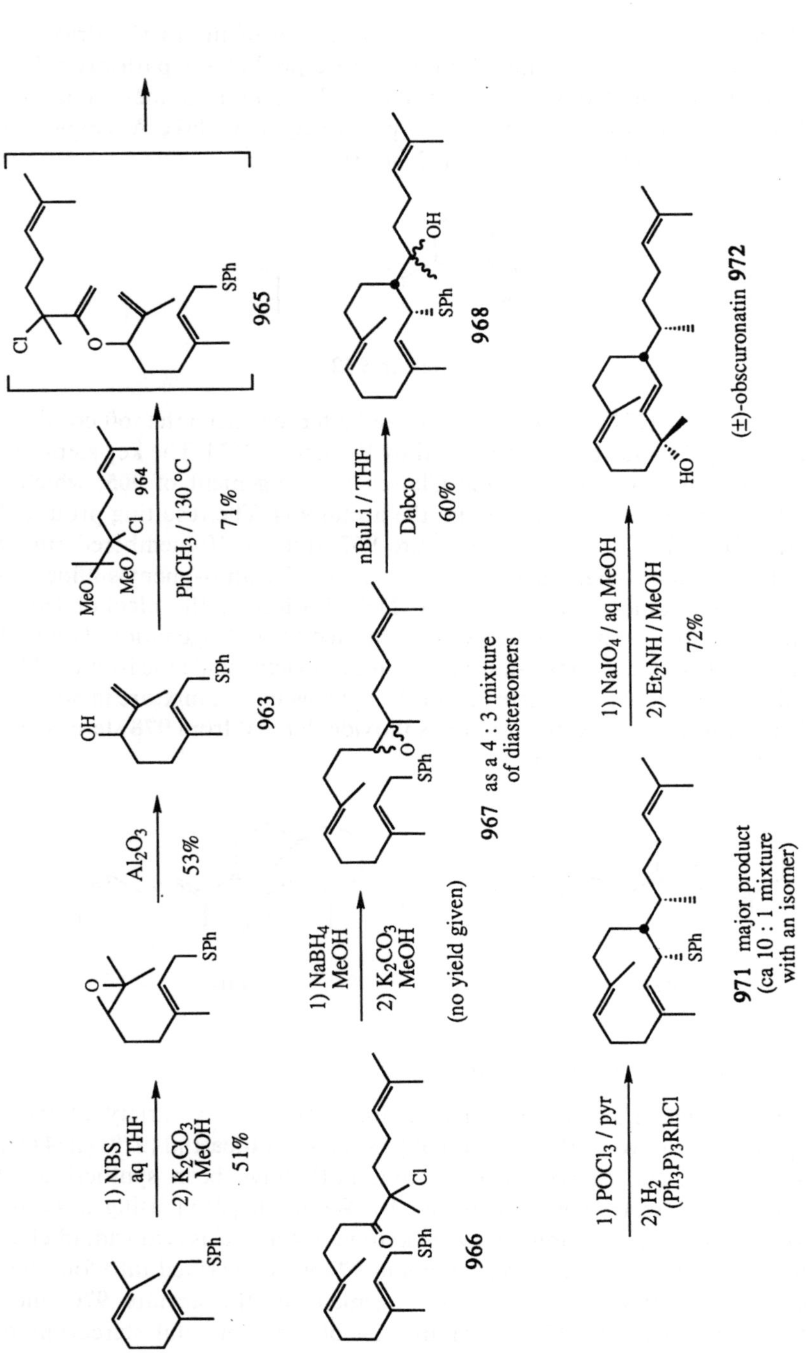

Scheme 160. Kodama Synthesis of (±)-Obscuronatin

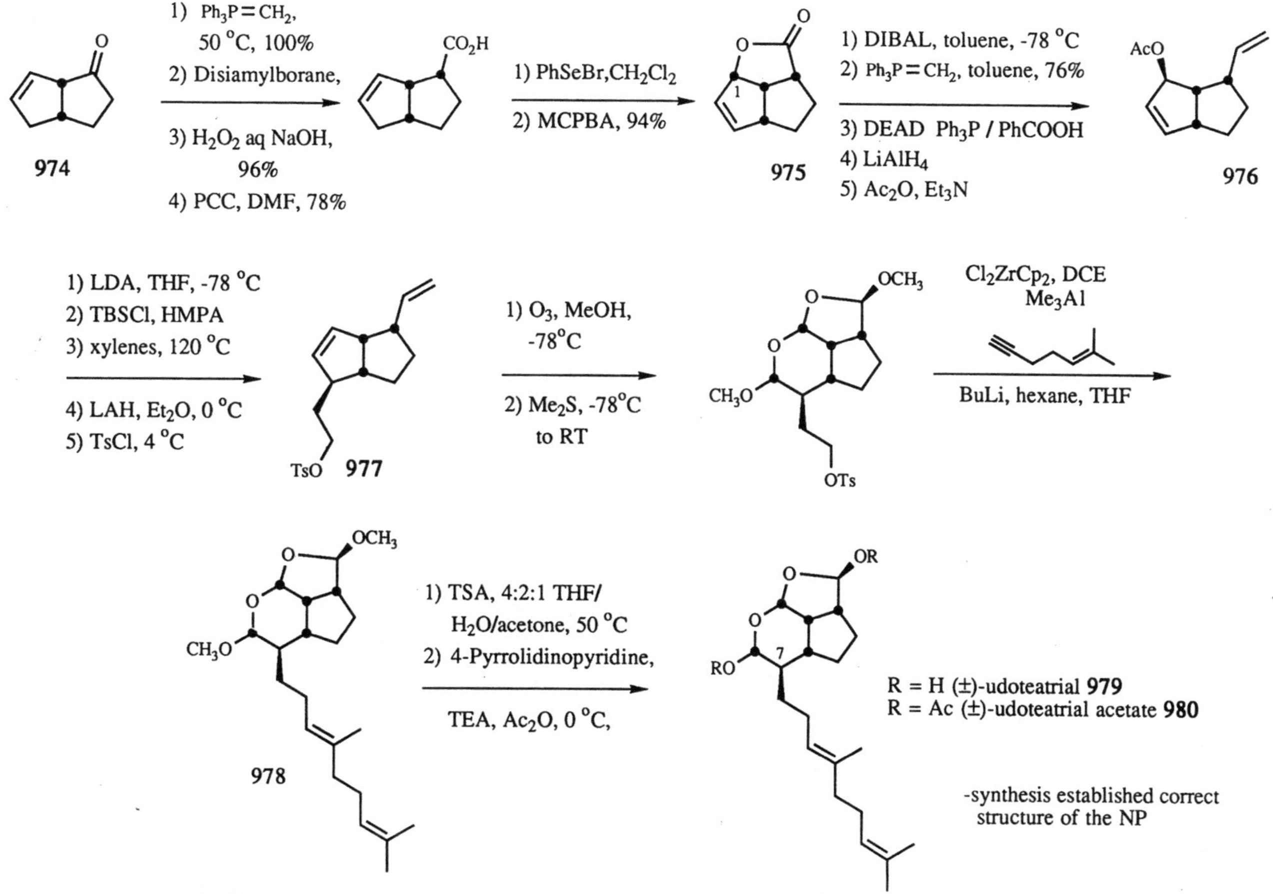

Scheme 161. Whitesell Synthesis of (±)-Udoteatrial.

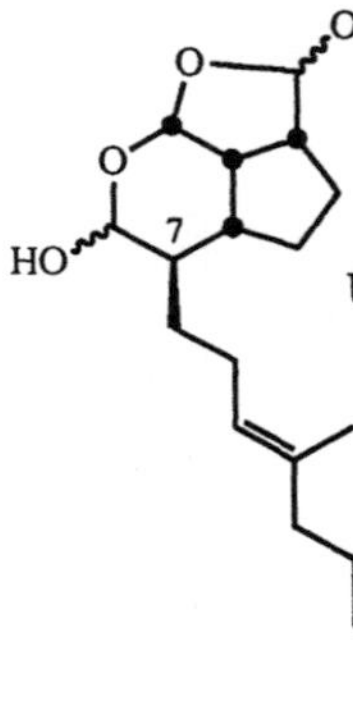

Udoteatrial **979**

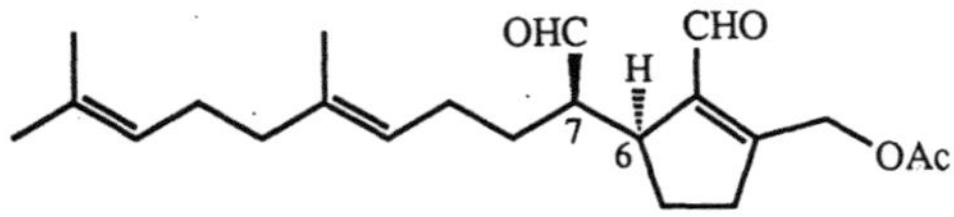

Petiodial **983**

with TsOH in aq acetone/THF gave udoteatrial which was analyzed as its diacetate (**980**) and found to be identical to material obtained by acetylation of natural udoteatrial. Whitesell originally prepared the C7 epimer of **979** by the same sequence of reactions without Mitsunobu inversion of the stereochemistry of the C1 alcohol. The end product of this sequence was not identical to naturally derived material and thus necessitated further study, eventually leading to reformulation of the natural product as **979**.

Petiodial has been synthesized by the Trost group in racemic form and by the Katsumura group as the (+)-enantiomer. Prior to this work no stereochemical details concerning the structure of **983** were available. Trost [256] converted *trans,trans*-farnesyl bromide to the allenic nitrile in 6 steps (Scheme 162). Palladium-catalyzed enallene cyclization resulted in the cyclopentenenitrile **981** in 85% yield. Functional group manipulation led to the hydroxyacetate **982**. Stereo and regioselective hydroboration-oxidation of the disubstituted olefin followed by Swern oxidation led to a major product which was identical to the natural product. Based on a predicted approach of the hydroboration reagent to the olefin, petiodial was postulated to have the relative configuration shown. The absolute configuration of natural (−)-petiodial was determined to be that shown in **983** (Scheme 163) by Katsumura [257] by synthesizing the (+)-isomer (**988**) beginning with (+)-genipin (**985**). This starting material possesses much of the intact carbon skeleton of **983** with the exception of the sidechain, which was attached via the geranyl sulfone **986**. Manipulation of this substance led to the hemiacetal **987** which was treated with DBU in benzene at reflux. Presumably this established an equilibrium between ring-opened aldehydes and hemiacetals. Once opened, isomerization of the olefin to the 2,3-position occurred and the ring-closed hemiacetals were no longer the favored forms at equilibrium and thus the two dialdehydes **988** and (**983**) could be isolated. The 6*S*,7*R*-isomer **988** was shown to be enantiomeric to natural petiodial, thus completing the synthesis and structural assignment.

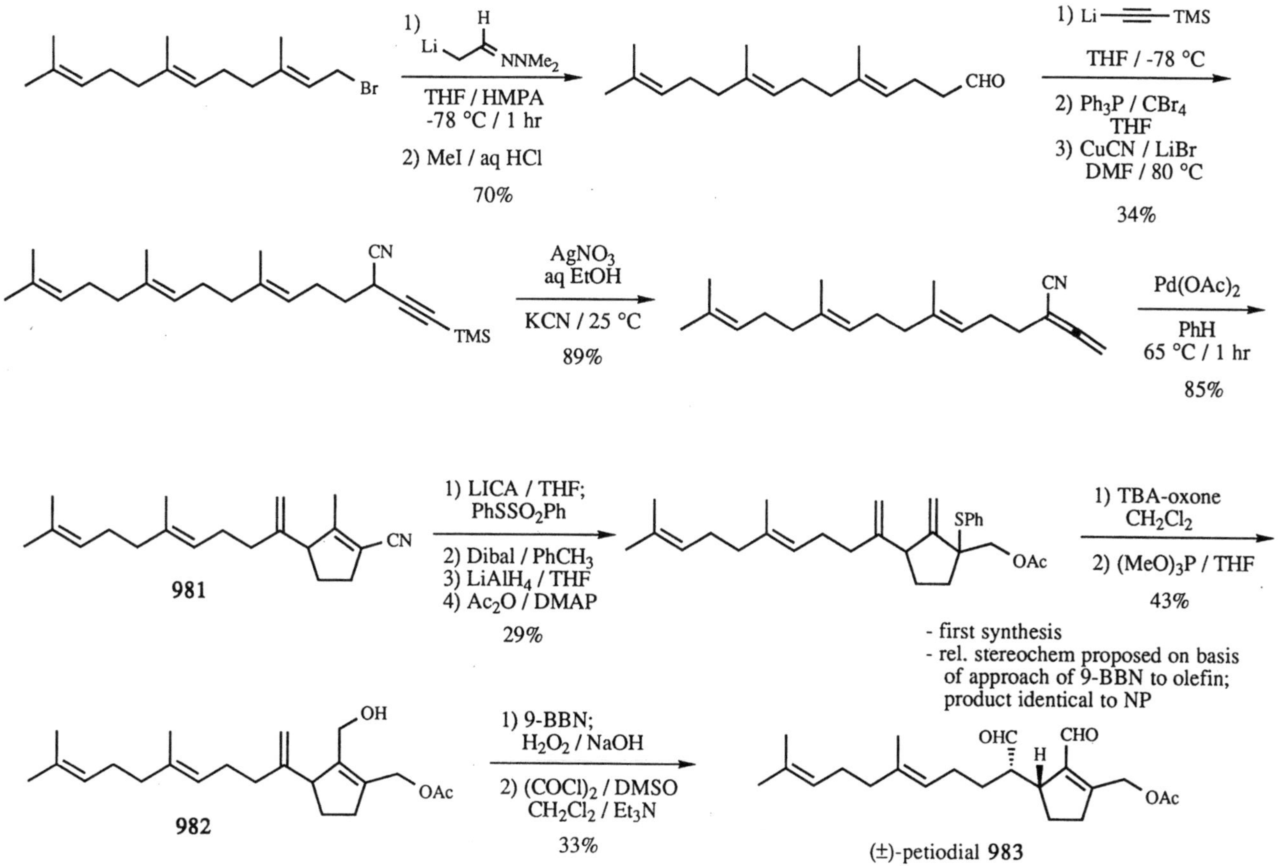

Scheme 162. Trost Synthesis of (±)-Petiodial

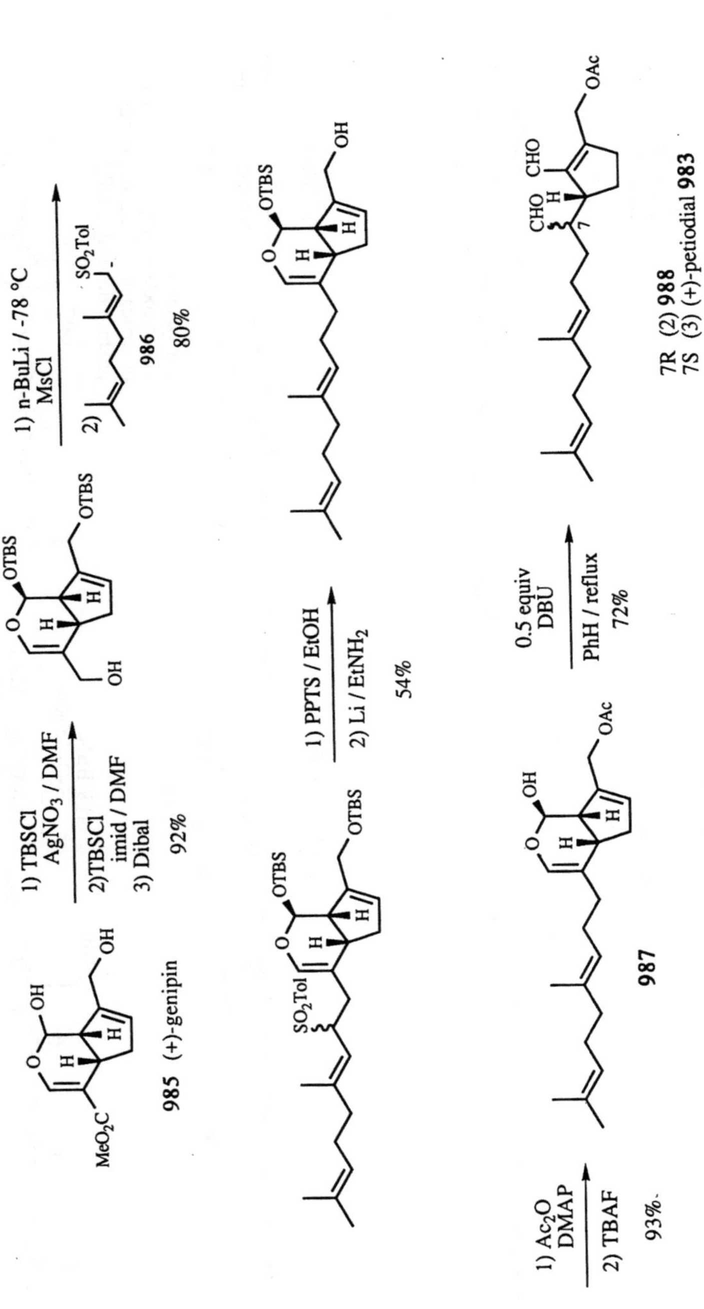

Scheme 163. Katsumura Synthesis of (±)-Petiodial

2.3.8.3 Reiswigin A

The absolute stereochemistry of reiswigin A [258] was worked out by Snider [259] in the only reported synthesis of this substance (Scheme 164). Allylsilane conjugate addition to enone **989** gave the mixture **990**. The ketone was protected and the cyclohexene oxidatively cleaved and aldol cyclized to the cyclopentene aldehyde **991**, still as a mixture. Formation of the imine **993** from the amino acid **992** was a prelude to Grignard conjugate addition and trapping to provide the aldehyde **994** stereogenically pure at three stereocenters. Treatment of **994** with dimethylaluminum chloride effected closure of the 7-membered ring. PCC oxidation and treatment with TsOH delivered (−)-reiswigin A and its C13 epimer in a 3:2 ratio in 84% from **994**. Having both C13 epimers availabale for spectroscopic study allowed the tentative assignment of relative stereochemistry at C13 based on ^{1}H and ^{13}C NMR spectroscopy.

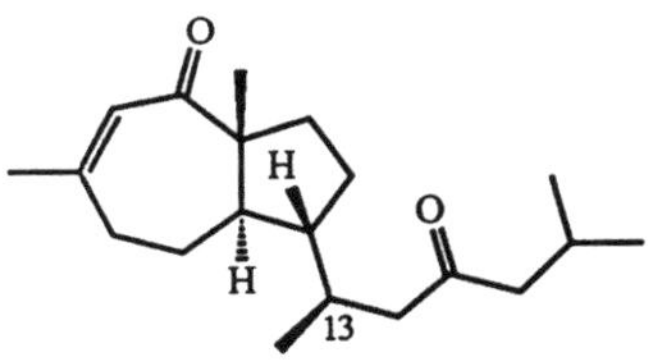

(-)-Reiswigin A **995**

2.3.8.4 Mayolide A

Mayolide A was the first secocembranoid isolated from soft coral (*Sinularia mayi*) [260]. Yamada used the chiral pool element **996** to provide the (+)-enantiomer (Scheme 165) [261]. Conjugate addition of lithio *t*-butylacetate occurred on the face of lactone **996** away from the protected alcohol group giving **997**. Functional group manipulation led to aldehyde **998**. Carbonyl addition of isopropenylmagnesium bromide gave a diastereomeric mixture of alcohols. Two successive Claisen rearrangements gave the dienic aldehyde **999**. Reasonable *E* over *Z* selectivity was observed in each rearrangement. The chain was extended by two carbons to give the methyl ketone **1000**. Attention was next turned to preparing the α-methylene portion of the molecule. Because the three alcohols in **1000** were differentially protected the selective demasking of each

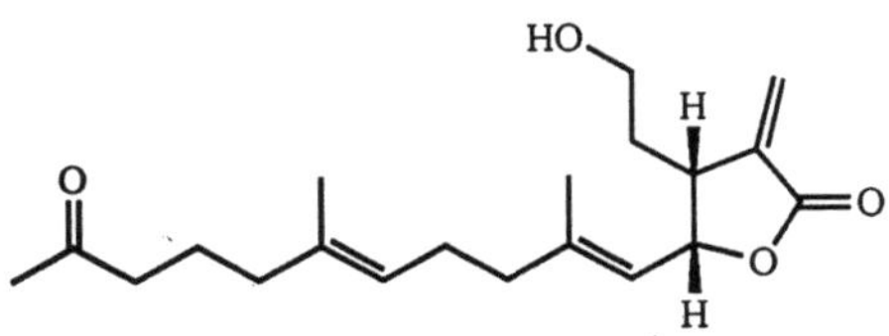

(+)-Mayolide A 1002

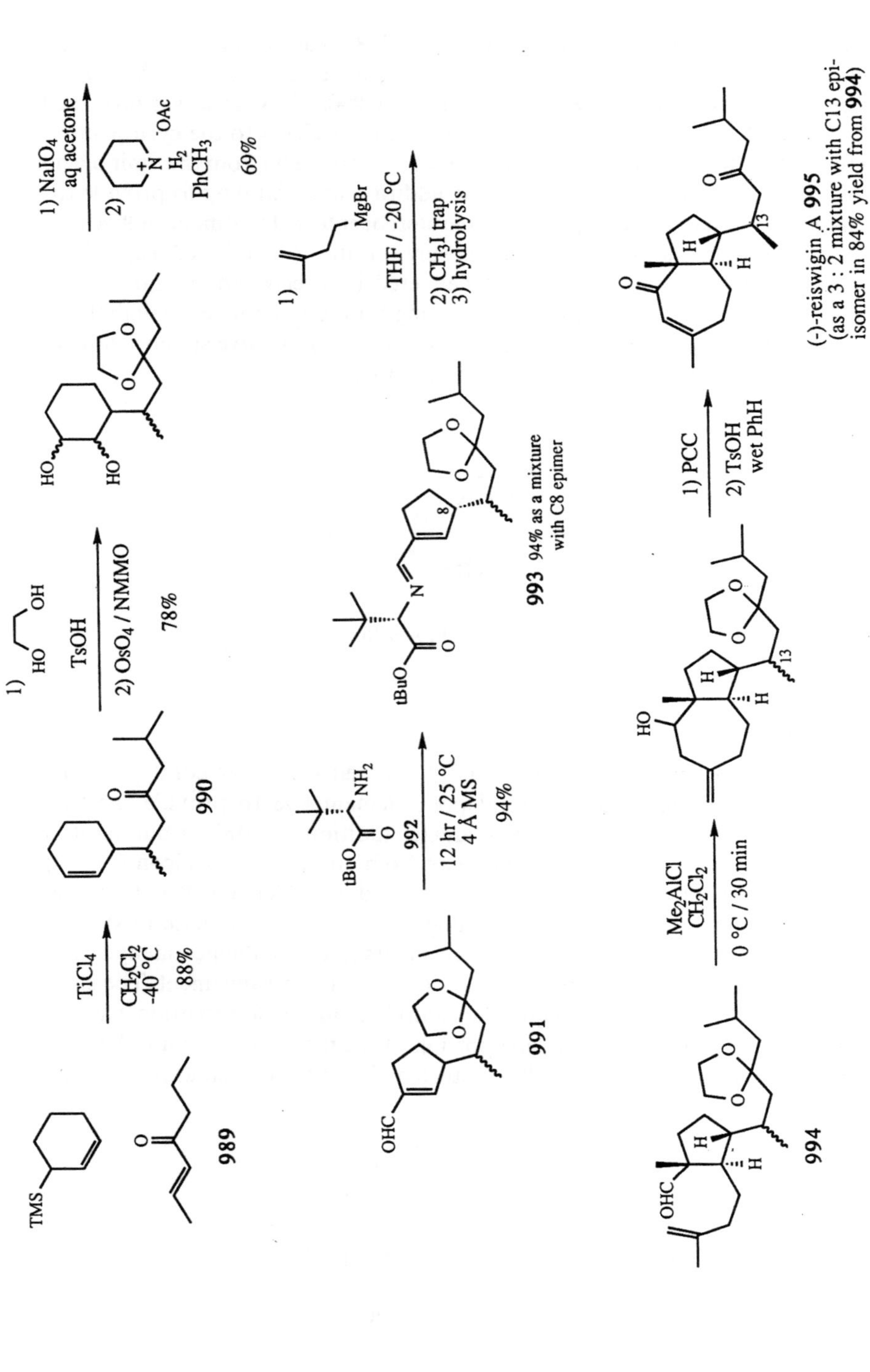

Scheme 164. Snider Synthesis of (−)-Reiswigin A

was apparently a straightforward task eventually resulting in lactone **1001**. α-Methylenation and final functional group manipulation gave (+)-mayolide which was opposite in configuration to naturally-derived material.

2.3.8.5 *Ambliols A and B*

Ambliol A and B [262] from the marine sponge *Dysidea amblia* have each been synthesized. A biomimetic cationic cyclization of the furanotriene was employed by Nishizawa [263] (Scheme 166) in which the furanotriene **1004** was treated with mercuric triflate resulting in 4 major products (**1005–1008**). The monocyclic organomercurial **1007** was reductively demercurated and desulfurized to racemic ambliol A (**1009**). Ambliol B was prepared via a non-biomimetic approach in Scheme 167 by Piers [264]. Vinyl cuprate addition to enone **1010** and trapping established stereocenters at C8 and C9. Methylenecyclohexene annulation was carried out using technology developed in the Piers lab involving the germanium-containing cuprate reagent **1012** eventually resulting in **1013**. Cyclopropanation allowed establishment of the gem-dimethyl group at C4 and extension of the sidechain at C9 into a furanoethyl group culminated in racemic ambliol B.

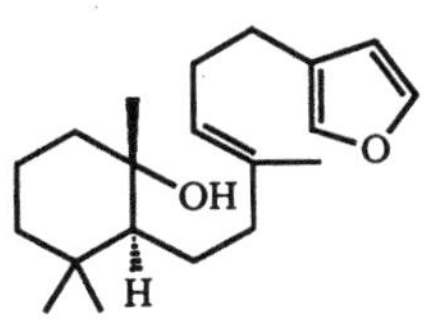

(±)-Ambliol A **1009**

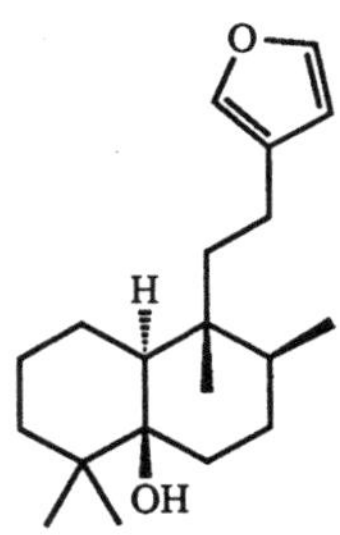

(±)-Ambliol B **1014**

2.3.8.6 *Taonianone*

The diterpene furan taonianone is an unusual metabolite isolated from brown algae (*Taonia atomaria*) and is unrelated to any other *Taonia* metabolite [265]. Two syntheses of the (+)-enantiomer have been reported, both of which showcase the oxidative cleavage and recyclization of a carvone derivative to establish the cyclopentenone ring of **1020**. Yoshikoshi [266] described the

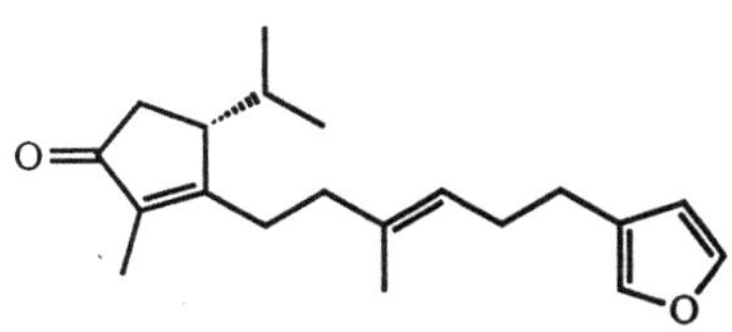

Taonianone **1020**

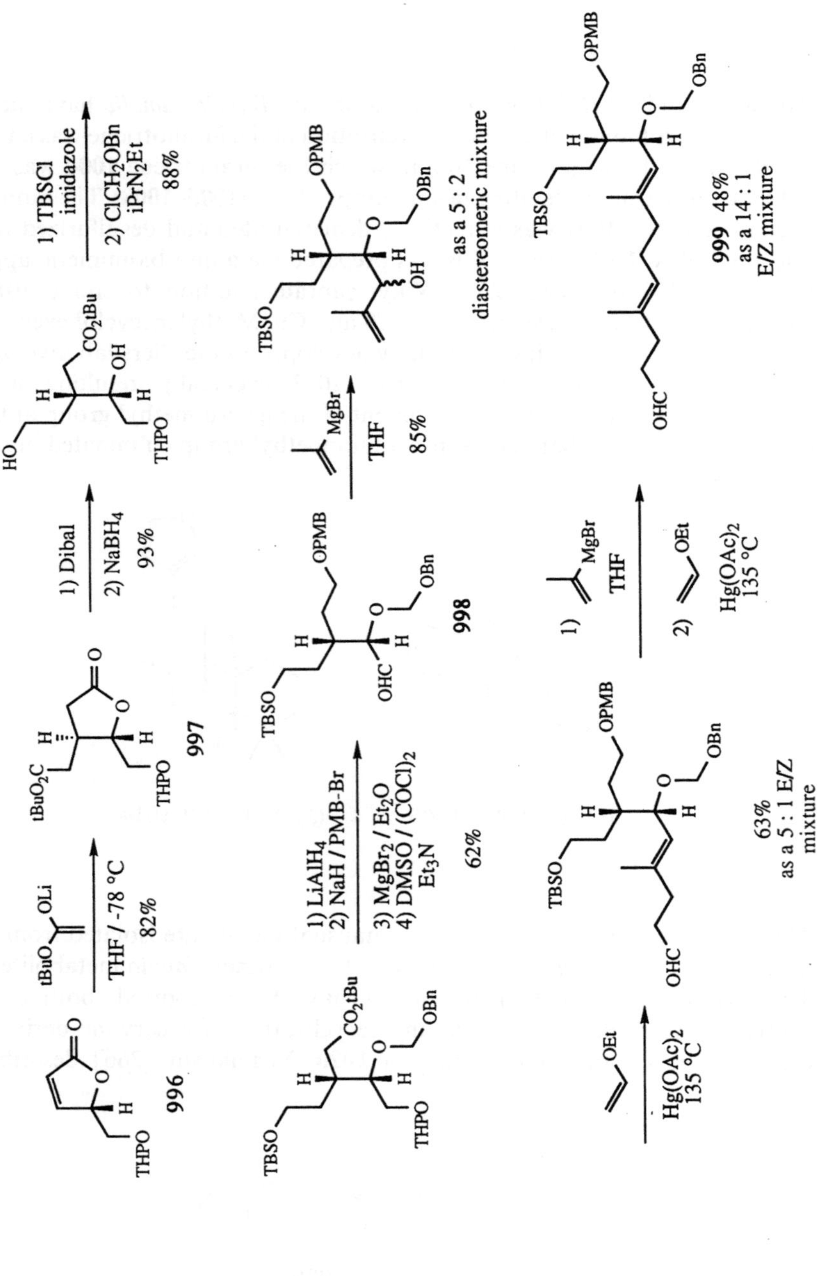

996
tBuO OLi
THF / -78 °C
82%
997
1) Dibal
2) NaBH4
93%
1) TBSCl
imidazole
2) ClCH2OBn
iPrN2Et
88%
1) LiAlH4
2) NaH / PMB-Br
3) MgBr2 / Et2O
4) DMSO / (COCl)2
Et3N
62%
998
MgBr
THF
85%
as a 5 : 2
diastereomeric mixture
OEt
Hg(OAc)2
135 °C
63%
as a 5 : 1 E/Z
mixture
1) MgBr
THF
2) OEt
Hg(OAc)2
135 °C
999 48%
as a 14 : 1
E/Z mixture

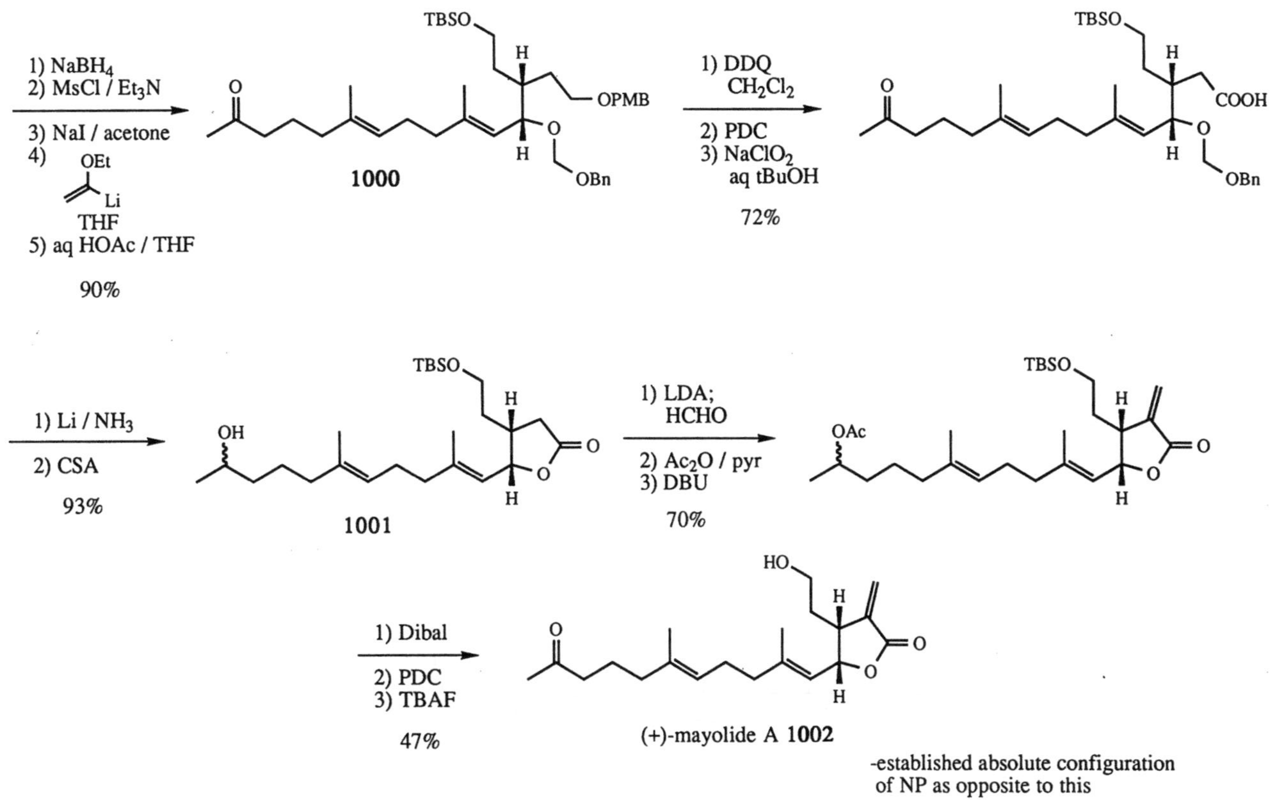

Scheme 165. Yamada Synthesis of (+)-Mayolide A

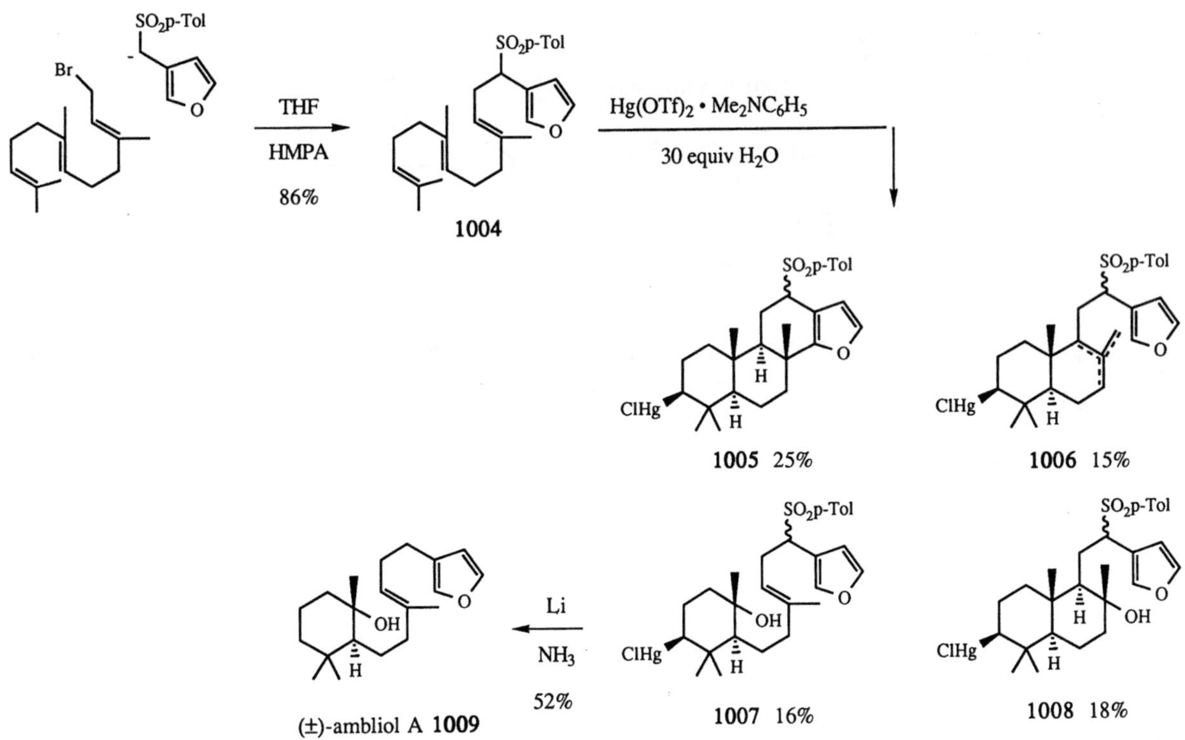

Scheme 166. Nishizawa Synthesis of (±)-Ambliol A

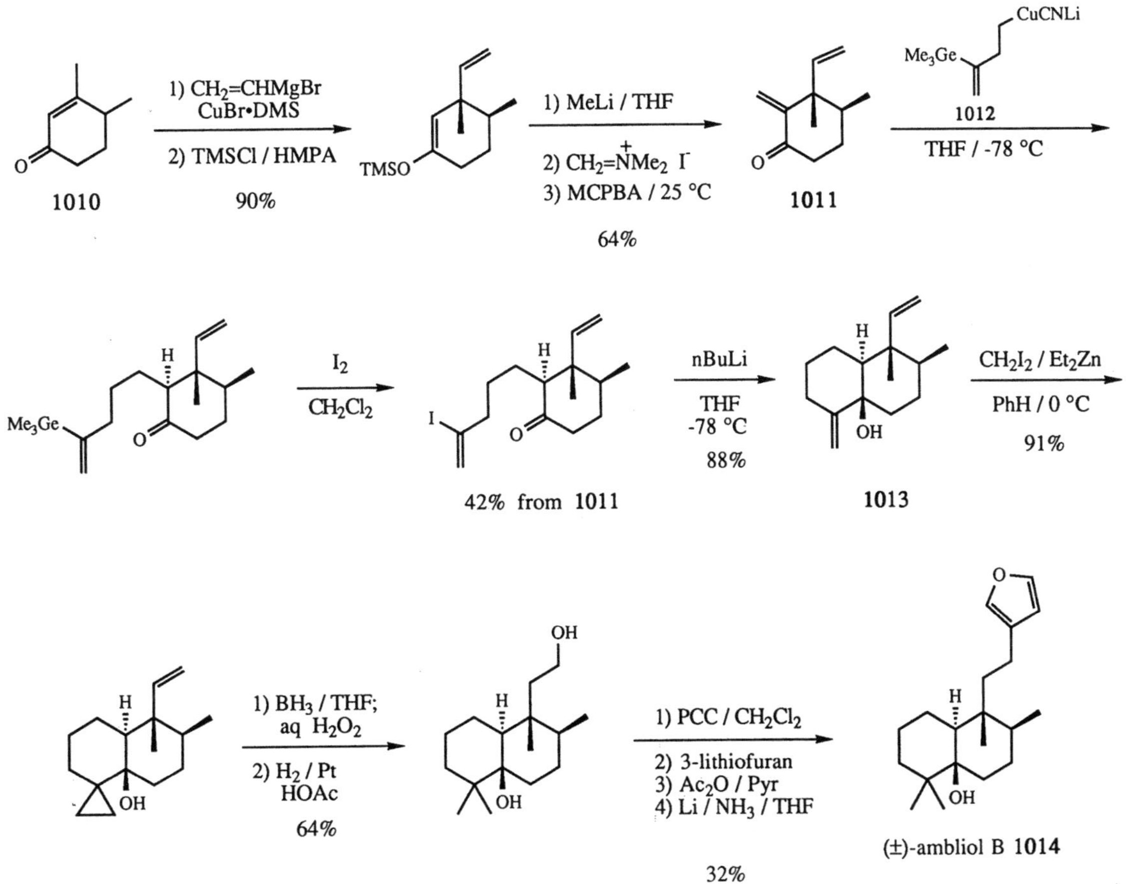

Scheme 167. Piers Synthesis of (±)-Ambliol B

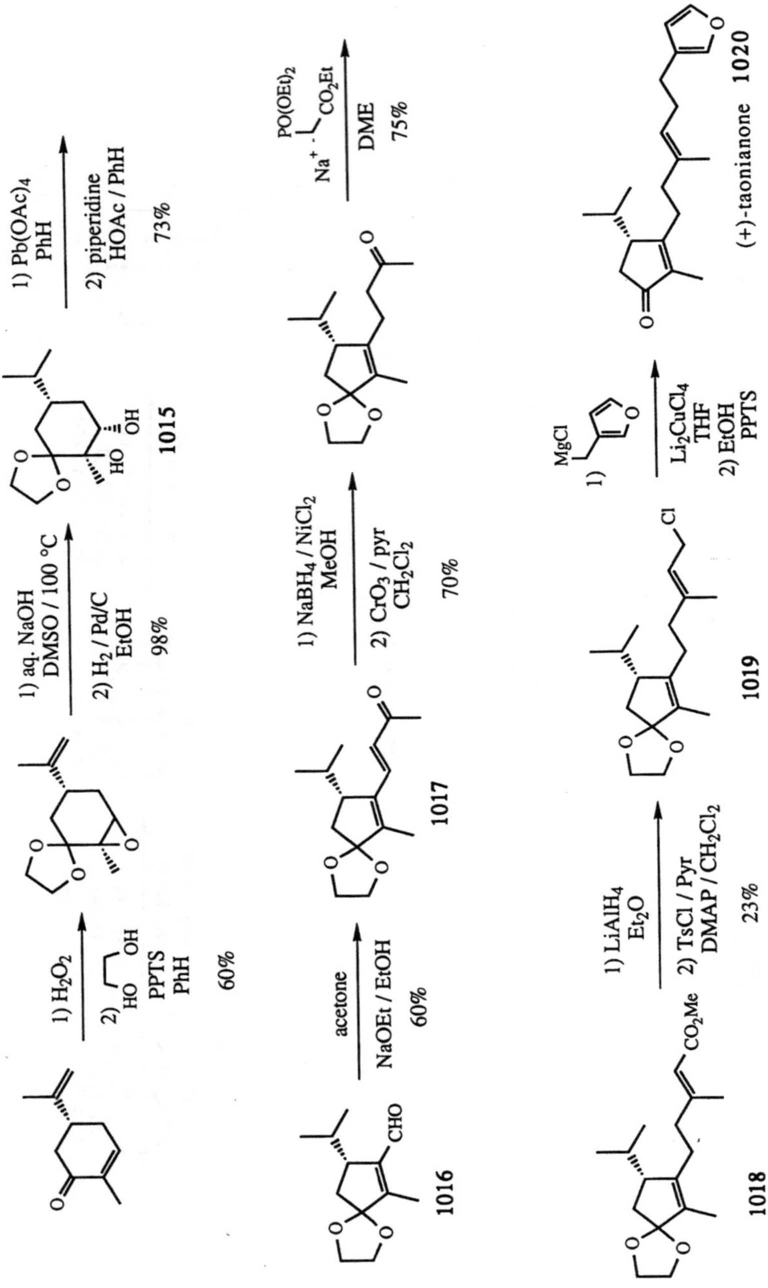

Scheme 168. Yoshikoshi Synthesis of (+)-Taonianone

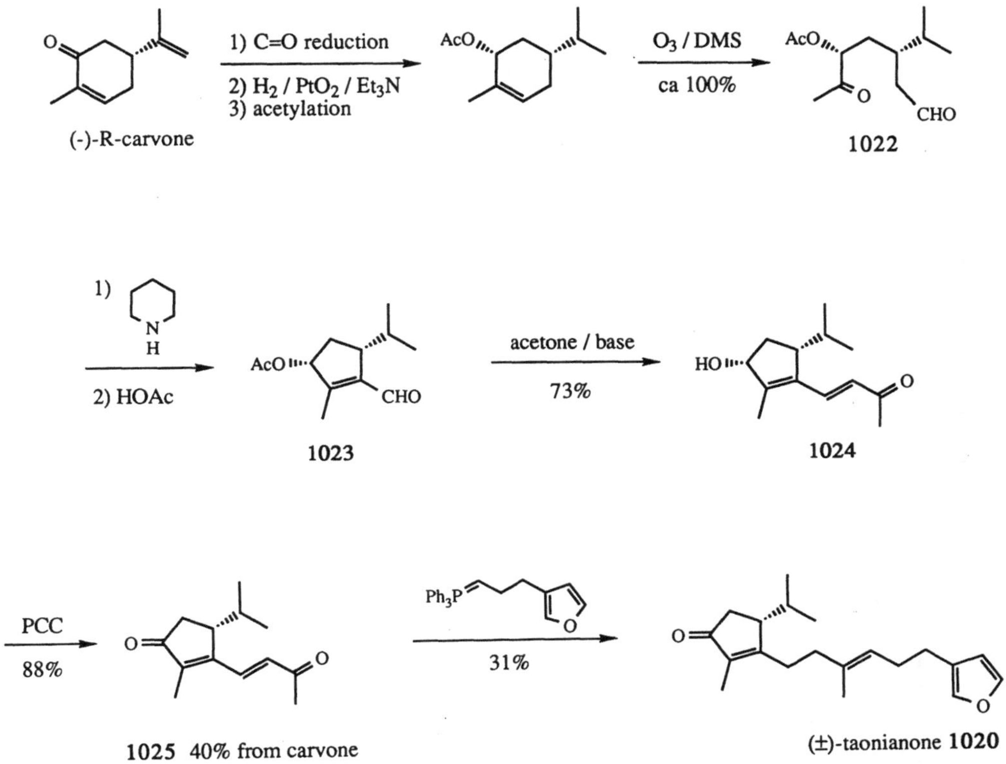

Scheme 169. Kreiser Synthesis of (+)-Taonianone

conversion of (−)-carvone to the aldehyde **1016** (Scheme 168) by cleavage of the *trans*-diol **1015** to a dialdehyde which was cyclized with piperidinium acetate. Extension of the side chain then took place in three phases, with initial 3-carbon homologation by aldol reaction with acetone giving **1017**. Next, a 2-carbon homoglation via the Wardsworth-Emmons reaction established the geometry of the trisubstituted olefin (**1018**). Conversion of **1018** to the allylic chloride **1019** and displacement with 2-furylmagnesium chloride gave the final skeleton of (+)-taonianone.

Direct ozonolytic cleavage of dihydrocarveol acetate leading to intermediate ketoaldehyde **1022** was used by Kreiser [267] (Scheme 169) in a synthesis which was quite similar to that of Yoshikoshi. Cyclization with base gave the aldehyde acetate **1023**. Chain extension by aldol reaction with acetone (as in the Yoshikoshi synthesis) occurred with acetate hydrolysis giving **1024** which was oxidized to diketone **1025**. Wittig reaction of **1025** occurred in 31% yield to give (+)-taonianone.

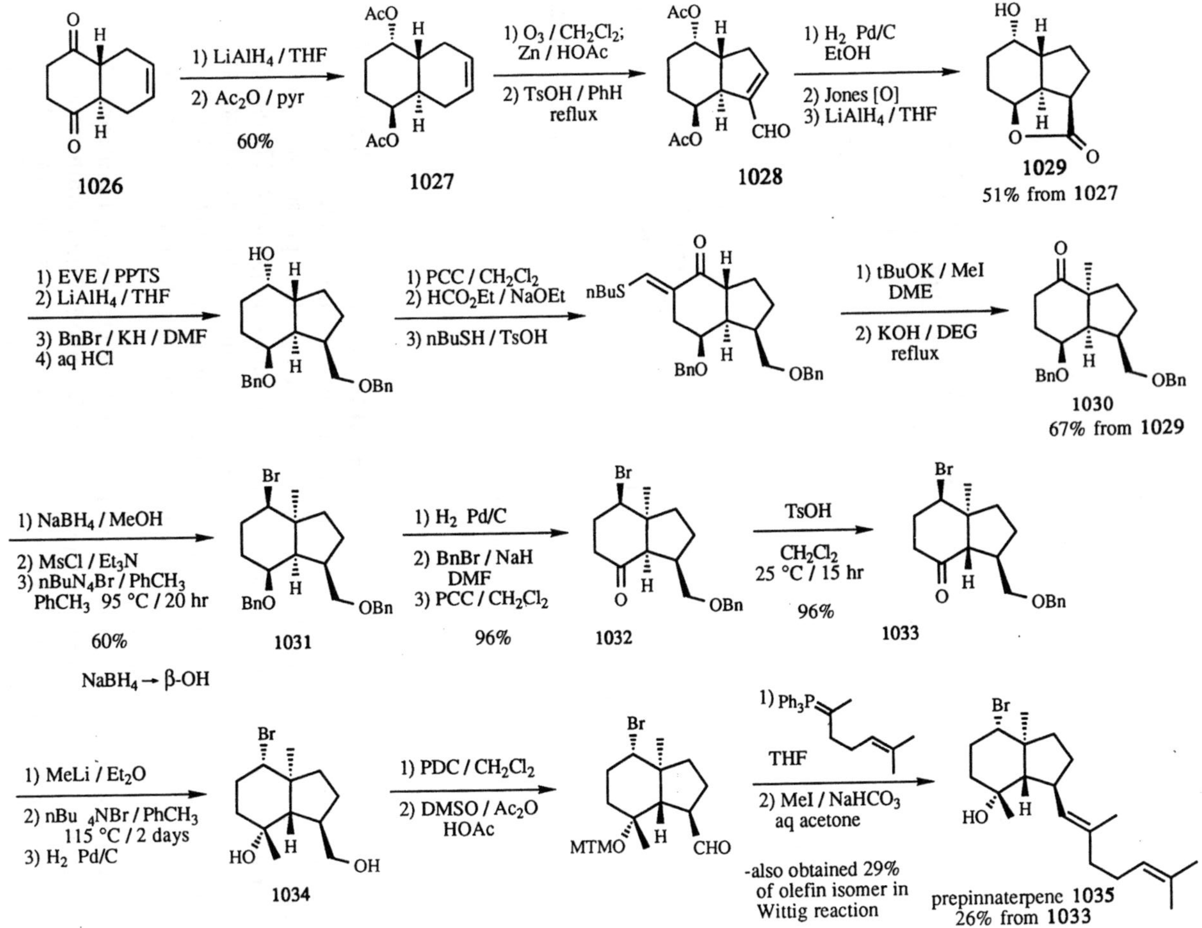

Scheme 170. Masamune Synthesis of Prepinnaterpene (**1035**)

2.3.8.7 Prepinnaterpene

The bromoditerpene prepinnaterpene [268] is related to the sesquiterpene oppositol, a synthesis of which was discussed earlier. Prepinnaterpene has been synthesized by Masamune via a 30 step process (Scheme 170) [269]. The (6,5) carbocyclic ring system was prepared from the bicyclic dione **1026** by ozonolysis of diacetate **1027** and aldol cyclization under acidic conditions. A routine series of transformations led from **1028** to the ketone **1030**. The C1 bromine was now installed by reduction of the ketone to the β-alcohol and mesylation. Treatment of the mesylate with tetrabutylammonium bromide in toluene at 95 °C for 20 h produced the β-bromide **1031** as the major product along with the α-bromide and an olefinic byproduct. The (incorrect) orientation of the bromine is equatorial in this intermediate. The synthetic pathway led to bromoketone **1032** which was epimerized to the trans-fused system with TsOH. The bromine now assumes an axial position. Methyllithium addition to the ketone produced the correct relative stereochemistry at C4. Inversion of C1 stereochemistry was now accomplished in a prolonged reaction producing **1034** containing all the stereocenters aligned properly. Extension of the C6 side chain was performed via Wittig coupling to provide racemic prepinnaterpene.

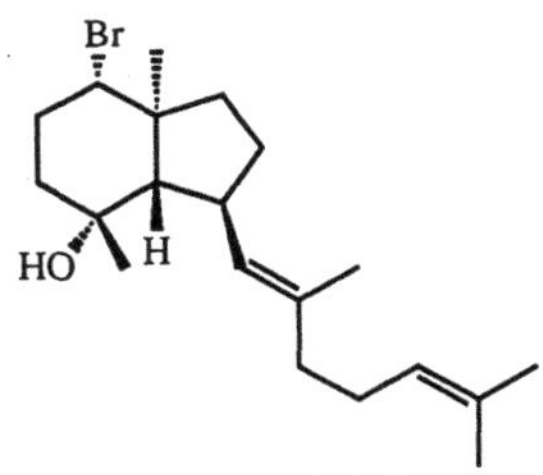

Prepinneterpene 1035

2.3.8.8 Sanadaol

Sanadaol (**1050**) has been found [270] in the brown algae *Pachydictyon coriaceum* and *Dictyota crenulata.* This diterpene possesses a unique bicyclo

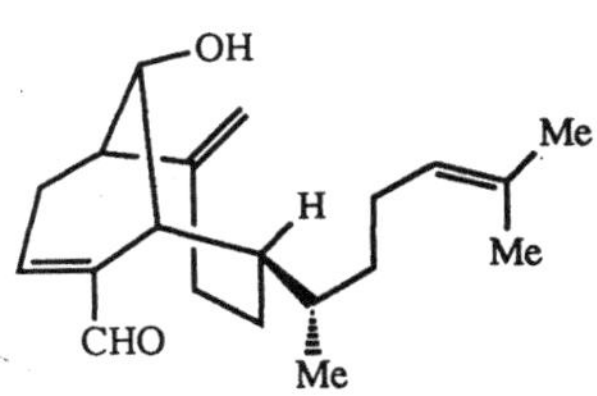

Sanadaol **1050**

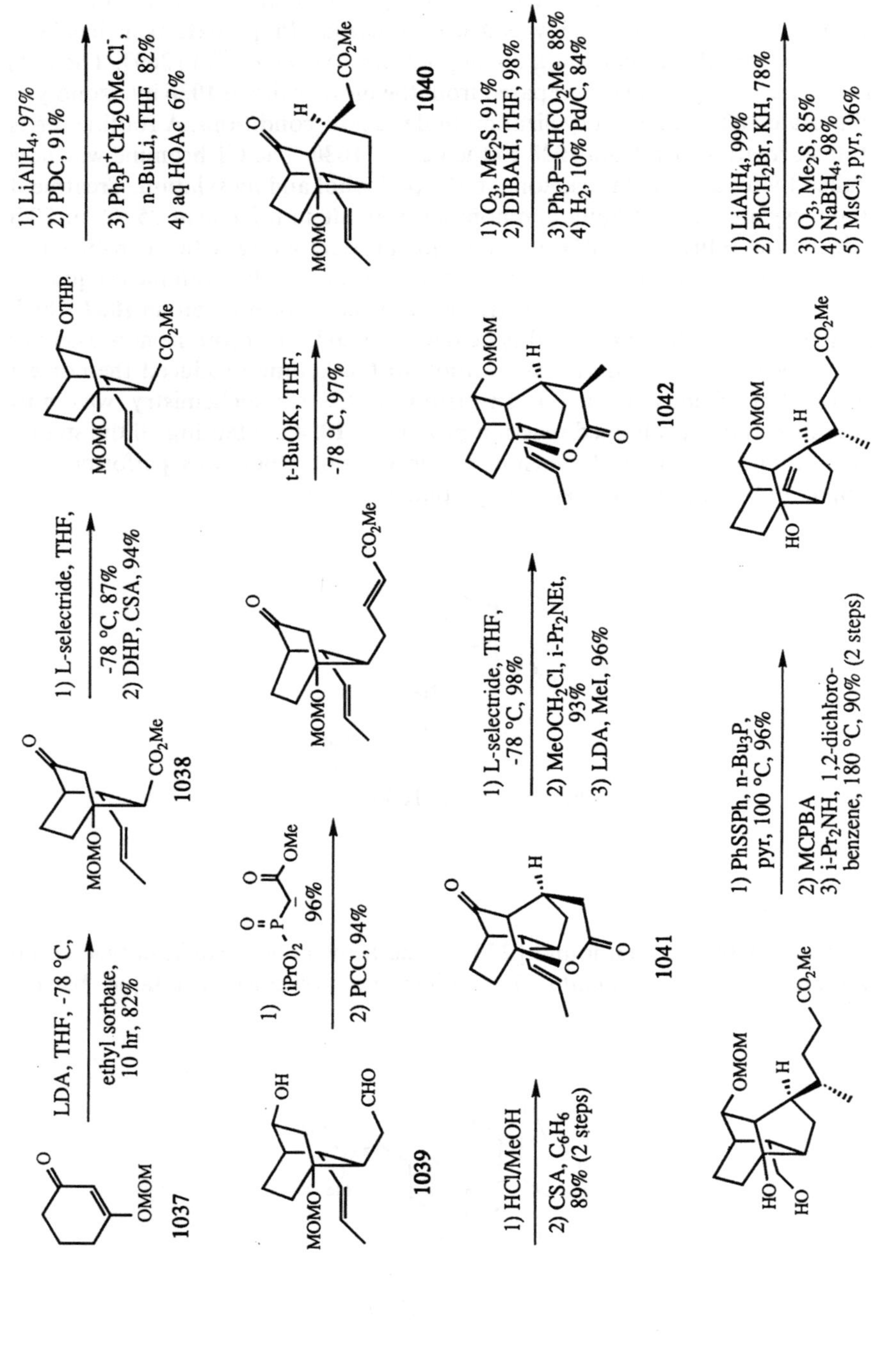
OMOM
1037
LDA, THF, -78 °C,
ethyl sorbate, 10 hr, 82%
MOMO
CO2Me
1038
1) L-selectride, THF, -78 °C, 87%
2) DHP, CSA, 94%
OTHP
1) LiAlH4, 97%
2) PDC, 91%
3) Ph3P+CH2OMe Cl-, n-BuLi, THF 82%
4) aq HOAc 67%
OH
CHO
1039
1) (iPrO)2P(O)CH2C(O)OMe 96%
2) PCC, 94%
t-BuOK, THF, -78 °C, 97%
H
1040
1) HCl/MeOH
2) CSA, C6H6 89% (2 steps)
1041
1) L-selectride, THF, -78 °C, 98%
2) MeOCH2Cl, i-Pr2NEt, 93%
3) LDA, MeI, 96%
OMOM
1042
1) O3, Me2S, 91%
2) DIBAH, THF, 98%
3) Ph3P=CHCO2Me 88%
4) H2, 10% Pd/C, 84%
HO
1043
1) PhSSPh, n-Bu3P, pyr, 100 °C, 96%
2) MCPBA
3) i-Pr2NH, 1,2-dichlorobenzene, 180 °C, 90% (2 steps)
1044
1) LiAlH4, 99%
2) PhCH2Br, KH, 78%
3) O3, Me2S, 85%
4) NaBH4, 98%
5) MsCl, pyr, 96%

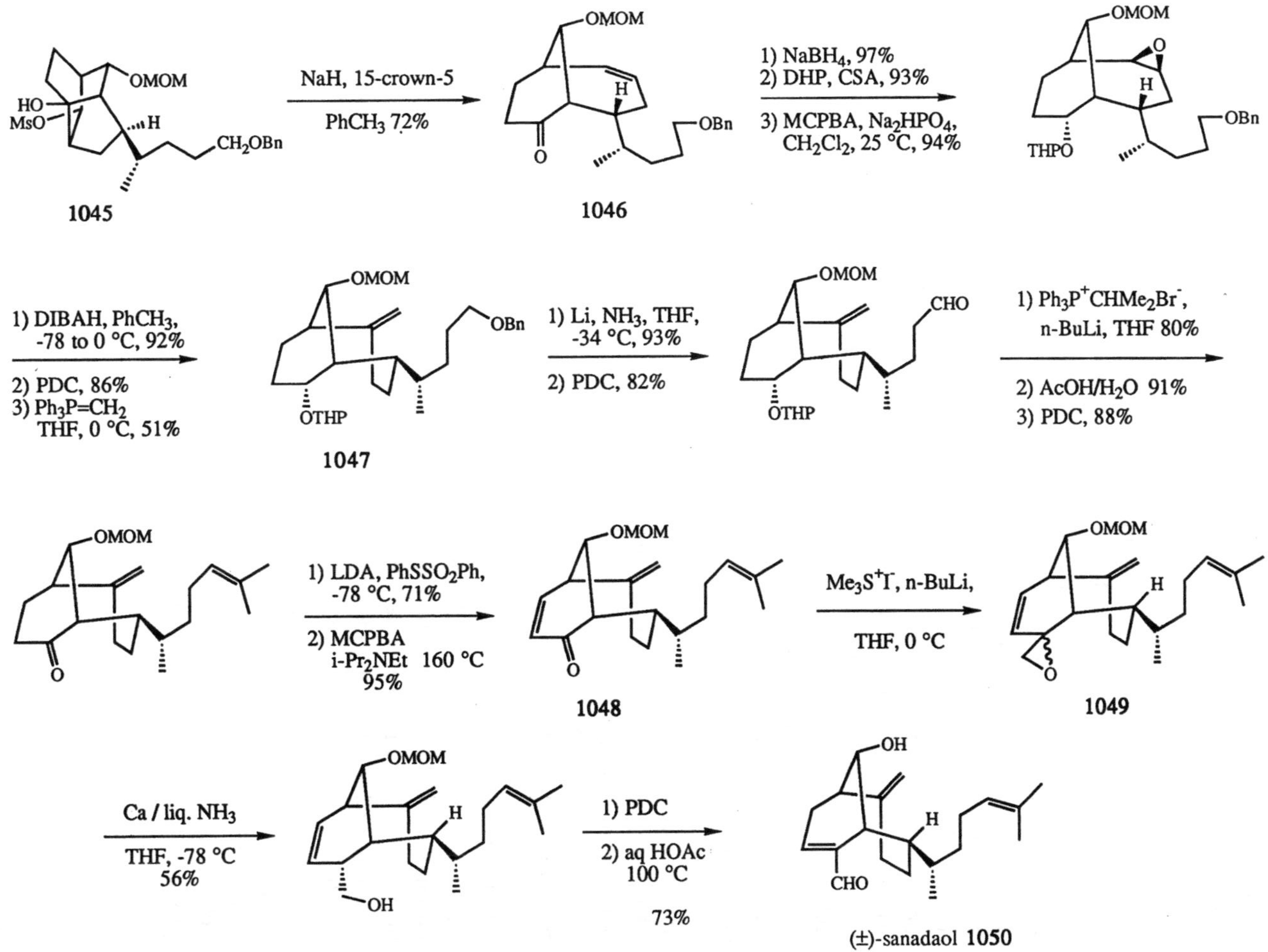

Scheme 171. Yamada Synthesis of (±)-Sanadaol

[4.3.1]decane ring system with five contiguous chiral centers. Confirmation of the structure of **1050** was obtained through a rather lengthy total synthesis by Yamada [271] (Scheme 171). The tandem Michael-intramolecular Michael reaction between cyclohexanone derivative **1037** and ethyl sorbate provides bicyclic ketone **1038** in 82% yield. Of the four chiral centers in the initial intermediate **1038**, only one is carried on to the product. A series of standard transformations results in minor refunctionalization of **1038** along with a one-carbon homologation to give aldehyde **1039**. A further set of standard transformations finishing with lactonization produces tetracyclic lactone **1041**, thus providing three of the final five chiral centers. The remaining two chiral centers are obtained via stereoselective reduction of the ketone and methylation of the lactone to give compound **1042**. At this point in the synthesis, intermediate **1042** contains thirteen of the twenty carbons present in sanadaol. Wittig olefination of the lactol from **1042** followed by elimination of the sulfoxide prepared from the primary hydroxyl of **1043** gives **1044** with fifteen of the needed twenty carbons. Standard transformations provide γ-hydroxymesylate **1045** which undergoes sodium hydride induced fragmentation to yield ketone **1046**, having the same ring system as the natural product. Selective DIBAL reduction of the epoxide formed from **1046** followed by oxidation and Wittig olefination provides the exocyclic methylene **1047** needed in the natural product. A series of standard functional group interconversions along with a Wittig olefination produces α,β-unsaturated ketone **1048**, containing nineteen of the final twenty carbons of **1050**. Treatment of **1048** with dimethylsulfonium methylide affords a diastereomeric mixture of epoxides **1049** which is converted to sanadaol **1050** in three steps. An extremely high average yield of 87% *per* step allows the preparation of **1050** in 44 steps with an overall yield of 0.27%.

An enantiospecific synthesis of sanadaol has also been developed by Yamada [272] (Scheme 172). The overall strategy followed is very similar to that used in the racemic route (Scheme 171). The synthesis begins with a Michael-intramolecular Michael reaction of optically pure ketal **1051**, available from D-mannitol [273], with dienolate **1052** gives **1053**. A series of standard transformations over fifteen steps provides ketone **1054**, which is nearly identical to intermediate **1040**, in the previous synthesis. Lactonization and methylation of **1054** provides intermediate **1055**, containing all five of the required chiral centers. Reduction of the lactone, Wittig olefination and hydrogenation produces diol **1056** which is converted into common intermediate **1046** using the methodology illustrated in Scheme 202. Continuation of the synthesis in an identical fashion as before provides the unnatural enantiomer (−)-sanadaol in 50 steps from ketal **1052**. The (+)-enantiomer was prepared in a similar manner.

2.3.8.9 Dictymal

As part of a program to synthesize terrestrial di- and sesterterpenes possessing a (5,8,5)-tricyclic ring system, Kato synthesized the related *Dictyota dichotoma*

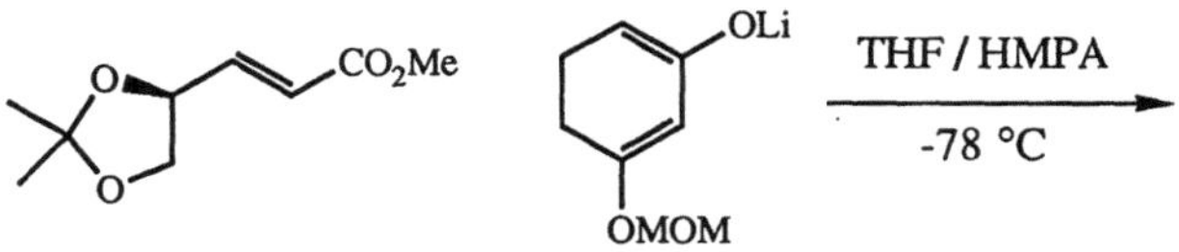

1051
(optically pure)

1052

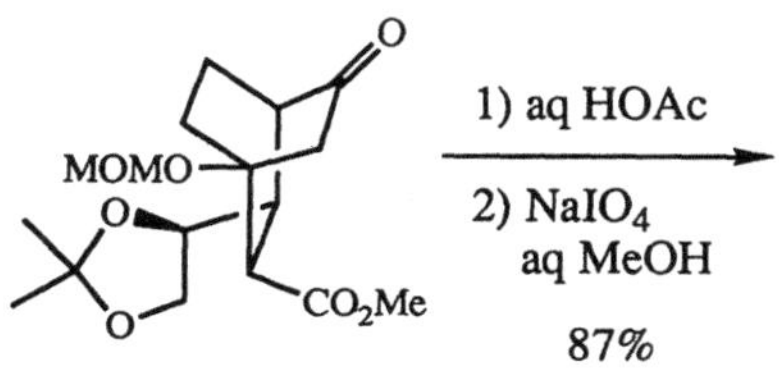

(-)-**1053**

major cycloadduct

metabolite dictymal (**1065**) [274] (Scheme 173) [275]. The strategy involved coupling of the two iridoids **1058** and **1059** prepared by a photochemical route including resolution of diastereomeric menthyl esters. Cr(II) chloride-mediated coupling of the allylic chloride **1059** to the aldehyde **1058** gave the α-alcohol **1060** as the major product. To prepare for the next key reaction the adduct **1060** was converted to the 1,5-diene **1061** which when heated to 180 °C was converted via Cope rearrangement to **1062** in good yield. Comparison of the structure of **1062** to that of the target molecule reveals that this advanced intermediate possesses much if not more structural complexity than **1065**. Six successive steps led to the diol **1063**. Reduction of the tetrasubstituted olefin with Na/tBuOH/HMPA led to a mixture of isomers from which **1064** (β-methyl) was separated and carried on. Five functional group manipulations were used to complete the synthesis providing (+)-dictymal and confirming the absolute configuration of this metabolite.

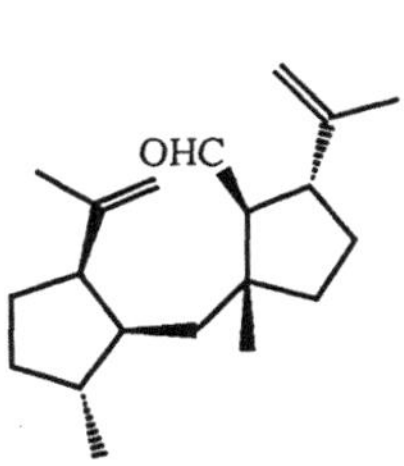

Dictymal **1065**

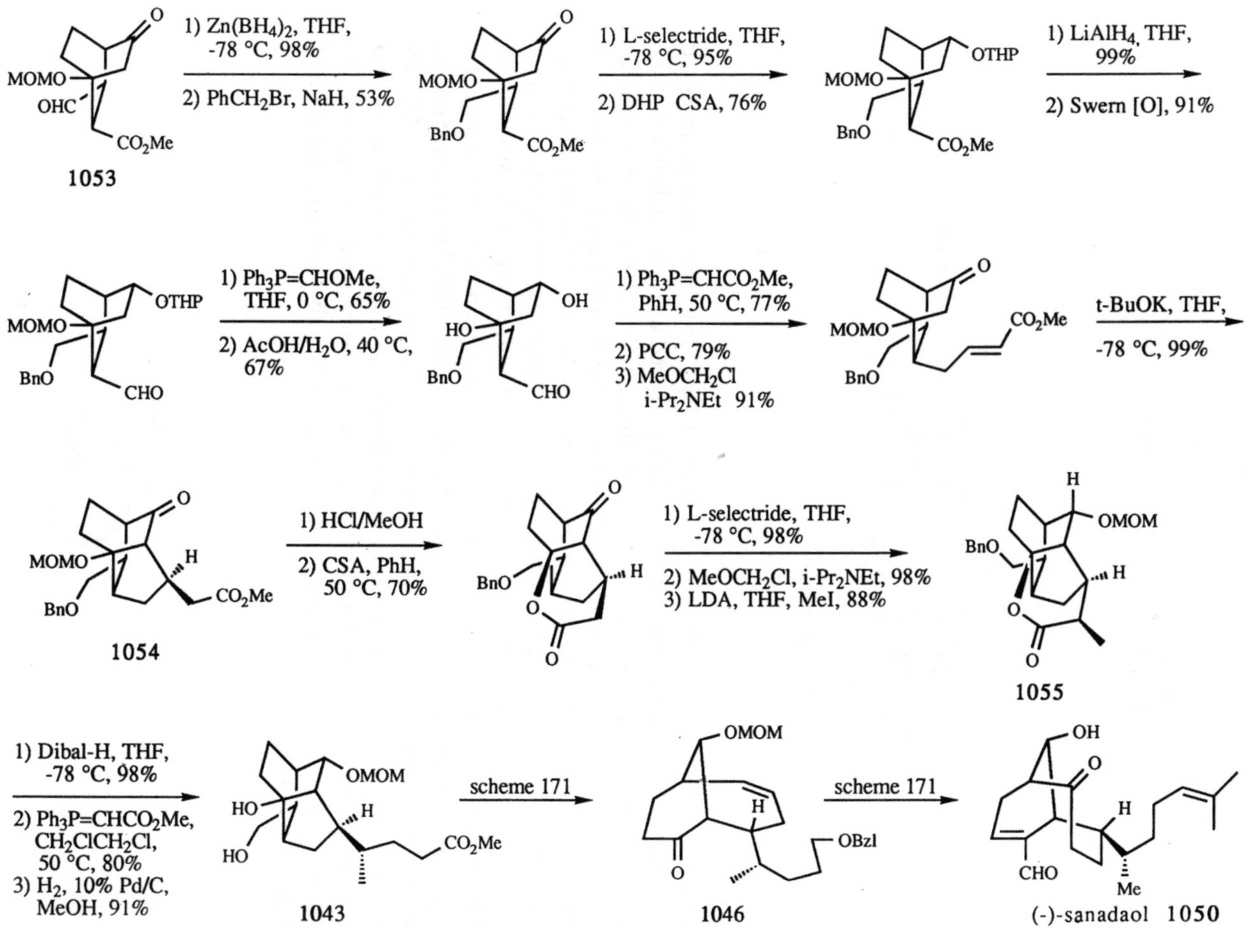

Scheme 172. Yamada Synthesis of (−)-Sanadaol

2.4 Sesterterpenoids

2.4.1 Manoalide and Analogs

Manoalide (**1073**), *seco*-manoalide (**1078**) and *E*- and *Z*-neomanoalide have been isolated [276] from the sponge *Lufferiella variabilis*. All four compounds exhibit antibacterial activity, while manoalide exhibits anti-inflammatory properties through the inhibition of phospholipases A_2 and C [277].

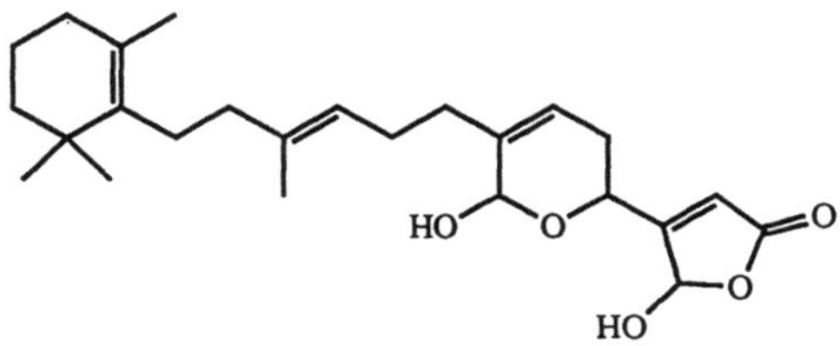

Manoalide **1073**

The first synthesis of manoalide (**1073**) is that of Isoe [278] shown in Scheme 174. Conversion of β-ionone-derived ester **1066** to the allylic bromide followed by alkylation with the lithium anion formed from the hydrazone of pyruvaldehyde dimethylacetal affords ketoacetal **1067**. Reaction of acetal **1067** with the enolate of *t*-butyl trimethylsilylacetate gives triene **1068** with 95% *Z*-selectivity at the newly formed double bond. Conversion of **1068** to the chloride and alkylation with the substituted furyllithium **1070** yields the furan derivative **1071**. In a key transformation, the γ-hydroxybutenolide is obtained by photosensitized oxygenation of the furan without oxidation of the tri- and tetrasubstituted double bonds. Hydrolysis of the acetal and the hydroxyl protecting groups with acetic acid provides manoalide in 9 steps and 23% overall yield. If 2 N HCl in THF is used in the hydrolysis of **1072**, *seco*-manoalide (**1078**) is obtained in place of manoalide (**1073**).

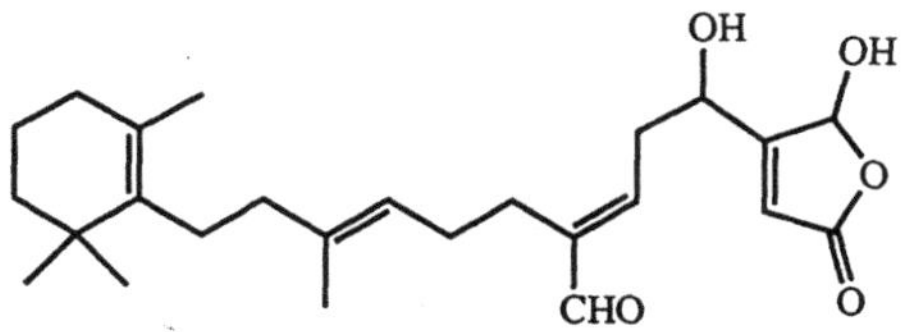

seco-Manoalide **1078**

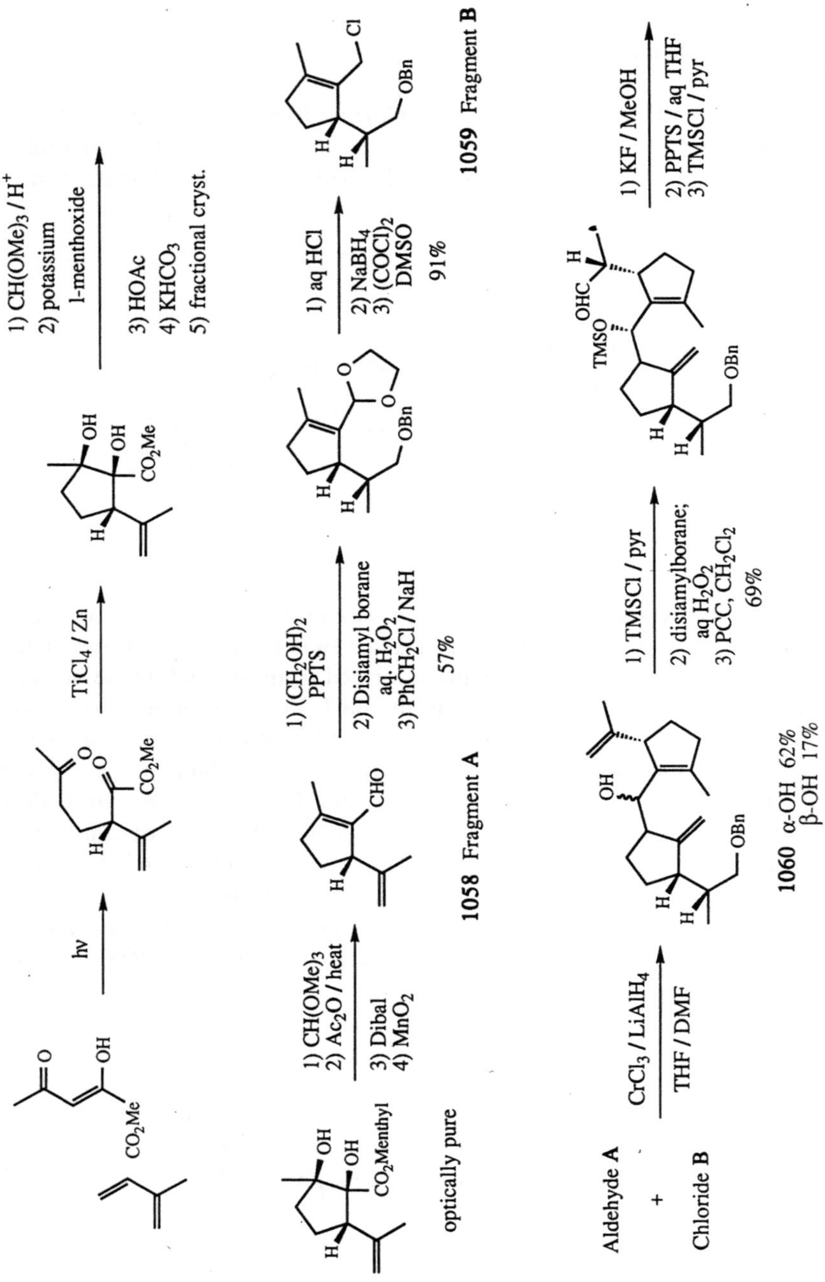
hν
TiCl4 / Zn
1) CH(OMe)3 / H+
2) potassium l-menthoxide
3) HOAc
4) KHCO3
5) fractional cryst.
optically pure
1) CH(OMe)3
2) Ac2O / heat
3) Dibal
4) MnO2
1058 Fragment A
1) (CH2OH)2 PPTS
2) Disiamyl borane aq. H2O2
3) PhCH2Cl / NaH
57%
1) aq HCl
2) NaBH4
3) (COCl)2 DMSO
91%
1059 Fragment B
Aldehyde A
+
Chloride B
CrCl3 / LiAlH4
THF / DMF
1060 α-OH 62%
β-OH 17%
1) TMSCl / pyr
2) disiamylborane; aq H2O2
3) PCC, CH2Cl2
69%
1) KF / MeOH
2) PPTS / aq THF
3) TMSCl / pyr

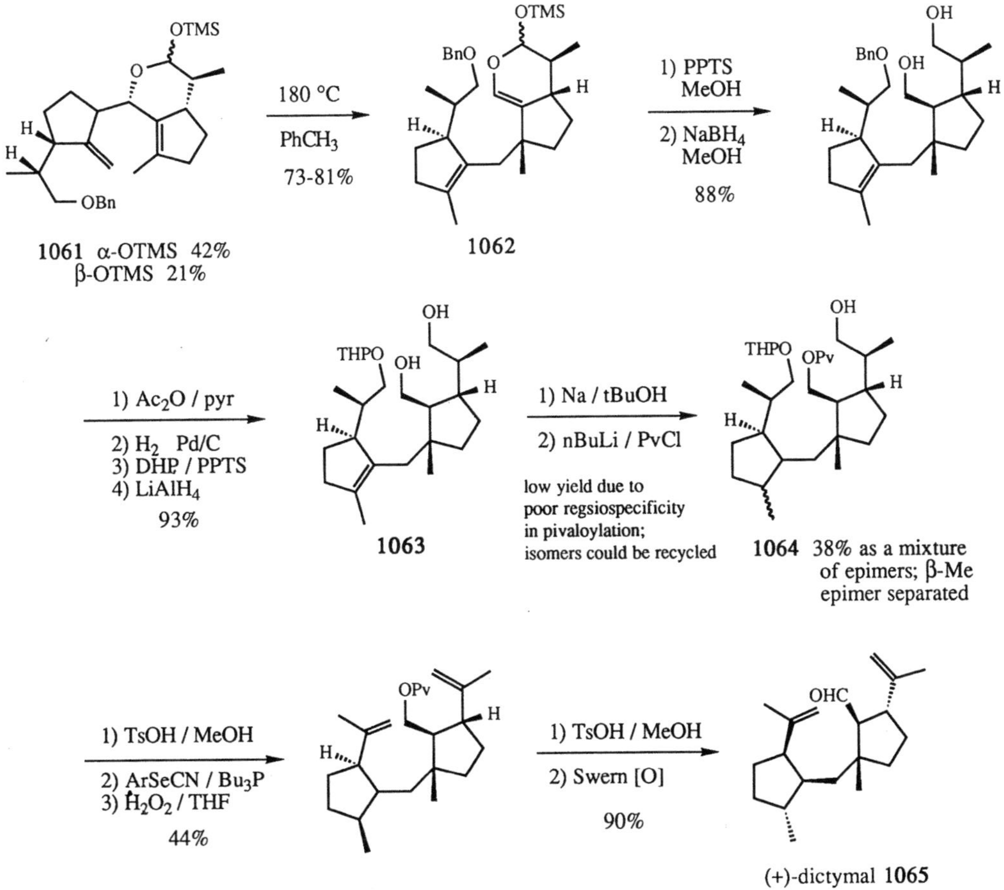

Scheme 173. Kato Synthesis of (+)-Dictymal

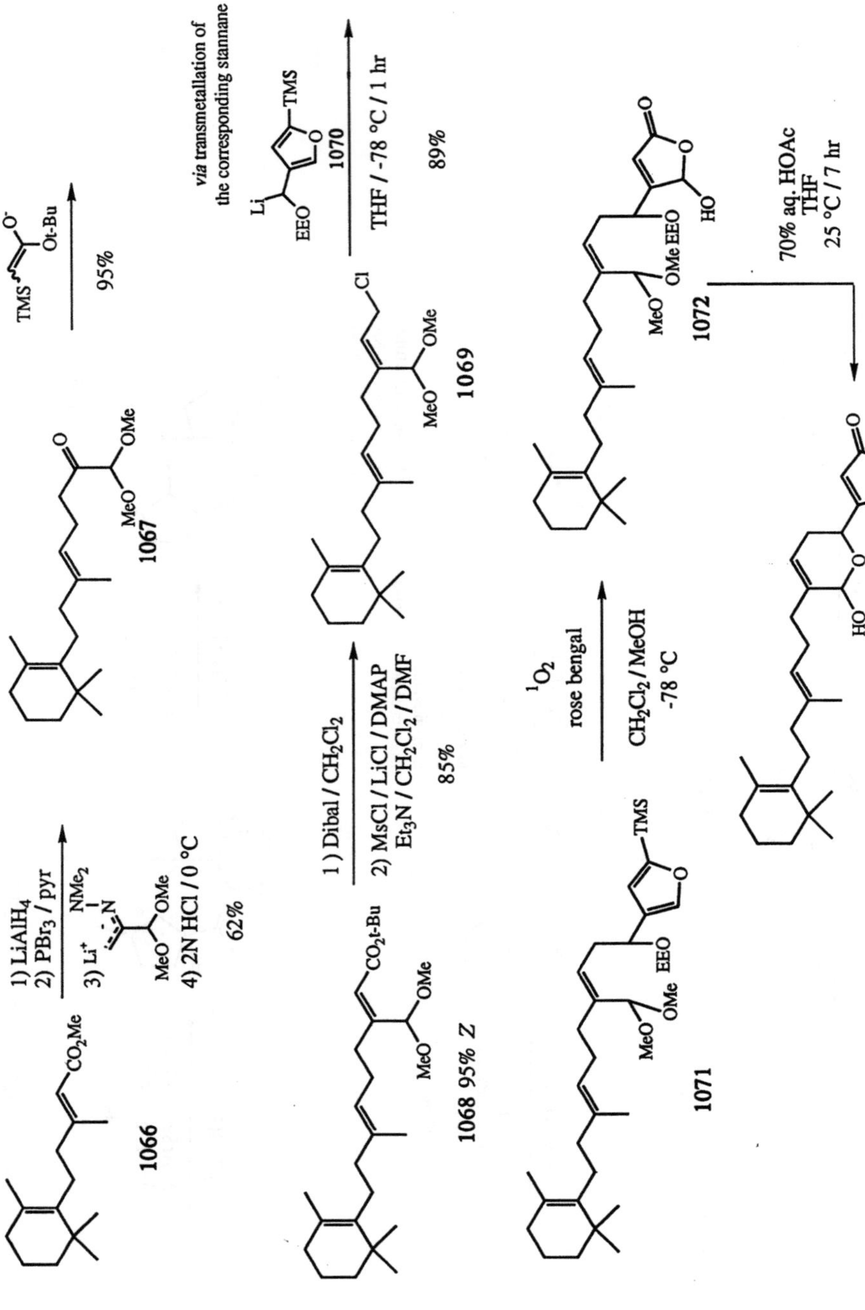

Scheme 174. Isoe First Synthesis of Manoalide

Alternative methodology developed by Isoe [279] for the completion of the manoalide synthesis is shown in Scheme 175. Palladium-catalyzed carbonylation and coupling of stannyl furan **1075** with allylic chloride **1069** affords ketone **1076** as a mixture of *E*- and *Z*-isomers (Scheme 175). LAH reduction of **1076**, protection of the alcohol and photosensitized oxygenation of the furan yields butenolide **1077** (a common intermediate). Hydrolysis and deprotection as before yields *seco*-manoalide (**1078**). Photochemical cyclization under the conditions developed by Scheuer [280] provides manoalide in quantitative yield from **1078**. An overall yield of 27% in 12 steps is obtained.

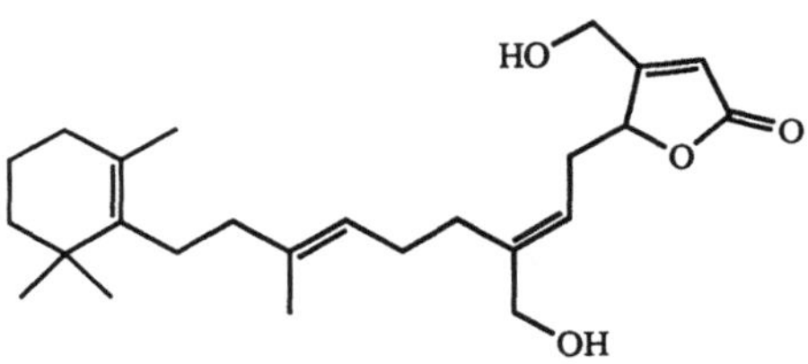

E-Neomanoalide **1083**

This strategy can be modified for the preparation [281] of *E*-neomanoalide (**1083**) (Scheme 176). Palladium-catalyzed coupling of allylic chloride **1069** and α-stannylfuran **1079** affords acetal **1080** as a 1:1 mixture of *E*- and *Z*-isomers. Equilibration to the *E*-olefin, reduction and silylation produces furan **1081**. Photosensitized oxidation of furan **1081** to hydroxybutenolide **1082** followed by reduction and hydrolysis provides *E*-neomanoalide in 13 steps and 17% overall yield.

A strategy very similar to that of Isoe was followed by Garst [282] in his synthesis of **1073** (Scheme 177). Conjugate reduction of β-ionone followed by Wittig olefination gives a 1:1 mixture of *E*- and *Z*-isomers **1085**. Conversion of alcohol **1085** to the bromide via the tosylate and alkylation with acyl anion equivalent **1086** provides aldehyde **1087**, along with 30–50% recovered starting material. Silylation of **1087** provides the TMS dienol ether **1088**. At this point, the original strategy called for a hetero Diels-Alder reaction between diene **1088** and aldehyde **1089** to construct the pyran ring **1091**. As this was not possible, **1091** was obtained via a vinylogous Mukaiyama aldol reaction to obtain **1090** and photochemical cyclization to obtain pyran **1091**. Photooxygenation of **1091** provides manoalide (**1073**) in 9 steps and approximately 1% yield overall.

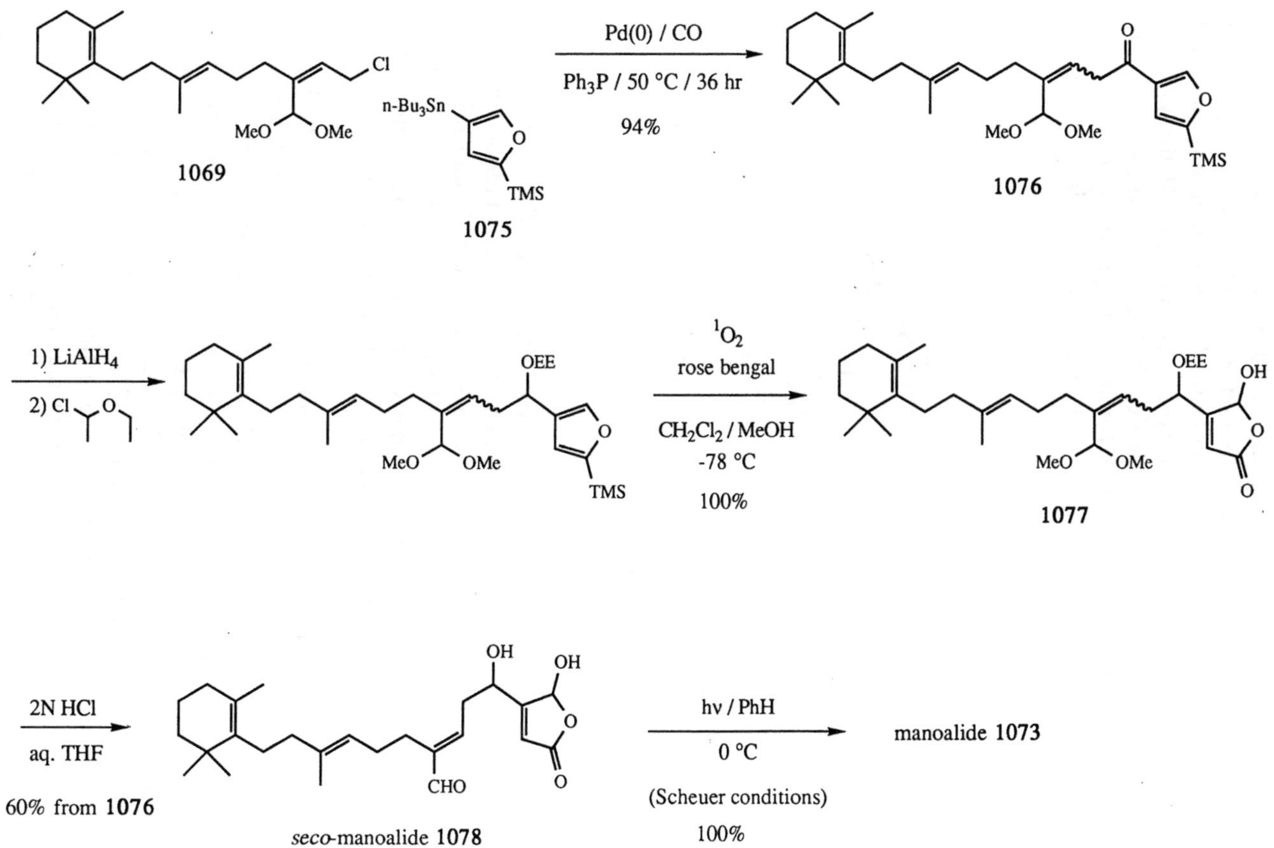

Scheme 175. Isoe Syntheses of Manoalide and *seco*-Manoalide

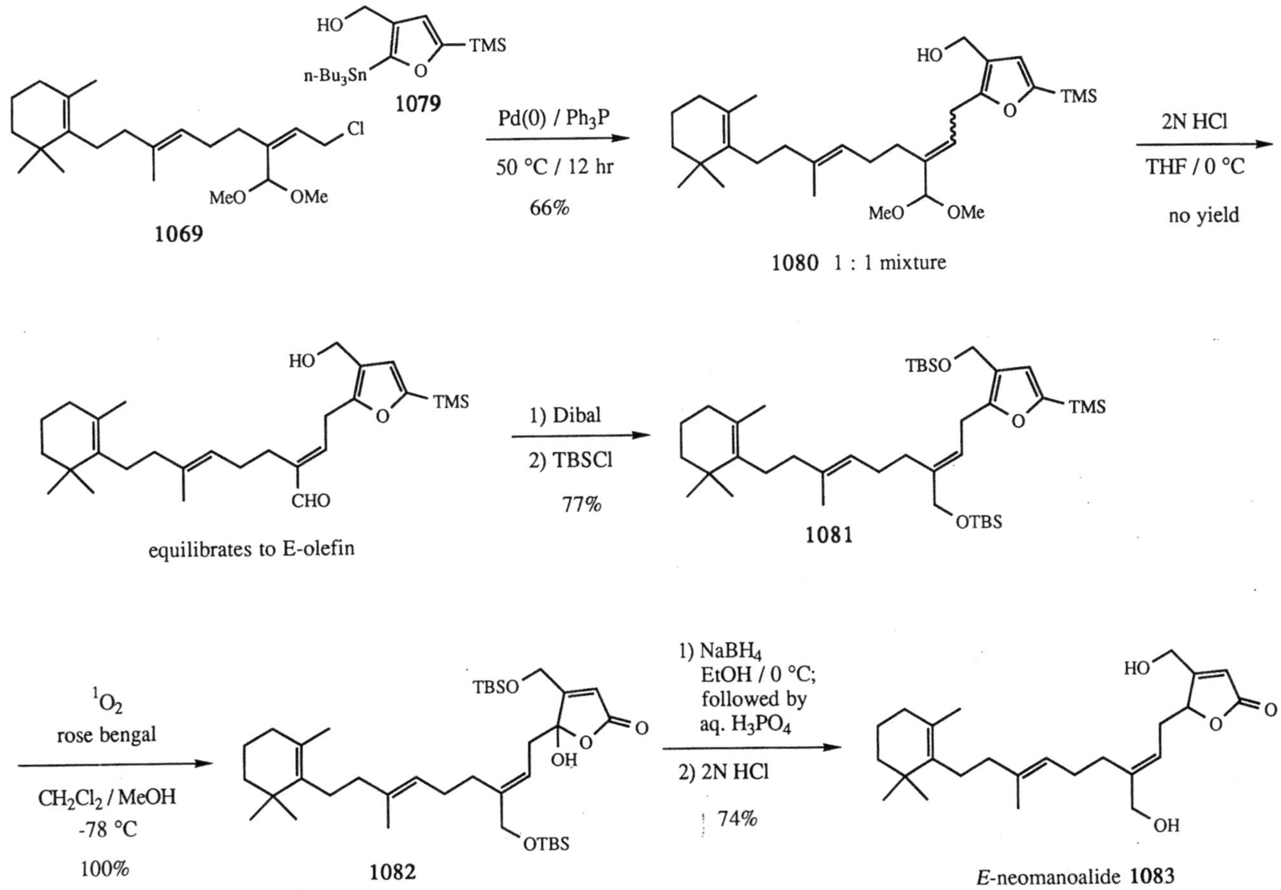

Scheme 176. Isoe Synthesis of of *E*-Neomanoalide

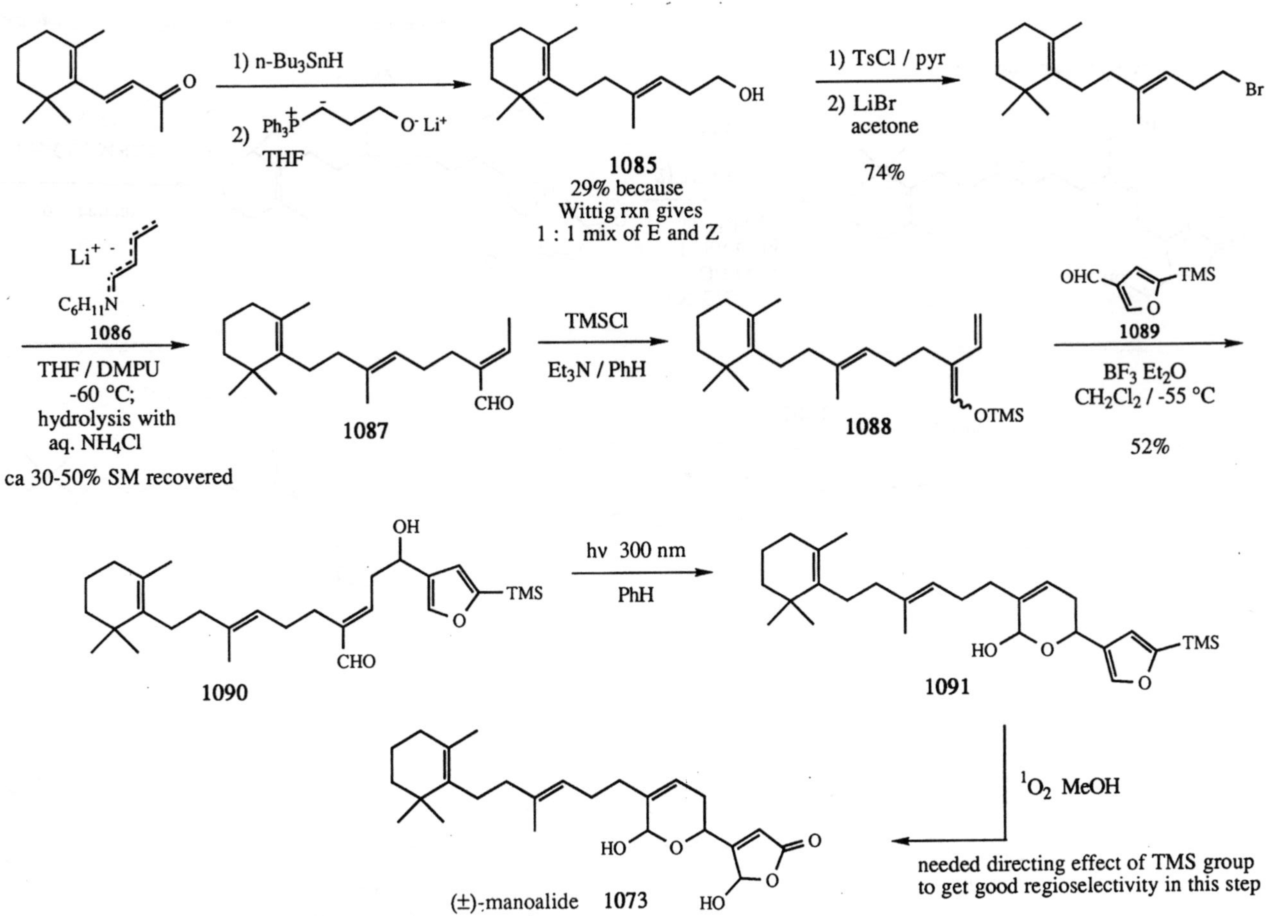

Scheme 177. Garst Synthesis of (±)-Manoalide

2.4.2 Miscellaneous Sesterterpenoids

2.4.2.1 Ircinianin

Ircinianin is a sesterterpene isolated [283] from the sponge *Ircinia wistarii.* It is structurally unique among other products isolated from the same species in that it contains a tricyclic spirotetronic acid system. It has been suggested [284] that the biosynthesis of ircinianin (**1100**) proceeds via an intramolecular Diels-Alder reaction. The synthetic strategy developed by Yoshii [285] for the synthesis of **1100** is patterned after such a biosynthetic model (Scheme 178). The strategy involves a series of homologation processes to build up to the final intermediate **1099** in which all olefins have *E*-geometry. The stereochemistry at all five stereocenters can then be set by the intramolecular Diels-Alder reaction. Coupling of allylic chloride **1093** (available from aldehyde **1092** in four steps) with 2-(3-furyl)ethylmagnesium bromide in the presence of Li_2CuCl_4 followed by desilylation provides furyl alcohol **1094**. Chlorination and conversion to sulfone **1095** allows coupling with aldehyde **1096** giving all *E*-triene **1097** after elimination and reductive removal of the sulfone moiety. Treatment of nitrile **1097** with DIBAL reveals the "protected" aldehyde **1098** which is coupled with the lithium enolate of methyl 2-methyltetronate to afford all *E*-pentaene **1099** after elimination. Demethylation of **1099** and heating in refluxing benzene initiates the intramolecular Diels-Alder reaction to produce ircinianin **1100** as the sole product. A yield of 0.8% is obtained over 15 steps.

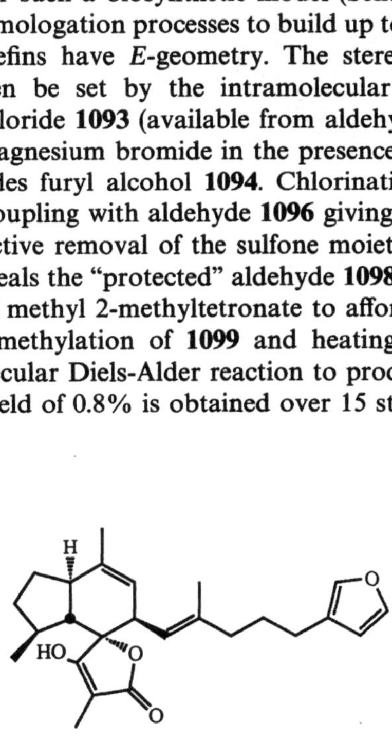

Ircinianin 1100

2.4.2.2 Dysideapalaunic Acid

Dysideapalaunic acid (**1109**) is a sesterterpene, isolated [286] from *Dysidea* sp. in the Palauan sea, which acts to inhibit aldose reductase. There are three points that must be addressed in order to develop an efficient synthesis of **1109**. These are construction of the carbon skeleton, obtention of the correct relative stereochemistry at the three chiral centers, and preparation of the

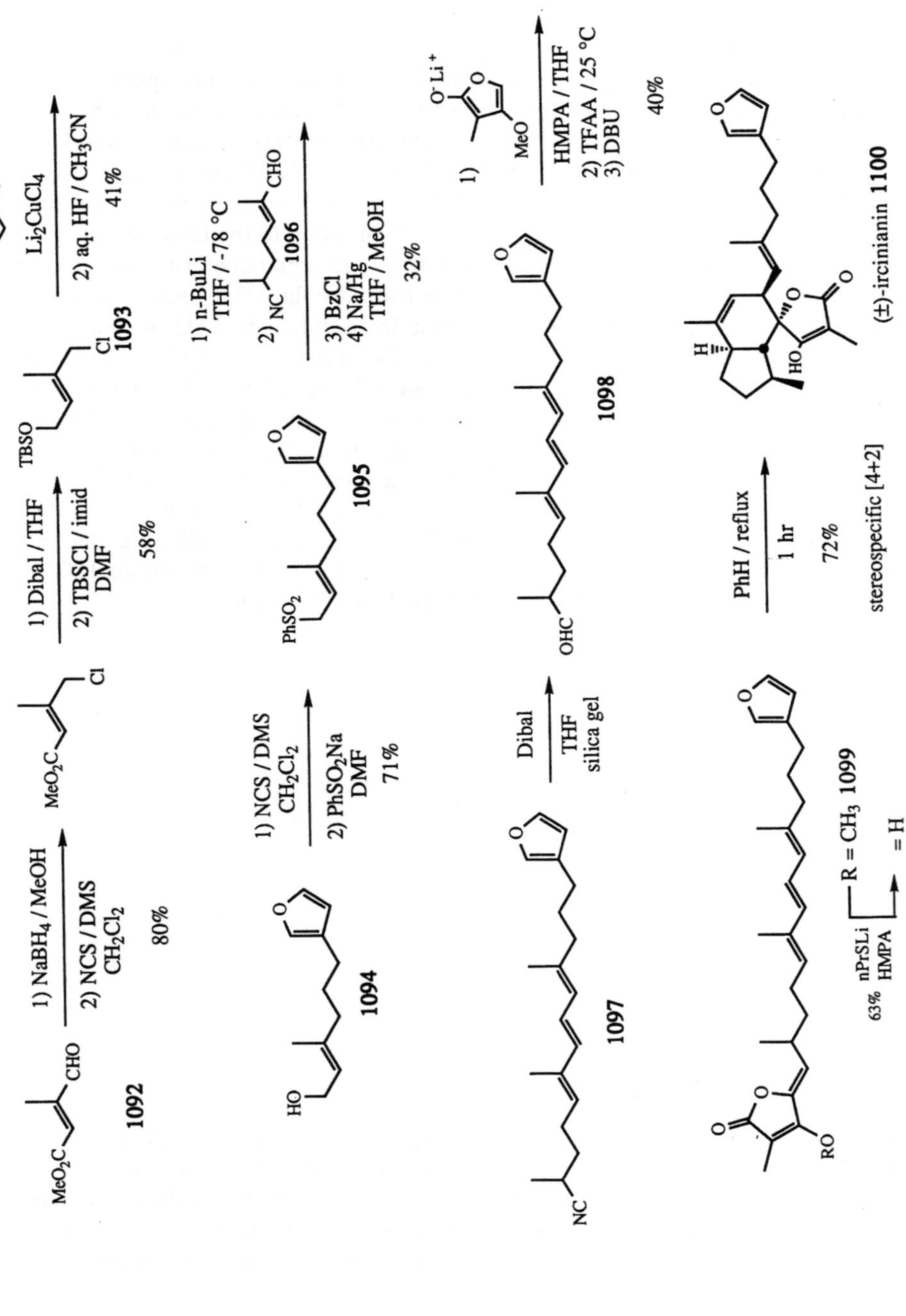

Scheme 178. Yoshii Synthesis of (±)-Ircinianin

(*E*)-α,β-unsaturated acid functionality with the correct geometry. An enantiospecific synthesis of **1109** which addresses all three of these points, and serves to establish the absolute configuration of **1109** has been developed by Hagiwara [287] (Scheme 179). The optically active ene-dione **1102** (80% ee) is selectively protected by acetalization and reductively alkylated by treatment with lithium in ammonia and allyl bromide to produce ketone **1103**, in which the correct stereochemistry at all three stereocenters has been obtained. The carbonyl functionality is removed by a standard sequence to afford ketal **1104**. Hydroboration-oxidation of **1104** followed by Swern oxidation and Wittig olefination provides **1105** and serves to complete attachment of the lower sidechain with correct stereochemistry. After deprotection of the ketone, a carbomethoxy functionality is added to prevent polymethylation, thus providing keto-ester **1106**. Methylation of **1106** and decarboxylation gives compound **1107**. The upper sidechain is attached by addition of the ultrasonically generated Grignard reagent of 4-bromobutan-2-one ethylene acetal to provide **1108** as a mixture of isomers. Dehydration and deprotection of **1108**, followed by condensation with the Horner-Emmons reagent provides (+)-dysideapalaunic acid, after saponification. A yield of 4.5% is obtained over 17 steps.

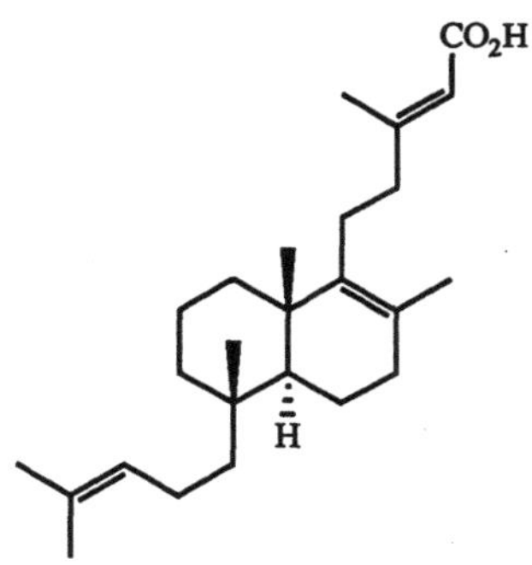

Dysideapalaunic acid **1109**

2.4.2.3 *Palauolide*

Palauolide is an antimicrobial sesterterpene isolated [288] from a mixture of three unknown sponges collected from Palau in the Western Caroline Islands in 1982. It is structurally unique, although the rearranged drimane skeleton of the bicyclo[4.4.0] portion of the molecule is fairly common among sponge metabolites. It possesses four contiguous chiral centers and an (*E*,*E*,*E*)-conjugated

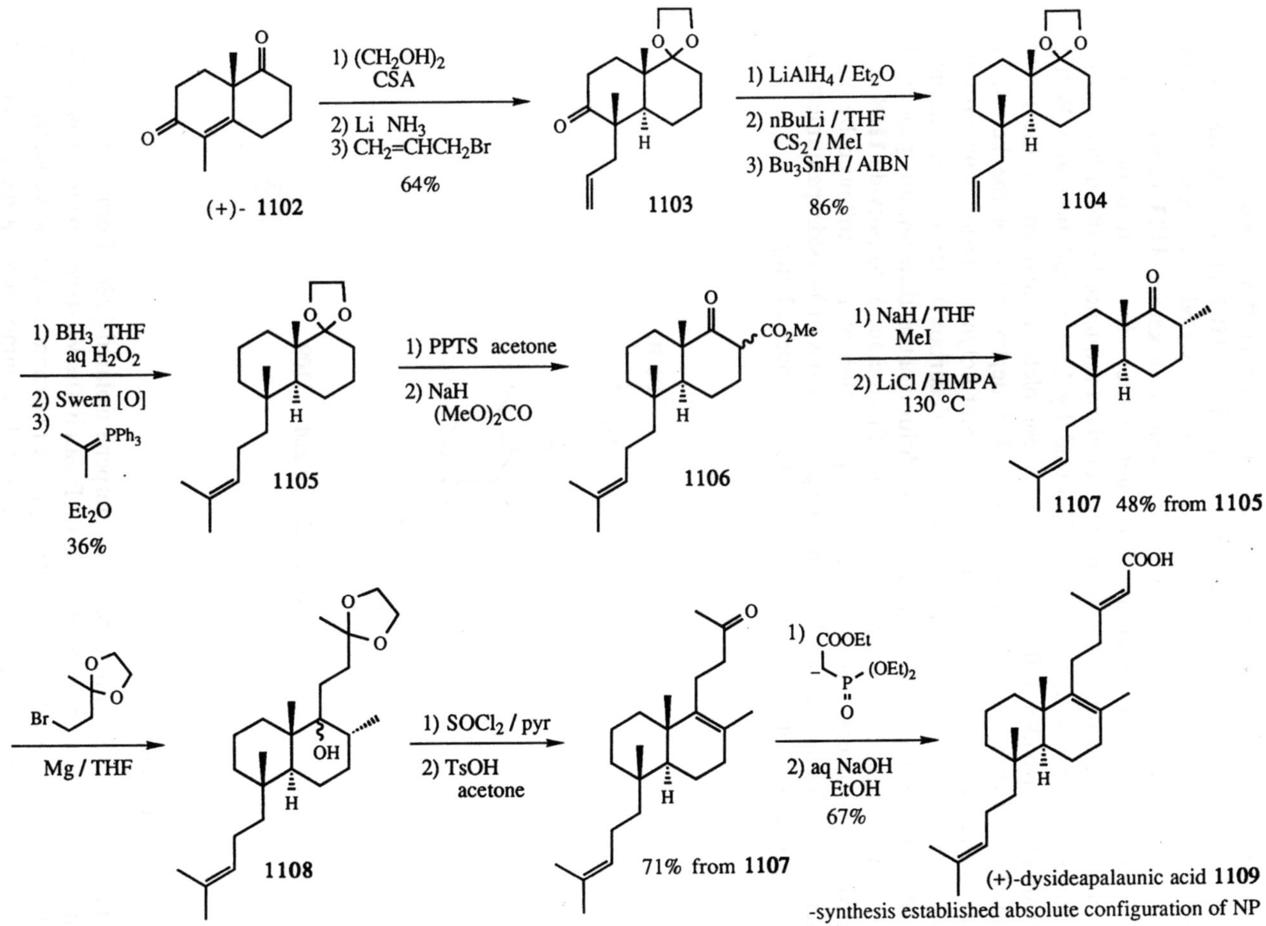

Scheme 179. Hagiwara Synthesis of (+)-Dysideapalaunic Acid

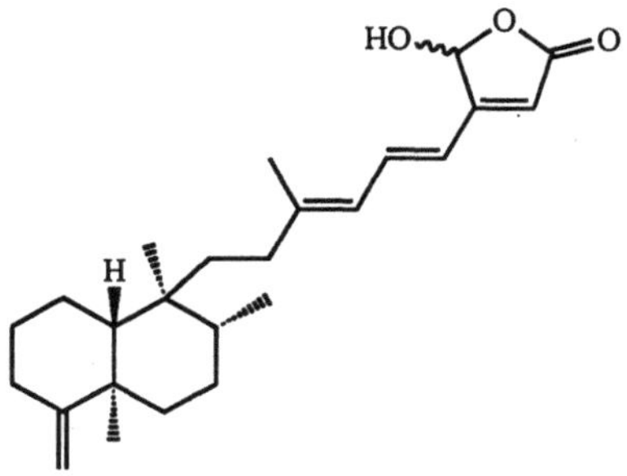

Palauolide 1119

triene system. The synthesis of **1119** by Piers is outlined in Scheme 180 [289]. Copper-catalyzed conjugate addition of Grignard reagent **1111** to ketone **1110** followed by intramolecular alkylation by treatment with alkoxide provides bicyclic ketone **1112** with three of the four chiral centers present in the product. Conversion of **1112** to a mixture of diastereomeric nitriles and alkylation with LDA and $I(CH_2)_3OMOM$ provides exclusively nitrile **1113**. The nitrile functionality is then converted to the primary alcohol, and further reduced by conversion to the phosphorodiamidate and treatment with lithium and ammonia to give compound **1114**. Deprotection of the primary alcohol and oxidation with PCC gives an aldehyde that is converted to methyl ketone **1115** by addition of methyllithium and PCC oxidation. Olefination of **1115** with the Horner-Emmons reagent provides a 10:1 *E*/*Z* mixture of olefins. Separation of the olefin mixture, DIBAL reduction and allylic oxidation yields aldehyde **1116**. Addition of sulfonyl stabilized anion **1117** to aldehyde **1116** followed by benzoylation and reductive elimination with sodium amalgam provides furyl triene **1118** in 51% yield. Photosensitized oxidation of the trimethylsilyl substituted furan **1118** yields palauolide in 18 steps and 5% overall yield.

2.5 Triterpenoids

2.5.1 Mokupalide

Mokupalide (**1129**) was isolated [290] from an unknown dark green sponge collected at Enewetak atoll in the Marshall islands. Mokupalide is the first example of this C_{30} isoprenoid skeleton, containing six isoprene units linked head-to-tail. A stereoselective synthesis of mokupalide has been reported by Weiler [291] (Scheme 181). Any efficient approach to **1129** must address the problems of controlling the stereochemistry of the three methyl-bearing trisubstituted olefins and of constructing the mokupalide skeleton efficiently. The key

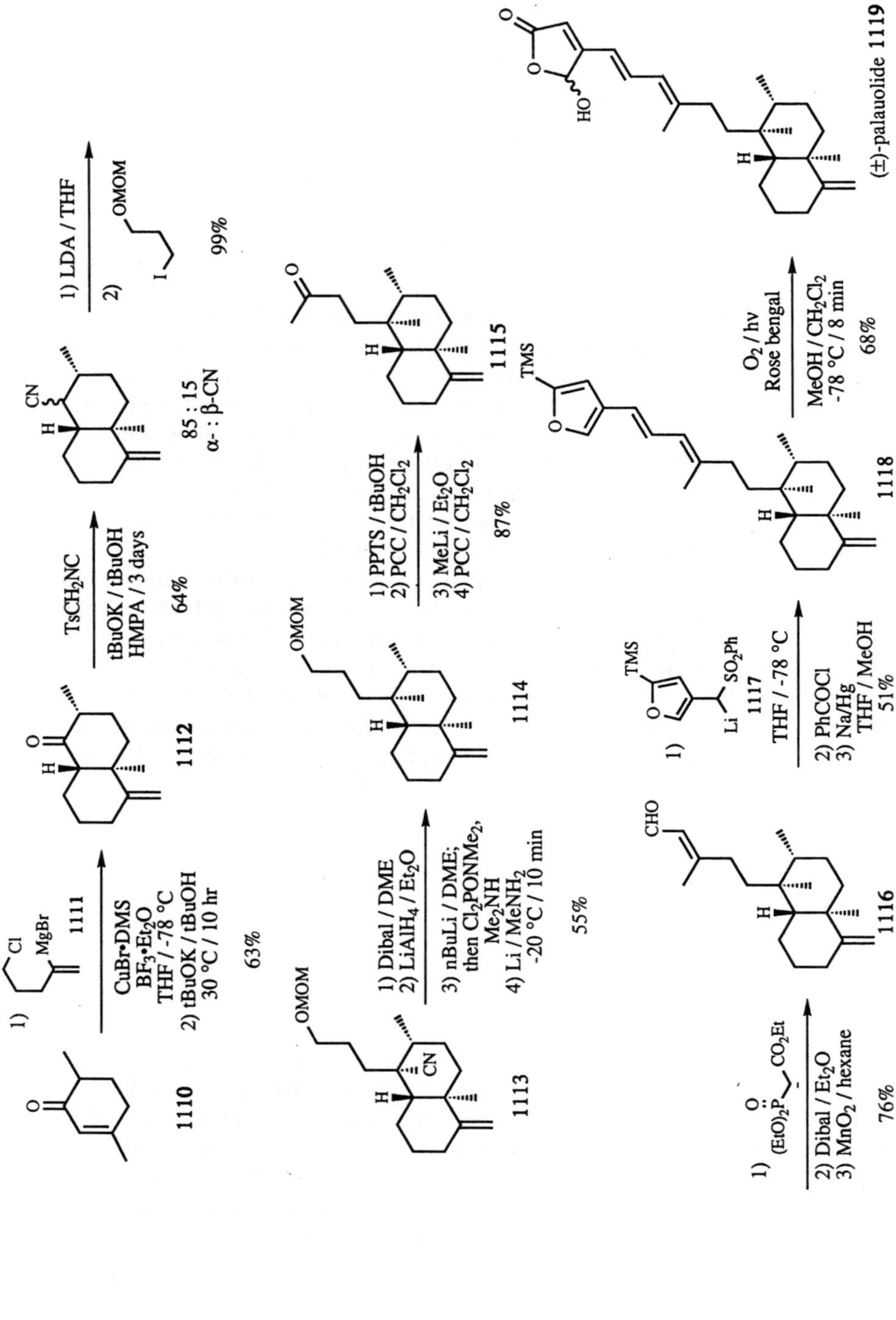

Scheme 180. Piers Synthesis of (±)-Palauolide

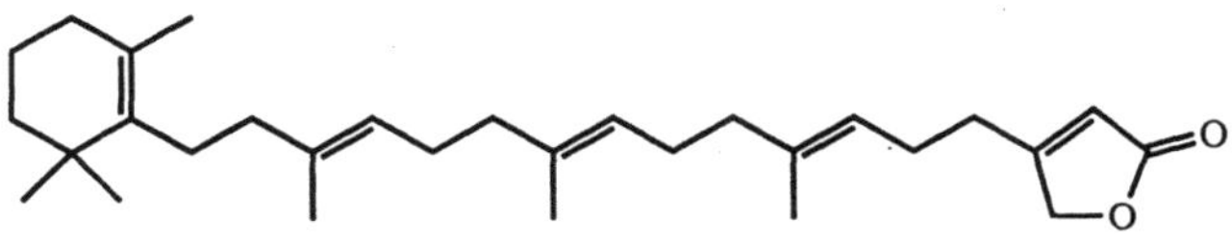

Mokupalide **1129**

starting material in Weiler's synthesis of **1129** is ester **1120** which can be prepared in three steps from ethyl acetoacetate. Allylic oxidation affords allylic alcohol **1121** in 38% yield. Mesylation of **1121** and treatment with lithium thiophenoxide gives allylic sulfide **1122**. Reduction of the ester functionality, to the allylic alcohol, and protection with DHP affords THP ether **1123**. Deprotonation α- to the phenyl sulfide, alkylation with bromide **1124** and hydrogenolysis of the sulfide gives tetraene **1125** in 70% yield. Hydrolysis of the THP group, conversion of the alcohol to the allylic bromide **1126** and coupling with the sulfone-stabilized butenolide anion **1127** gives phenylsulfonyl substituted mokupalide **1128**. Hydrogenolysis of **1128** with sodium amalgam in methanol yields mokupalide (**1129**) in 11 steps from **1120** and approximately 10% overall yield.

2.5.2 Squalene-10,11-Oxide

In 1982, Yamada isolated (+)-squalene-10,11-oxide (**1133**) from the red alga *Laurencia okamurai*, collected at Goza, Mie Prefecture, Japan [292]. Interestingly, its enantiomer has been isolated from the green alga *Caulerpa prolifera* by Fattorusso [293]. The absolute configuration of (+)-**1133** has been established by Yamada [294] through total synthesis (Scheme 182). Sharpless asymmetric epoxidation of *trans,trans*-farnesol provides epoxide **1130** with an enantiomeric excess of 96%. Tosylation of **1130** and treatment with sodium iodide in acetone yields the primary iodide. Bishomologation of the C-15 skeleton is achieved by reaction of the iodide with *t*-butyl lithioacetate to afford the ester, which upon reduction with DIBAL produces aldehyde **1131**. Wittig olefination of the aldehyde **1131** with phosphorane **1132** produces (+)-squalene-10,11-oxide as a 4:3 mixture of *E*/*Z* isomers. After separation of the isomers on silver nitrate impregnated silica an overall yield of 13% is obtained in 6 steps.

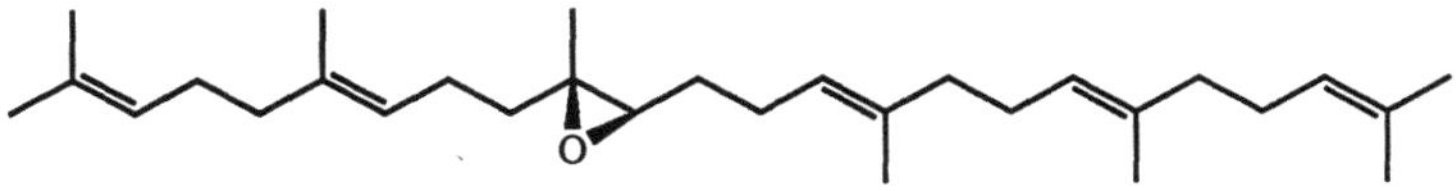

(+)-Squalene-10,11-oxide **1133**

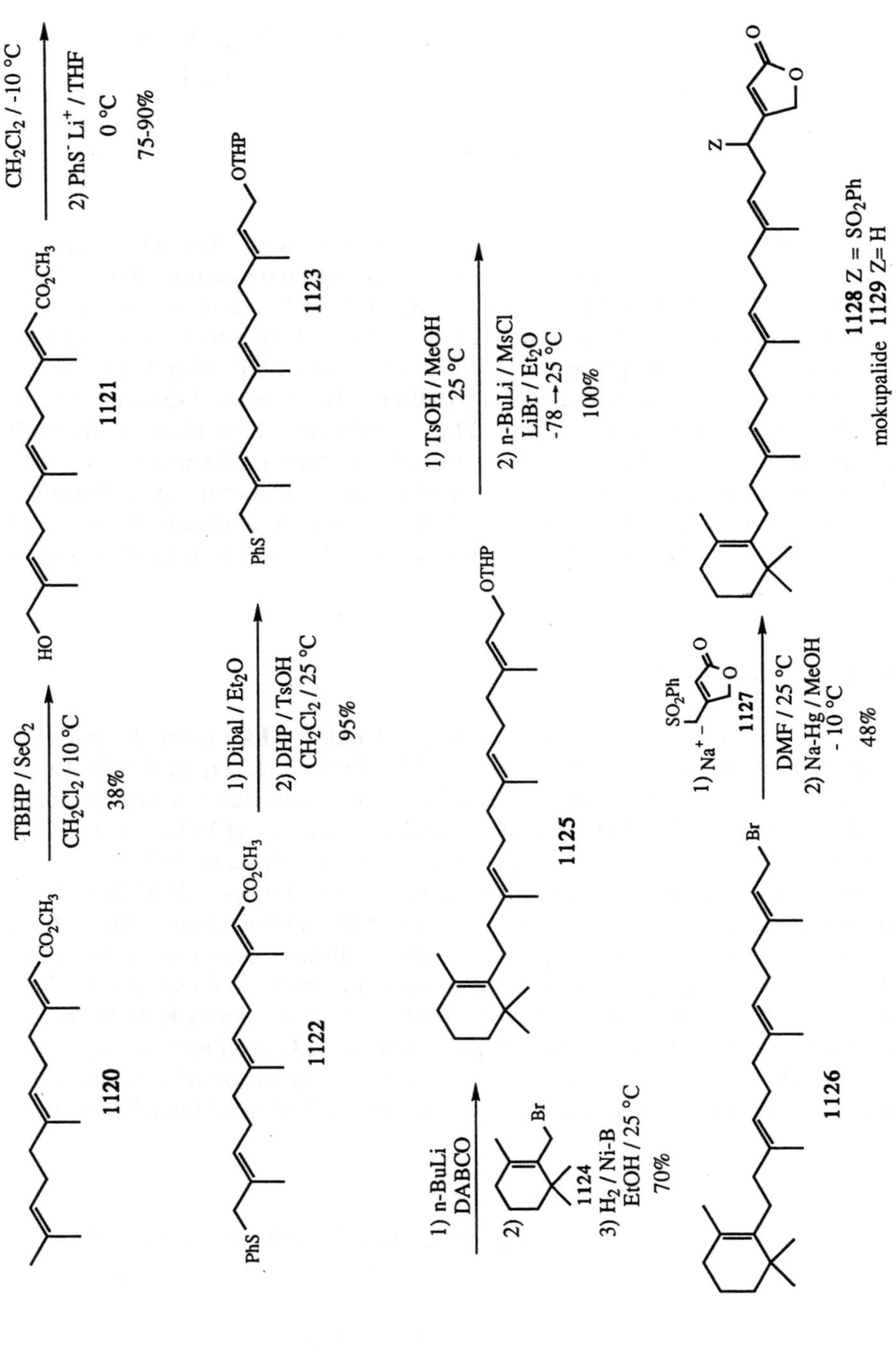

Scheme 181. Weiler Synthesis of Mokupalide

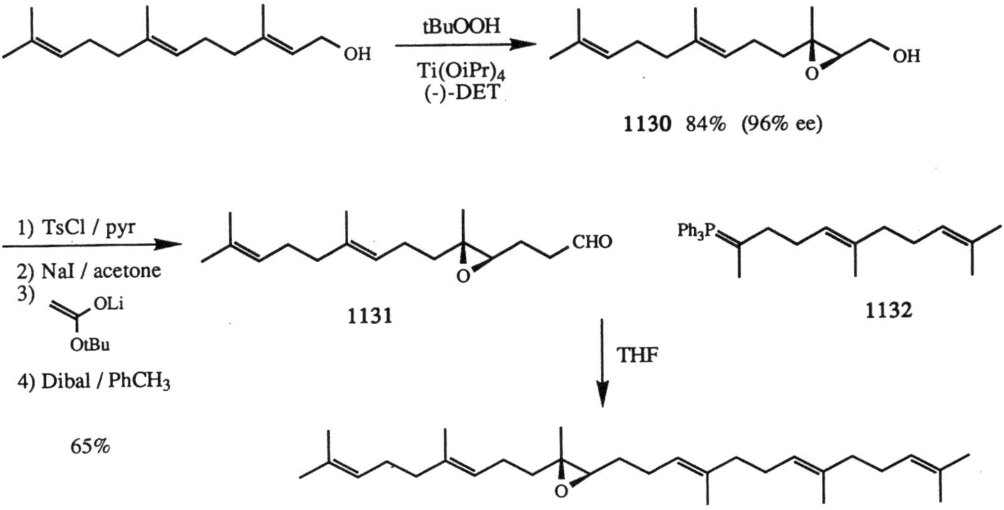

(+)-squalene-10,11-oxide **1133**
as a 4 : 3 mixture of
E/Z isomers

Scheme 182. Yamada Synthesis of (+)-Squalene-10,11-oxide from *Laurencia okamurai*

2.5.3 Teurilene

Teurilene (**1147**) is a *meso*-triterpene ether isolated [295] from the red alga *Laurencia obtusa*. It possesses three linked tetrahydrofuran units in the center of the molecule and eight chiral centers arranged such that the molecule has C_s symmetry. The strategy utilized by Shirahama [296] in the synthesis of **1147** (Scheme 183) is interesting in that optically pure fragments are used to provide an optically inactive product. The synthesis is divergent/convergent, in that a single starting material provides two enantiomers (**1135** and **1136**) that are coupled (**1142** → **1143**) after one of them, **1136** has undergone significant modification. Sharpless asymmetric epoxidation of geraniol using (−)-DIPT produces an epoxide (88% ee) which is rearranged via standard methodology to terminal epoxide **1136** in 46% overall yield. In a similar fashion, geraniol is converted into the enantiomeric epoxide **1135**. Ring opening of epoxide **1136** by the sulfur-stabilized lithioanion of **1137** followed by reductive removal of the

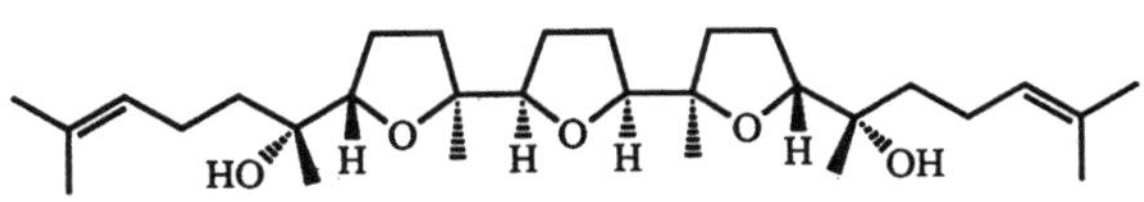

Teurilene **1147**

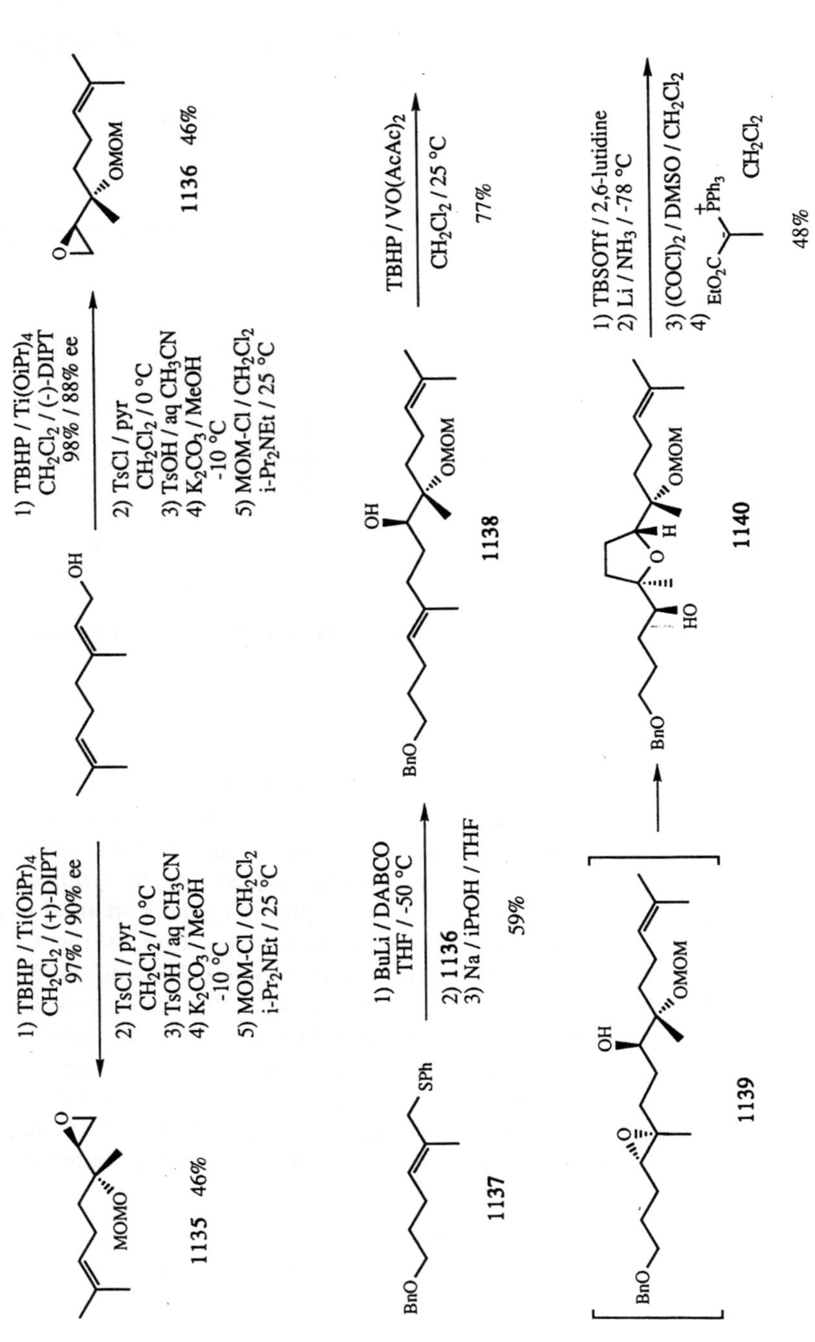
1) TBHP / Ti(OiPr)4
CH2Cl2 / (+)-DIPT
97% / 90% ee
2) TsCl / pyr
CH2Cl2 / 0 °C
3) TsOH / aq CH3CN
4) K2CO3 / MeOH
-10 °C
5) MOM-Cl / CH2Cl2
i-Pr2NEt / 25 °C
MOMO
1135 46%
OH
1) TBHP / Ti(OiPr)4
CH2Cl2 / (-)-DIPT
98% / 88% ee
2) TsCl / pyr
CH2Cl2 / 0 °C
3) TsOH / aq CH3CN
4) K2CO3 / MeOH
-10 °C
5) MOM-Cl / CH2Cl2
i-Pr2NEt / 25 °C
OMOM
1136 46%
BnO
SPh
1137
1) BuLi / DABCO
THF / -50 °C
2) 1136
3) Na / iPrOH / THF
59%
BnO
OH
OMOM
1138
TBHP / VO(AcAc)2
CH2Cl2 / 25 °C
77%
BnO
OH
OMOM
1139
BnO
HO
H
OMOM
1140
1) TBSOTf / 2,6-lutidine
2) Li / NH3 / -78 °C
3) (COCl)2 / DMSO / CH2Cl2
4) EtO2C PPh3
CH2Cl2
48%

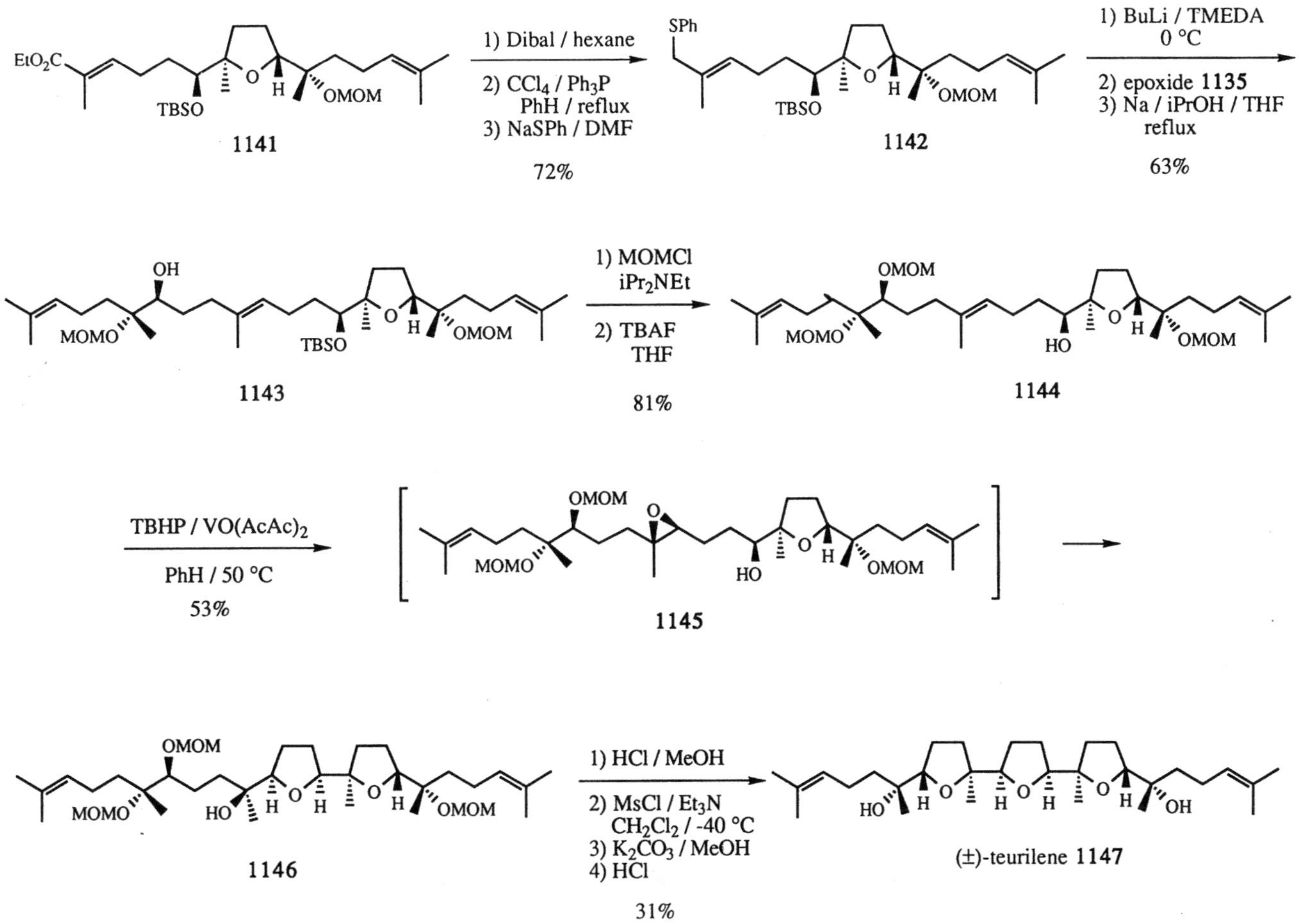

Scheme 183. Shirahama Synthesis of Teurilene

phenylthio group provides alcohol **1138** as one enantiomer. Vanadium(IV) catalyzed epoxidation of **1138** proceeds with 75% stereoselectivity yielding intermediate epoxide **1139**, which under the reaction conditions cyclizes to tetrahydrofuran **1140**. The free hydroxyl group is protected as its TBS ether and the benzyl group removed to provide the primary alcohol. Swern oxidation of the primary alcohol followed by Wittig olefination yields ester **1141**. DIBAL reduction of the ester, formation of the primary allylic chloride and displacement with thiophenoxide yields allylic sulfide **1142**. This now sets the stage for coupling of **1142** with the other optically active epoxide fragment **1135**. Treatment of sulfide **1142** with *n*-butyllithium in the presence of TMEDA followed by addition of **1135** yields a β-hydroxysulfide. Reaction with sodium in THF/*iso*-propanol reductively cleaves the phenylthio moiety to provide compound **1143**. Protection of the secondary alcohol **1143** as its MOM ether and treatment with fluoride affords bishomoallylic alcohol **1144**. Vanadium(IV)-catalyzed epoxidation of **1144** stereospecifically yields the bistetrahydrofuran compound **1146**, through the intermediacy of epoxide **1145**. At this point all protecting groups are removed by treatment with HC1 in methanol and the secondary hydroxyl group is selectively converted to a mesylate. Successive treatment of the mesylate with potassium carbonate in methanol and with HC1 provides teurilene **1147** in 23 steps and 0.61% overall yield. Although the product is racemic, the correct relative sterochemistry was obtained at all eight stereocenters.

Shirahama [297] has also developed a second strategy for the preparation of **1147** (Scheme 184) via an alternate preparation of an advanced intermediate from the previous synthesis (Scheme 183). Opening of epoxide **1135** with the sulfur-stabilized lithio anion of **1149** followed by protection of the secondary alcohol and reductive removal of the phenylthio moiety affords benzyl protected alcohol **1150**. Deprotection of the primary alcohol **1150**, conversion to the chloride and displacement with thiophenoxide produces sulfide **1151**. At this point, **1151** can be coupled with the enantiomer of epoxide **1135**. Treatment of allylic sulfide **1151** with *n*-butyllithium in the presence of TMEDA, followed by addition of epoxide **1136** yields the coupled product. Protection of the secondary alcohol as its MOM ether and reductive removal of the phenylthio group gives tetraene **1152**. Vanadium(IV) catalyzed epoxidation of **1152** in the presence of acetic acid proceeds via a series of deprotection-oxidation-cyclization steps to give bis-tetrahydrofuran **1146**, a key intermediate in Shirahama's previous synthesis of teurilene (**1147**).

2.5.4 Haloethers

2.5.4.1 Thyrsiferol and Venustatriol

The brominated tetracyclic squalene-derived polyethers thyrsiferol [298] and venustatriol [299] have been isolated from red algae of the genus *Laurencia*. Several members of this class of compounds have been shown to exhibit strong

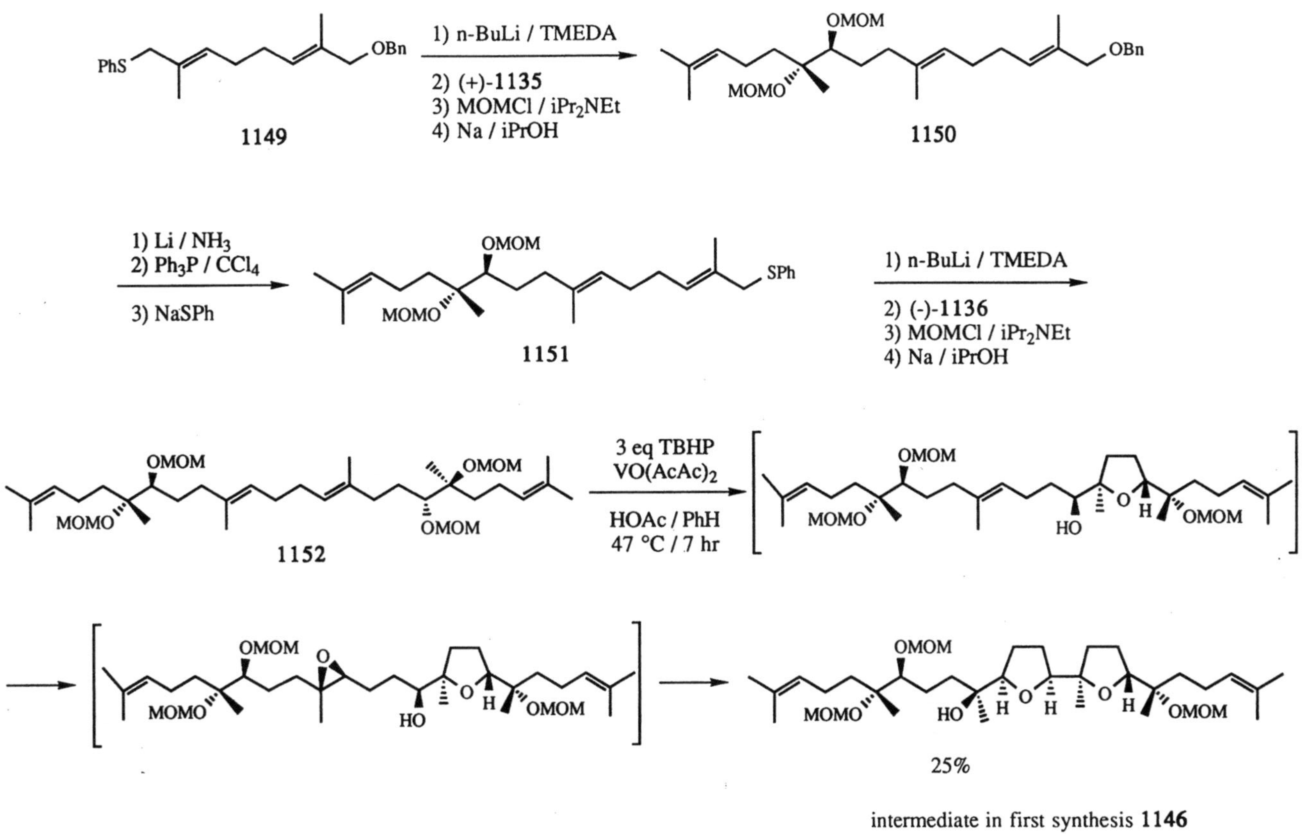

Scheme 184. Shirahama's Second Route to Teurilene Intermediate **1146**

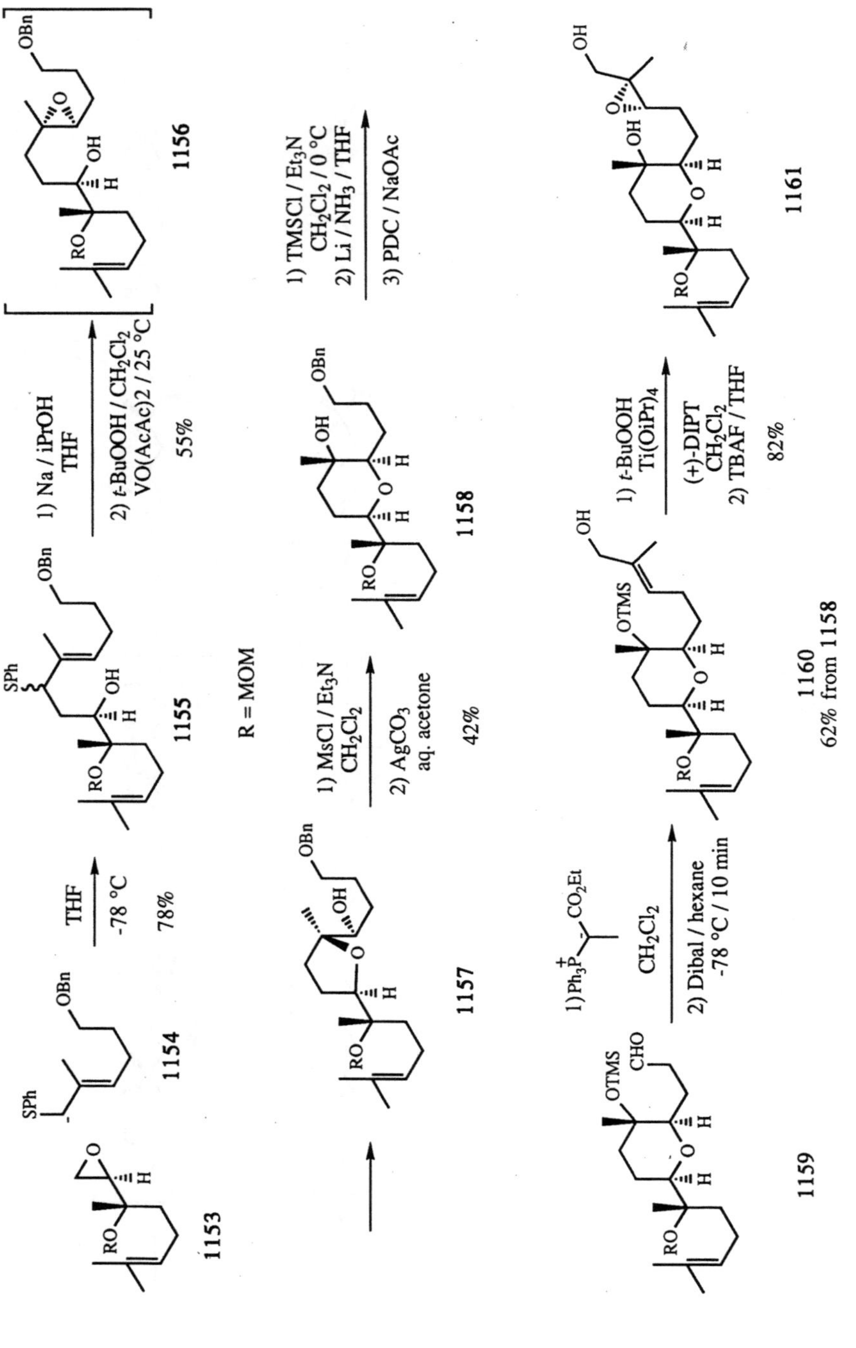
1153
1154
THF
-78 °C
78%
1155
R = MOM
1) Na / iPrOH
THF
2) t-BuOOH / CH2Cl2
VO(AcAc)2 / 25 °C
55%
1156
1157
1) MsCl / Et3N
CH2Cl2
2) AgCO3
aq. acetone
42%
1158
1) TMSCl / Et3N
CH2Cl2 / 0 °C
2) Li / NH3 / THF
3) PDC / NaOAc
1159
CH2Cl2
2) Dibal / hexane
-78 °C / 10 min
1160
62% from 1158
1) t-BuOOH
Ti(OiPr)4
(+)-DIPT
CH2Cl2
2) TBAF / THF
82%
1161

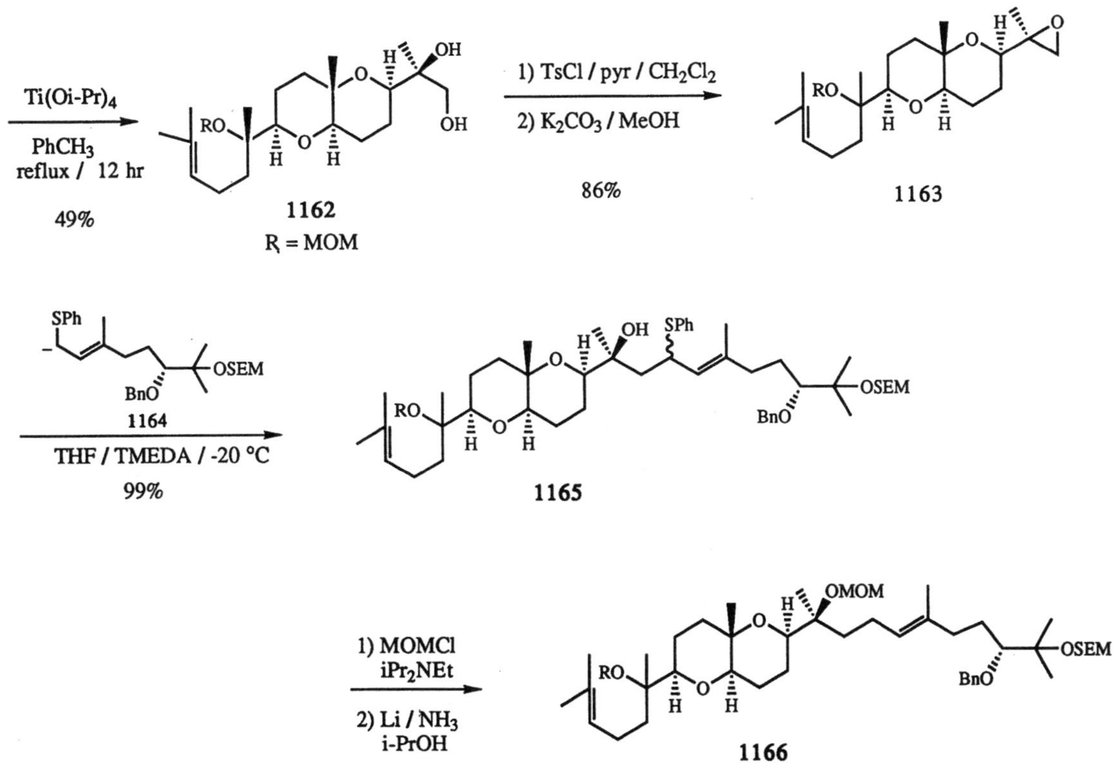

Scheme 185. Shirahama Synthesis of (+)-Venustatriol and (+)-Thysiferol

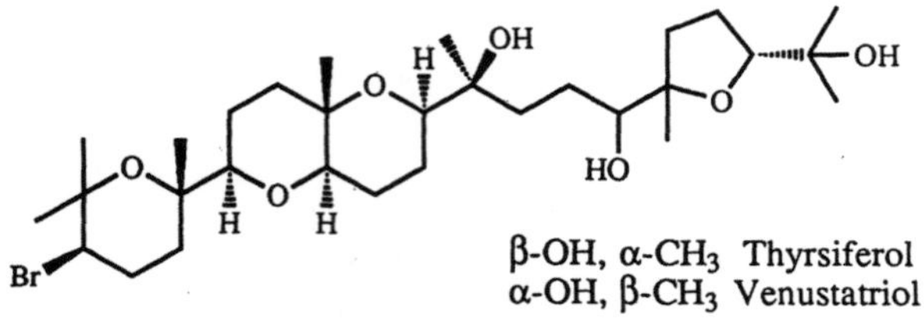

β-OH, α-CH₃ Thyrsiferol **1169**
α-OH, β-CH₃ Venustatriol **1170**

cytotoxicity against P388 [300] or significant anti-viral activity [301]. Thyrsiferol and venustatriol have been prepared in a common synthesis by Shirahama [302], as shown in Scheme 185. The synthesis is initiated by the coupling of optically active epoxide **1153** with the sulfide lithio anion **1154** to give **1155**. Reductive removal of the phenylthio group and Sharpless epoxidation provides tetrahydrofuran **1157** via intermediate epoxide **1156**. Mesylation of the α-hydroxytetrahydrofuran and treatment with silver carbonate induces ring expansion to the tetrahydropyran **1158**, thus providing the B-ring. The C-ring is appended in an eight-step process. Protection of the secondary alcohol, removal of the benzyl group, and PDC oxidation affords aldehyde **1159**. Wittig olefination with a carboethoxymethylidene phosphorane and DIBAL reduction of the resulting ester stereospecifically provides allylic alcohol **1160**. Sharpless asymmetric epoxidation of **1160** and desilylation gives epoxide **1161**, which is converted to bicylic ether **1162** by treatment with $Ti(OiPr)_4$. Further homologation and construction of the D-ring proceeds as follows. Epoxide **1163** is formed by selective tosylation of diol **1162** and treatment with potassium carbonate. Epoxide opening with sulfur-stabilized anion **1164** gives compound **1165** in 99% yield. Protection of the secondary alcohol functionality as its MOM ether and reductive removal of the phenylthio group gives the multiply protected diene **1166**. Sharpless epoxidation of **1166** followed by hydrolysis of the protecting groups gives a 4:1 mixture of diastereomeric tetrahydrofurans **1167** and **1168** (Scheme 186). Separation of the diastereomers and bromonium ion-initiated cyclization of each by treatment with TBCO produces thyrsiferol (**1169**) and venustatriol (**1170**) in low yielding final steps. Thyrsiferol and venustatriol are obtained after 21 steps.

A somewhat different approach towards the synthesis of venustatriol was utilized by Corey [303] as shown in Scheme 187. The general strategy followed is convergent and involves construction of the four rings of venustatriol in the order C → B → A → D. The C- and B-rings are prepared via acid-catalyzed cyclization processes (**1175** → **1176** and **1177** → **1178**). The A-ring with its attendant bromine substituent is prepared via bromonium ion initiated cyclization (**1178** → **1179**). Finally, the D-ring is obtained by attachment of a separately prepared tetrahydrofuran unit **1186** (**1180** → **1187**). The synthesis is initiated by the Sharpless asymmetric epoxidation of (*E,E,E*)-farnesol to provide alcohol **1172**. Collins oxidation of **1172** followed by Wittig olefination with a carbomethoxymethylidene phosphorane yields **1173** in 74% yield from farnesol.

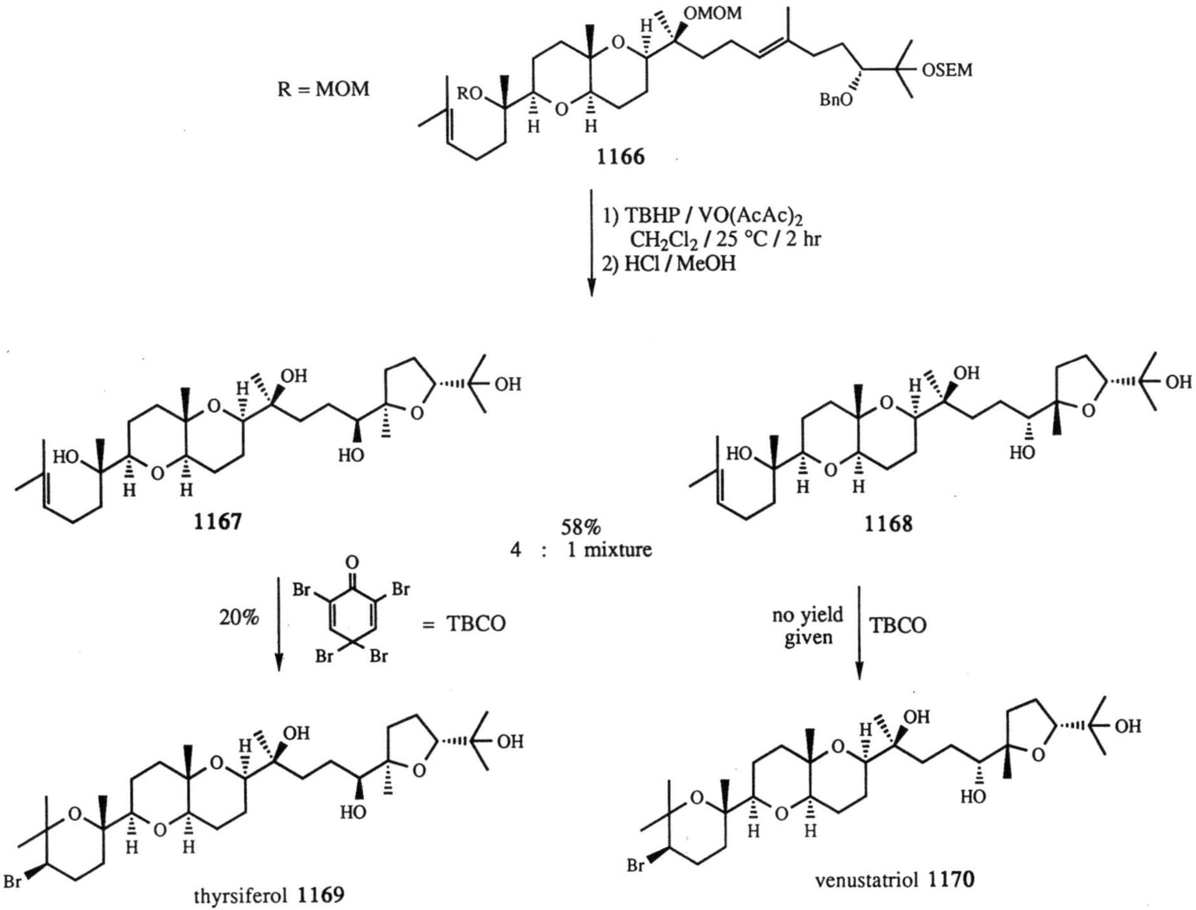

Scheme 186. Conversion of Intermediate **1166** to Venustatriol and Thyrsiferol

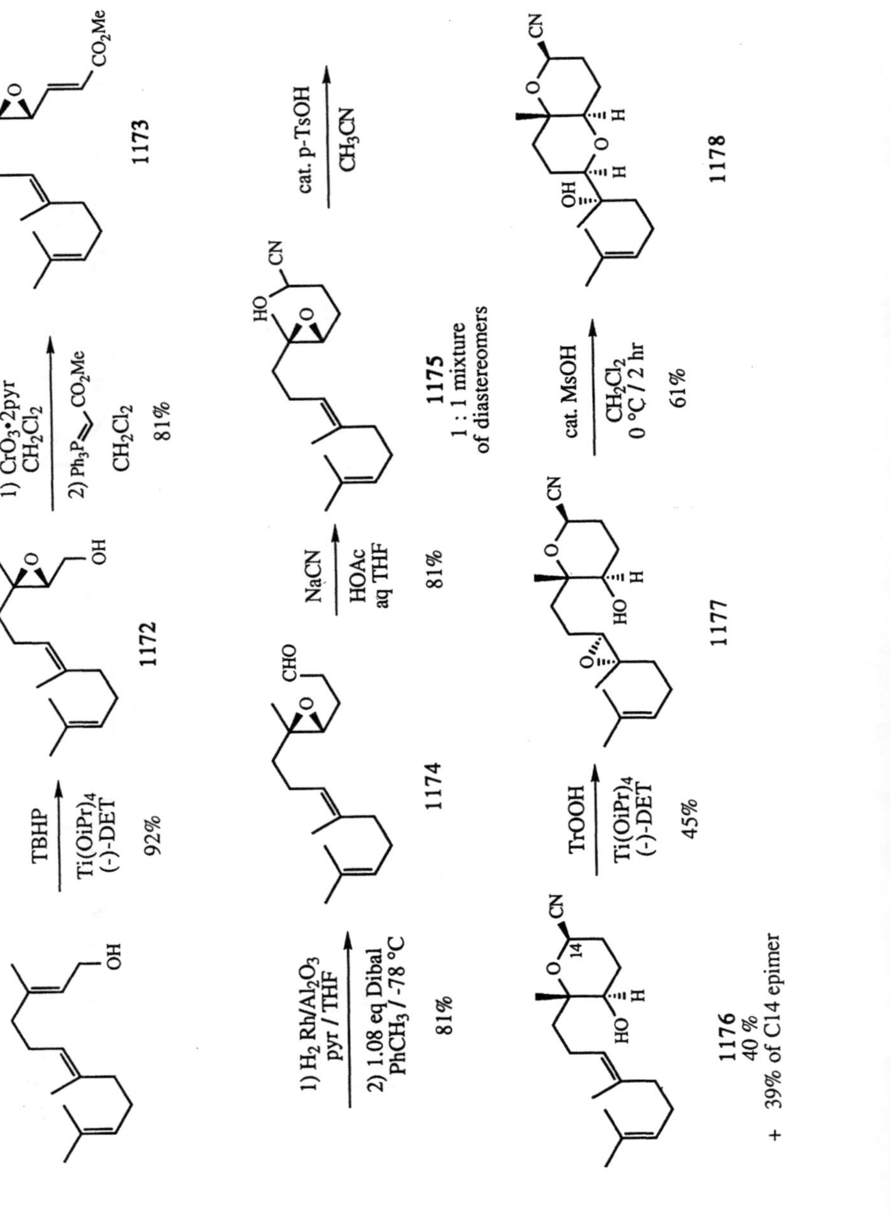
OH
TBHP
Ti(OiPr)$_4$
(-)-DET
92%
OH
1172
1) CrO$_3$•2pyr
CH$_2$Cl$_2$
2) Ph$_3$P=CO$_2$Me
CH$_2$Cl$_2$
81%
CO$_2$Me
1173
1) H$_2$ Rh/Al$_2$O$_3$
pyr / THF
2) 1.08 eq Dibal
PhCH$_3$ / -78 °C
81%
CHO
1174
NaCN
HOAc
aq THF
81%
HO
CN
1175
1 : 1 mixture
of diastereomers
cat. p-TsOH
CH$_3$CN
CN
14
HO
H
1176
40 %
+ 39% of C14 epimer
TrOOH
Ti(OiPr)$_4$
(-)-DET
45%
CN
HO
H
1177
cat. MsOH
CH$_2$Cl$_2$
0 °C / 2 hr
61%
CN
OH
H
H
1178

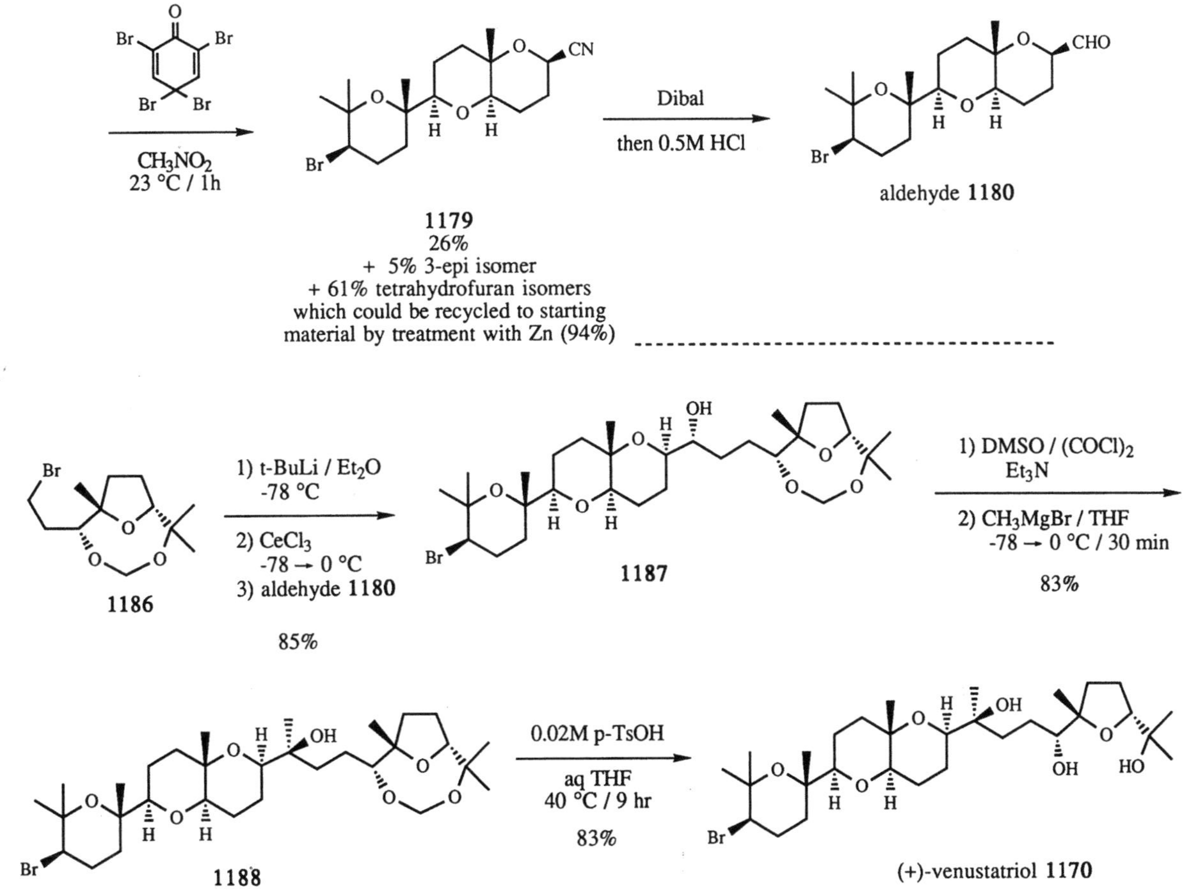

-last two C-C bond forming steps are chelation controlled additions

Scheme 187. Corey Synthesis of (+)-Venustatriol

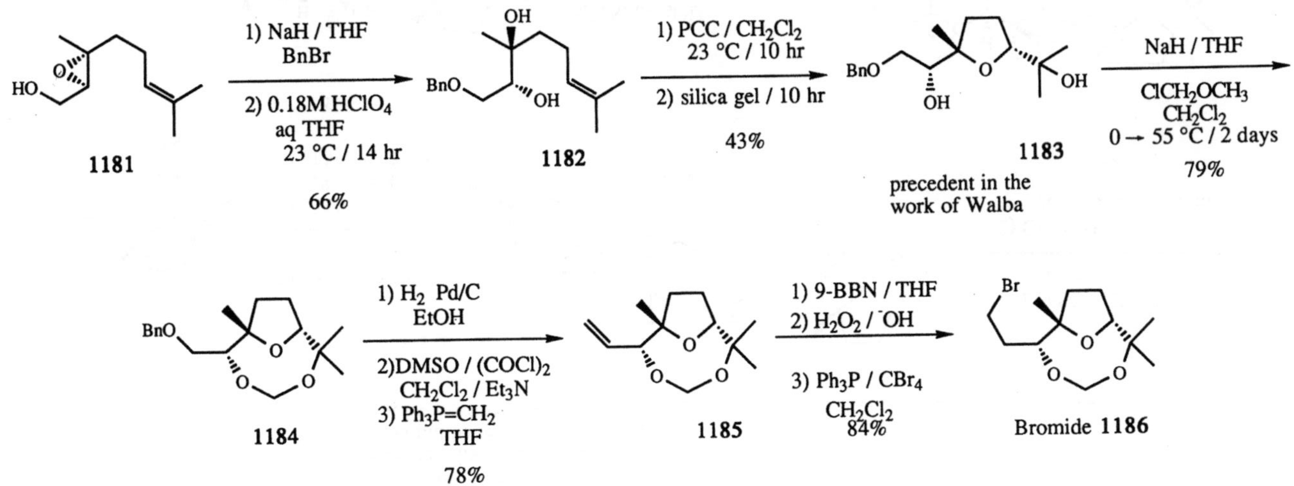

Scheme 188. Synthesis of Bromide **1186** (Corey)

Selective hydrogenation of the conjugated double bond over Rh/Al_2O_3 followed by DIBAL reduction gives aldehyde **1174**. Formation of the cyanohydrin of **1174** as a 1:1 mixture of diastereomers followed by treatment with TsOH gives the C ring in the form of nitrile **1176**. Separation of the 1:1 mixture of C14 epimeric nitriles and Sharpless epoxidation of the β-epimer stereoselectively provides epoxide **1177**. This is an unusual example of the stereoselective epoxidation of a trishomoallylic alcohol. Cyclization of **1177** to form the B-ring **1178** is effected by treatment with methanesulfonic acid in dichloromethane. Reaction of **1178** with tetrabromocyclohexadienone provides the A-B-C ring system of venustatriol in 26% yield in the form of nitrile **1179**. Also obtained were 5% of the epimeric bromide and 61% of a mixture of tetrahydrofuran isomers. These could be recycled back to starting material **1178** by reduction with zinc. DIBAL reduction of nitrile **1179** and hydrolysis affords aldehyde **1180**. The D-ring fragment **1186** was prepared as illustrated in Scheme 188. Epoxide **1181** (available from geraniol) is protected as its benzyl ether and subjected to hydrolysis with perchloric acid in aqueous THF. The resulting unsaturated diol **1182** is oxidized with PCC and cyclized to tetrahydrofuran **1183** on silica, as precedented in the work of Walba [304]. Reaction of diol **1183** with sodium hydride and MOMCl for two days affords the cyclic acetal **1184** in 79% yield. Debenzylation, Swern oxidation and Wittig methylenation yields vinyl ether **1185**. Hydroboration-oxidation and conversion of the resulting alcohol to the primary bromide gives the needed D-ring fragment in 11 steps and 15% overall yield. The venustatriol framework is obtained in a chelation-controlled coupling reaction by lithiation of bromide **1186**, transmetallation to form the cerium reagent, and addition of aldehyde **1180** to yield **1187**. Swern oxidation of the secondary alcohol **1187** and addition of methylmagnesium bromide produces tertiary alcohol **1188** stereoselectively. Hydrolysis of the acetal yields venustatriol (**1170**) in 15 steps and 0.82% overall yield.

2.6 Miscellaneous Terpenoids

2.6.1 Aplidiasphingosine

Aplidiasphingosine is an antimicrobial and antitumor terpenoid that can be regarded as a derivative of sphingosine. It was originally isolated [305] from an orange-flecked tunicate, identified as an *Aplidium* species. The major problem

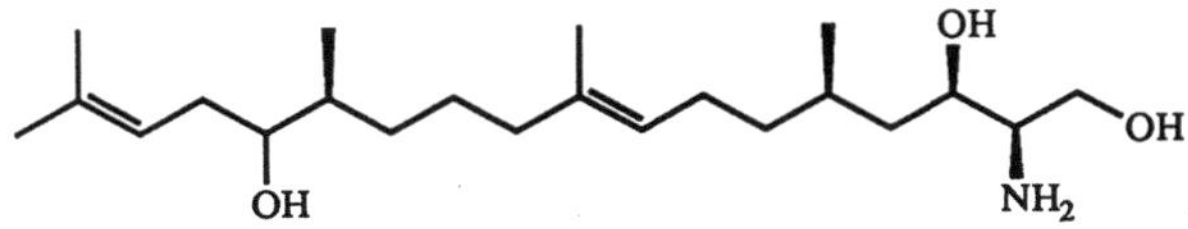

(+)-Aplidiasphingosine **1202**

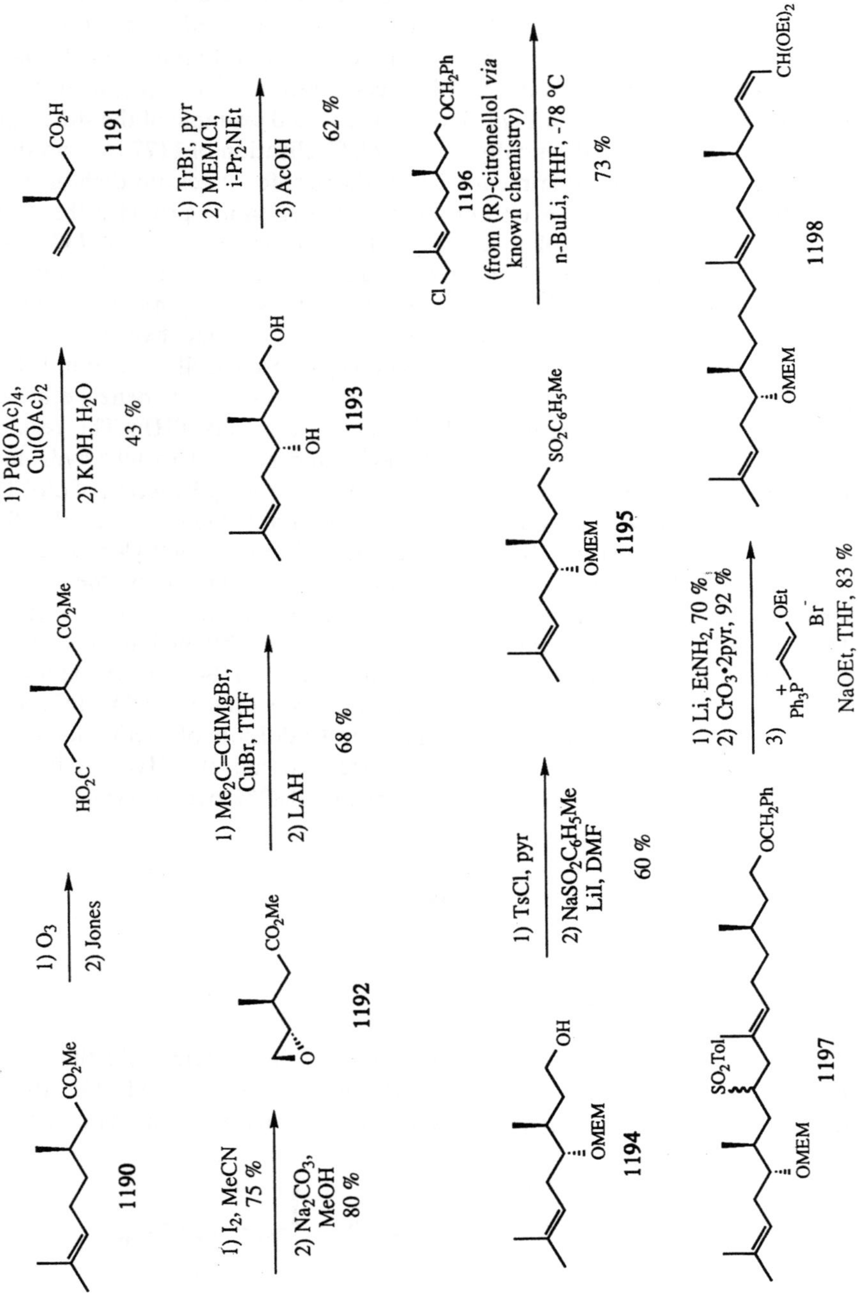
1190
1) O3
2) Jones
HO_2C
CO_2Me
1) Pd(OAc)4, Cu(OAc)2
2) KOH, H2O
43 %
CO_2H
1191
1) I2, MeCN
75 %
2) Na2CO3, MeOH
80 %
1192
1) Me2C=CHMgBr, CuBr, THF
2) LAH
68 %
OH
1193
1) TrBr, pyr
2) MEMCl, i-Pr2NEt
3) AcOH
62 %
OMEM
1194
1) TsCl, pyr
2) NaSO2C6H5Me
LiI, DMF
60 %
SO2C6H5Me
1195
Cl
OCH2Ph
1196
(from (R)-citronellol via known chemistry)
n-BuLi, THF, -78 °C
73 %
SO2Tol
1197
1) Li, EtNH2, 70 %
2) CrO3•2pyr, 92 %
3) Ph3P OEt Br
NaOEt, THF, 83 %
CH(OEt)2
1198

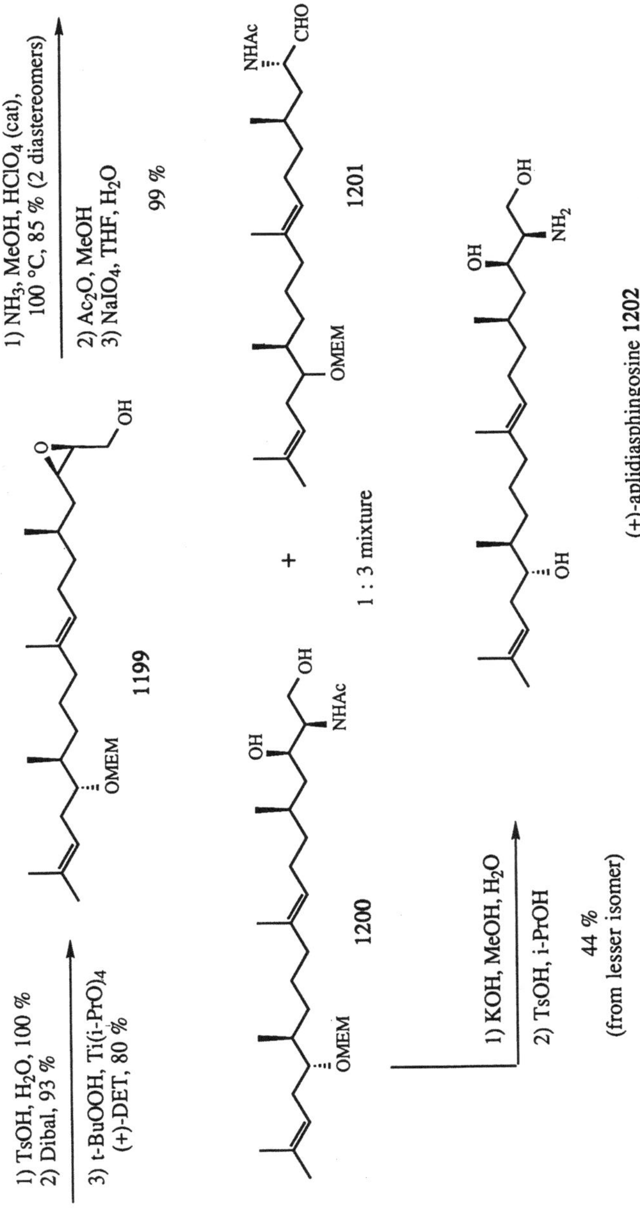

Scheme 189. Mori Synthesis of (+)-Aplidiasphingosine

that must be addressed in the synthesis of **1202** is control of the relative stereochemistry of the five stereocenters. Mori [306] solves this problem in an enantiospecific synthesis of **1202** by the coupling of two optically pure fragments to give three of the five stereocenters (Scheme 189). The remaining two centers are obtained via Sharpless asymmetric epoxidation. The left-hand fragment is prepared from methyl (+)-citronellate (**1190**) in a 12-step sequence. Ozonolysis of **1190** followed by Jones oxidation, oxidative decarboxylation and saponification affords the unsaturated acid **1191**. Iodolactonization followed by basic opening of the lactone stereospecifically provides epoxide **1192**. Grignard opening of the epoxide **1192** followed by reduction yields diol **1193**, which is selectively protected at the secondary alcohol via a three-step sequence to produce **1194**. Conversion of **1194** to the sulfone allows coupling of **1195** with allylic chloride **1196** (obtained from (*R*)-citronellol via known chemistry) to afford **1197**. Reductive removal of the sulfone moiety along with the benzyl protecting group followed by oxidation with Collins reagent and olefination of the resulting aldehyde gives the unsaturated acetal **1198**. Hydrolysis and reduction to the allylic alcohol followed by Sharpless assymmetric epoxidation affords epoxide **1199**. Opening with ammonia results in the formation of isomeric 1,2- and 1,3-glycols. Acylation of the amine and periodate oxidation gives the diol **1200** and aldehyde **1201** as a separable 1:3 mixture. The diol is converted to (+)-aplidiasphingosine **1202** by amide hydrolysis and removal of the MEM protecting group. The completely stereocontrolled stereosynthesis of **1202** requires 25 steps and gives an overall yield of 0.7%. Mori has also prepared aplidiasphingosine as a mixture of diastereomers in a non-stereocontrolled fashion [307].

2.6.2 *Prorocentrum* Metabolite

An extracellular metabolite produced in laboratory cultures of the dinoflagellate *Prorocentrum minimum* has been isolated and shown to be 1-(2,6,6-trimethyl-4-hydroxycyclohexenyl)-1,3-butanedione (**1206**). In vitro antibiotic activity against *Staphylococcus aureus* has been demonstrated [308]. The only synthesis of **1206** has been that of Constantino [309] shown in Scheme 190. The synthesis proceeds in a straightforward manner. Addition of the sodium monoanion of butadiyne to the known hydroxy ketone **1204** produces **1205** as a mixture of diastereomers. Dehydration of **1205** to the conjugated ene-diyne followed by

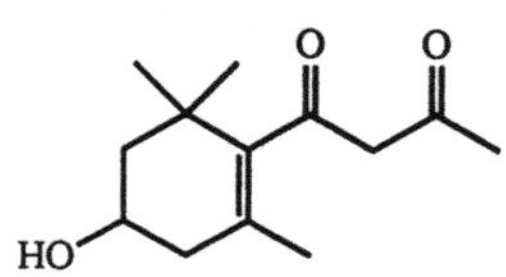

Prorocentrum Metabolite **1206**

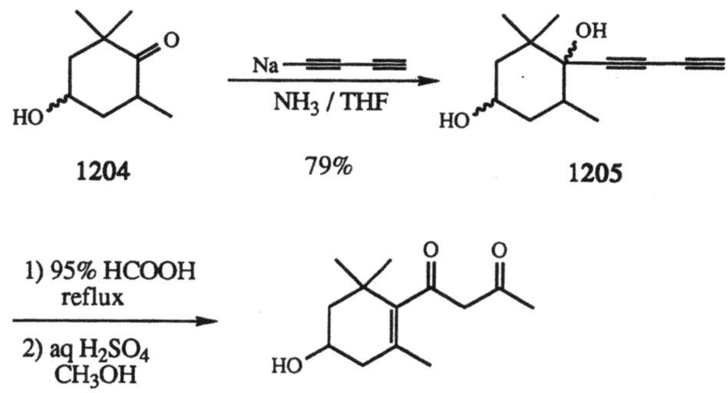

Prorocentrum metabolite 1206

50%

Scheme 190. Constantino Synthesis of *Prorocentrum* Metabolite **1206**

hydration with aqueous sulfuric acid gives the natural product in 29% yield over the 2 steps.

2.6.3 *Sigmosceptrella* Metabolite

The unnamed metabolite **1211** has been isolated [310] from the sponge *Sigmosceptrella laevis*, collected in the vicinity of Laing Island (Papua-New Guinea). The synthesis of **1211** by Sarma [311] (Scheme 191), begins from the functionalized decalone **1207**. Dissolving metal reduction of **1207** and alkylation of the regiospecifically generated enolate produces ester **1208**. Wittig olefination of **1208** and hydrogenation provides **1209** as a 4:1 (β-Me/α-Me) mixture of diastereomers. PCC oxidation of **1209** and saponification yields the keto-acid. Conversion of the acid to the acid chloride, treatment of with diazomethane and subsequent silver-catalyzed Wolff rearrangement in methanol results in Arndt-Eistert one-carbon homologation of the ester to give **1210**. Wittig olefination, saponification and addition of methyllithium to afford the methyl ketone produces the natural product **1211** in a 13 step process in 2.5% overall yield.

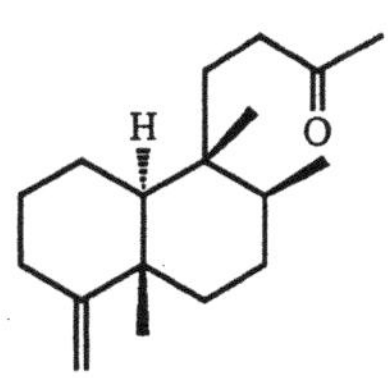

Sigmosceptrella Metabolite 1211

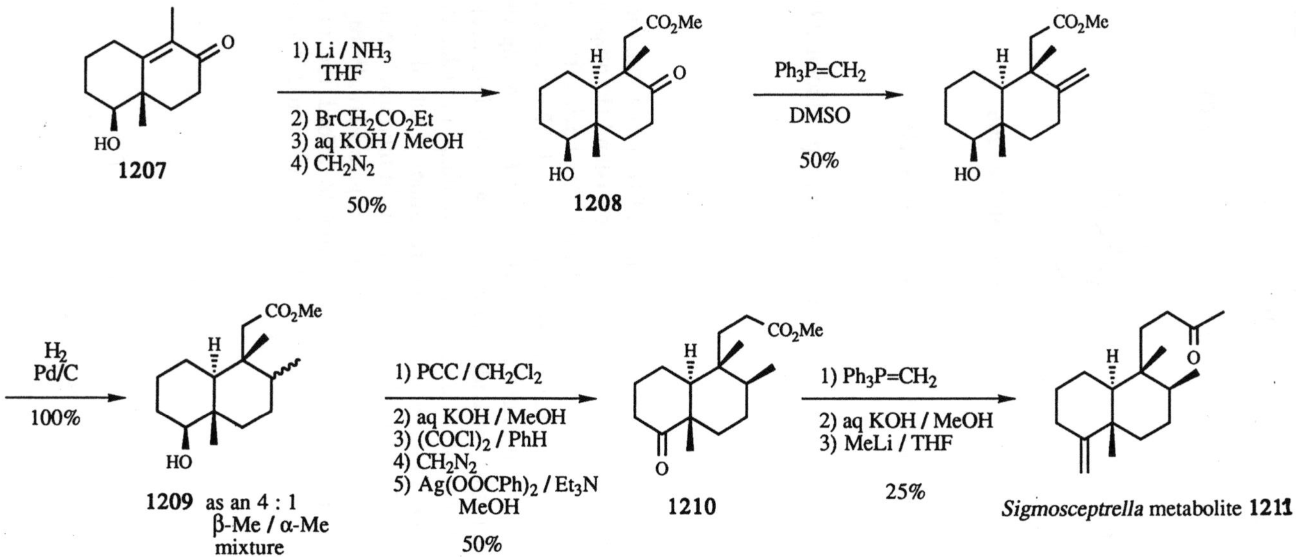

Scheme 191. Sarma Synthesis of *Sigmosceptrella* Metabolite **1211**

3 References

1. There are a number of reviews in this area covering various aspects of marine natural products chemistry. Probably the most comprehensive coverage of isolation are the Faulkner reviews: (a) Faulkner DJ (1984) Nat Prod Rep 1:251, (b) Faulkner DJ (1984) Nat Prod Rep 1:551, (c) Faulkner DJ (1986) Nat Prod Rep 3:1, (d) Faulkner DJ (1987) Nat Prod Rep 4:539, (e) Faulkner DJ (1988) Nat Prod Rep 5:613. In-depth coverage of a broader range of topics is covered in previous volumes of this series and prior to that in : (f) Scheuer PJ (ed) (1978–1983) Marine Natural Products: Chemical and Biological Perspectives. Academic, New York, vols 1–5, 1978–1983.
2. Our thanks to John Faulkner for supplying this information from his extensive database. Note that the isolation total for 1989 is incomplete and the number of syntheses shown for the year 1970 actually reflect 1960–70
3. A qualitative and selective review of marine natural products synthesis has been provided by Pietra F (1985) Gazz 115:443. Our review is more in the style of those which appear in the book series: ApSimon J (ed) (1973–1988). The Total Synthesis of Natural Products, Vol 1–7, Wiley-Interscience, New York with emphasis on convenient graphical presentation and critical analysis
4. Tius M (1988) Chem Rev 88:719
5. Kishi Y and co-workers (1989) J Am Chem Soc 111:7525 and 7530
6. Kishi Y (1989) Pure Appl Chem 61:313
7. Note that the synthesis numbers in the Scheme 1 graph do not include manuscripts in any of the aforementioned categories which we have omitted.
8. Paul VJ, McConnell OJ, Fenical W (1985) J Org Chem 45:3401
9. Zegarski J, Howard BM (1985) Tetrahedron Lett 26:1363
10. Masaki Y, Hashimoto K, Sakuma K, Kaji K (1984) Bull Chem Soc Japan 57:3466
11. Ichikawa N, Naya Y, Enomoto S (1974) The Abstract of the 18th Symposium of Terpene Essential Oils and Aromatic Chemistry 67
12. Yoshihara K, Hirose Y (1978) Bull Chem Soc Japan 51:653
13. Stierle DB, Wing RM, Sims JJ (1976) Tetrahedron Lett 4455
14. Williard PG, Grab LA, De Laszlo SE (1983) J Org Chem 48:1123
15. Williard PG, Grab LA (1984) Tetrahedron Lett 25:5009
16. Stierle DB, Sims JJ (1979) Tetrahedron 35:1261
17. Mynderse JS, Faulkner DJ (1974) J Am Chem Soc 96:6771
18. Williard PG, De Laszlo SE (1985) J Org Chem 50:3738
19. Sato A, Shindo T, Kasanuki N, Hasegawa K (1989) J Nat Prod 52:975
20. Pathirana C, Andersen RJ, Wright JCL (1989) Tetrahedron Lett 30:1487
21. (a) Irie T, Yasunari Y, Suzuki T (1965) Tetrahedron Lett 3619, (b) Irie T, Suzuki T, Ito S, Kurosawa E (1967) Tetrahedron Lett 3187, (c) Hall JG, Reiss JA (1986) Aust J Chem 39:1401
22. Srikrishna S, Sunderbabu G (1989) Tetrahedron Lett 30:3561
23. Taber DF, Anthony DF (1980) Tetrahedron Lett 21:2779
24. McMurry JE, Von Beroldingen LA (1974) Tetrahedron 30:2027
25. (a) Yamamura S, Hirata Y (1963) Tetrahedron 19:1485, (b) Yamada K, et al (1969) Tetrahedron 25:3509
26. Kazlauskas R, Murphy PT, Quinn RJ, Wells RJ (1976) Aust J Chem 29:2533
27. Irie T, Suzuki M, Kurosawa E, Masamune T (1970) Tetrahedron 26:3271
28. Yamada K, Yazawa H, Toda M, Hirata Y (1968) J Chem Soc Chem Commun 1342

29. Yamada K, Yazawa H, Uemura D, Toda M, Hirata Y (1969) Tetrahedron 25:3509
30. Goldsmith DJ, John TK, Kwong CD, Painter III GR (1980) J Org Chem 45:3989
31. For an earlier example of this type of tricothecane ring system formation see: Anderson WK, Lee GE (1980) J Org Chem 45:501
32. (a) Ronald RC (1976) Tetrahedron Lett 4413, (b) Ronald RC, Gewali MB, Ronald BP (1980) J Org Chem 45:2224
33. Ghosh A, Biswas S, Venkateswaran RV (1988) J Chem Soc Chem Commun 1421
34. Laronze J-Y, Boukili RE, Cartier D, Laronze J, Levy J (1989) Tetrahedron Lett 30:2229
35. Sheradsky T, Salemnick G, Nir Z (1972) Tetrahedron 28:3833
36. Sheradsky T (1966) Tetrahedron Lett 5225
37. Gewali MB, Ronald RC (1982) J Org Chem 47:2792
38. Feutrill GI, Mirrington RN, Nichols RJ (1973) Aust J Chem 26:345
39. (a) Irie T, Suzuki M, Kurosawa E, Masamune T (1966) Tetrahedron Lett 1837, (b) Irie T, Suzuki M, Kurosawa E, Masamune T (1970) Tetrahedron 26:3271
40. Ito S, Endo K, Yoshida T, Yatagai M, Kodama M (1967) J Chem Soc Chem Commun 186
41. Fenical WH, Howard BM (1976) Tetrahedron Lett 2519
42. Sims JJ, Rose AF, Izac RR (1978) In: Scheuer PJ (ed) Marine Natural Products: Chemical and Biological Perspectives, Vol I, Academic, New York, p 340.
43. Suzuki M, Kurosawa E (1978) Tetrahedron Lett 4805
44. (a) Suzuki M, Kowata N, Kurosawa E (1980) Tetrahedron 36:1551, (b) Suzuki M, Kurosawa E, Irie T (1970) Tetrahedron Lett 4995
45. Sakai R, Higa T, Jefford CW, Bernardinelli G (1986) Helv Chem Acta 69:91
46. Faulkner DJ (1976) Pure Appl Chem 48:25
47. Ichinose I, Kato T (1978) Chem Lett 61
48. (a) Martin JD, Palazon JM, Perez C, Ravelo JL (1986) Pure Appl Chem 58:395, (b) Martin JD, Perez C, Ravelo JL (1986) J Am Chem Soc 108:7801
49. Suzuki M, Segawa M, Suzuki T, Kurosawa E (1983) Bull Chem Soc Japan 56:3824
50. Niwa H, Yoshida Y, Hasegawa T, Yamada K (1985) Chem Lett 1687
51. Iwata C, Miyashita K, Koga Y, Shinoo Y, Yamada M, Tanaka T (1983) Chem Pharm Bull 31:2308
52. Iwata C, Akiyama T, Miyashita K (1988) Chem Pharm Bull 36:2872
53. Murai A, Kato K, Masamune T (1982) Tetrahedron Lett 23:2887
54. Cimino G, De Stefano S, Minale L, Trivellone E (1972) Tetrahedron 28:4761
55. Masaki Y, Hashimoto K, Serizawa Y, Kaji K (1984) Bull Chem Soc Japan 57:3476
56. Scott WJ, Crisp GT, Stille JK (1984) J Am Chem Soc 106:4630
57. Hochlowski JE, Walker RP, Ireland C, Faulkner DJ (1982) J Org Chem 47:88
58. Dunlop RW, Kazlauskas R, March G, Murphy PT, Wells RJ (1982) Aust J Chem 35:95
59. Akita H, Naito T, Oishi T (1979) Chem Lett 1365
60. Nakano T, Maillo MA (1981) Synth Comm 11:463
61. Nakano T, Maillo MA, Rojas A (1987) J Chem Soc Perkin Trans 1 2137
62. Ley SV, Mahon M (1981) Tetrahedron Lett 22:4747
63. Weinheimer AJ, Washecheck PH (1969) Tetrahedron Lett 3315
64. Kido F, Noda Y, Maruyama T, Kabuto C, Yoshikoshi A (1981) J Org Chem 46:4264
65. Bergstrom DE, Anantha Reddy P (1983) J Heterocyclic Chem 20:469
66. Cimino G, De Stefano S, Guerriero A, Minale L (1975) Tetrahedron Lett 1417, 1421 and 1425
67. For leading references see: Hochlowski JE, Walker RP, Ireland C, Faulkner DJ (1982) J Org Chem 47:89
68. Thompson JE, Walker RP, Wratten SJ, Faulkner DJ (1982) Tetrahedron 38:1865
69. Guella G, Guerriero A, Pietra F (1985) Helv Chim Acta 68:39
70. Matsumoto T, Usui S (1978) Chem Lett 105
71. Liotta D, Ott W (1987) Synth Comm 17:1655
72. Naispuri D, Das G (1979) J Chem Soc Perkin Trans 1 2776
73. Tanis SP, Herrinton PM (1983) J Org Chem 48:4572
74. Akita H, Oishi T (1981) Chem Pharm Bull 29:1580
75. Smith III AB, Mewshaw R (1984) J Org Chem 49:3685
76. Tius MA, Takaki KS (1982) J Org Chem 47:3166
77. Kurth MJ, Soares CJ (1987) Tetrahedron Lett 28:1031
78. Baker R, Sims RJ (1981) Tetrahedron Lett 22:161
79. Kazlauskas R, Murphy PT, Wells RJ (1978) Tetrahedron Lett 4949 and 4951
80. Hirota H, Kitano M, Komatsubara K-I, Takahashi T (1987) Chem Lett 2079

81. (a) Vaillancourt V, Agharahimi MA, Sundram U, Faulkner DJ, Albizati KF (1991) J Org Chem 55:378, (b) Richou O, Vaillancourt V, Faulkner DJ, Albizati KF (1989) J Org Chem 54:4729
82. (a) Schulte G, Scheuer PJ, McConnell OJ (1980) Helv Chim Acta 63:2159, (b) Hochlowski JE, Walker RP, Ireland WC, Faulkner DJ (1982) J Org Chem 47:88, (c) Cardellina II JH, Barnekow DE (1988) J Org Chem 53:882
83. Uyehara T, Sugimoto M, Suzuki I, Yamamoto Y (1989) J Chem Soc Chem Commun 1841
84. Tanis SP, Herrinton PM (1985) J Org Chem 50:3988
85. (a) Cimino G, De Stefano S, Minale L, Trivellone E (1975) Tetrahedron Lett 3727, (b) Cimino G, De Stefano S, Minale L, Trivellone E (1978) Experientia 34:1425
86. (a) Marshall JA, Conrow RE (1980) J Am Chem Soc 102:4274, (b) Marshall JA, Conrow RE (1983) J Am Chem Soc 105:5679
87. (a) Stallard MO, Fenical W, Kittredge JS (1978) Tetrahedron 34:2077, (b) Fenical W, Sleeper HL, Paul VJ, Stallard MO, Sun HH (1979) Pure Appl Chem 51:1865
88. Greene AE, Coelho F, Barreiro E, Costa PRR (1986) J Org Chem 51: 4250
89. Hendrickson JB, DeCapite PM (1985) J Org Chem 50:2112
90. Greene AE, Serra AA, Barreiro EJ, Costa PRR (1987) J Org Chem 52:1170
91. Hall SS, Faulkner DJ, Fayos J, Clardy J (1973) J Am Chem Soc 95:7187
92. Shizuri Y, Yamaguchi S, Terada Y, Yamamura S (1986) Tetrahedron Lett 26:57
93. Niwa M, Iguchi M, Yamamura S (1978) Tetrahedron Lett 4043
94. Izac RR, Poet SE, Fenical W, Van Engen D, Clardy J (1982) Tetrahedron Lett 23:3743
95. Paul VJ, Mc Connell OJ, Fenical W (1980) J Org Chem 45:3401
96. Martin M, Clardy J (1982) Pure Appl Chem 54:1915
97. (a) Ohsuka A (1962) J Chem Soc Japan 83:757, (b) Barnes CS, Loder JW (1962) Aust J Chem 15:322
98. (a) Kubo I, Lee Y-W, Pettei M, Pilkiewicz F, Nakanishi K (1976) J Chem Soc Chem Commun 1013
99. (a) Okuda RK, Scheuer PJ, Hochlowski JE, Walker RP, Faulkner DJ (1983) J Org Chem 48:1866, (b) Cimino G, De Rosa S, De Stefano S, Morrone R, Sodano G (1985) Tetrahedron 41:1093
100. Kato T, Suzuki T, Tanemura M, Kumanireng AS, Ototani N, Kitahara Y (1971) Tetrahedron Lett 1961
101. Nakanishi K, Tanis SP (1979) J Am Chem Soc 101:4398
102. (a) Howell SC, Ley SV, Mahon M, Worthington PA (1981) J Chem Soc Chem Commun 507, (b) Hollinshead DM, Howell SC, Ley SV, Mahon M, Ratcliffe N, Worthington PA (1983) J Chem Soc Perkin Trans 1 1579
103. Jallali-Naini M, Boussac G, Lemaitre P, Larcheveque M, Guillerm D, Lallemand J-Y (1981) Tetrahedron Lett 22:2995, (b) Jalali-Naini M, Guillerm D, Lallemand J-Y (1983) Tetrahedron 39:749
104. Mori K, Watanabe H (1986) Tetrahedron 42:273
105. (a) Fenical W, Sims JJ, Wing RM, Radlick PC (1972) Phytochemistry 11:1161, (b) Iguchi M, Niwa M, Yamamura S (1973) Bull Chem Soc Japan 46:2920
106. Andersen NH, Syrdal DD, Lawrence BM, Terhune SJ, Hogg JW (1973) Phytochemistry 12:827
107. Williams JR, Callahan JF, Lin C (1983) J Org Chem 48:3162
108. Ohta Y, Hirose Y (1972) Chem Lett 263
109. Fenical W, Sims JJ (1974) Tetrahedron Lett 1137
110. Sims JJ, Donnell MS, Leary JV, Lacy GH (1975) Antimicrob Ag Chemother 7:320
111. (a) Chen EY (1982) Tetrahedron Lett 46:4769, (b) Chen EY (1984) J Org Chem 49:3245
112. Ando M, Wada K, Takase K (1985) Tetrahedron Lett 26:235
113. Dischmann M, Ernst L, Wolf H (1989) Liebigs Ann Chem 727
114. Kikuchi H, Tsukitani Y, Yamada Y, Iguchi K, Drexler SA, Clardy J (1982) Tetrahedron Lett 23:1063
115. Kulkarni YS, Niwa M, Ron E, Snider BB (1987) J Org Chem 52:1568
116. Barton DHR, McCombie SW (1975) J Chem Soc Perkin Trans 1 1574
117. Fattorusso E, Magno S, Mayol L (1979) Gazz Chim Ital 109:589
118. Rogers C, Keay BA (1989) Tetrahedron Lett 30:1349
119. Faulkner DJ (1976) Phytochemistry 15:1992
120. Hoye TR, Kurth MJ (1979) J Org Chem 44:3461
121. Weinheimer AJ, Washecheck PH, Hossain MB, Van der Helm DJ (1968) Chem Soc Chem Commun 1070

122. Boeckman Jr RK, Silver SM (1973) Tetrahedron Lett 3497
123. Hellou J, Andersen RJ, Thompson JE (1982) Tetrahedron 38:1875
124. Ohta Y, Andersen NH, Liu C-B (1977) Tetrahedron 33:617
125. Armstrong RJ, Harris FL, Weiler L (1986) Can J Chem 64:1002
126. Shishido K, Tokunaga Y, Omachi N, Hiroya K, Fukumoto K, Kametani T (1989) J Chem Soc Chem Commun 1093
127. Wijnberg JBPA, Vader J, De Groot A (1983) J Org Chem 48:4380
128. Uijttewaal AP, Jonkers FL, Van der Gen A (1979) J Org Chem 44:3157
129. Tursch B, Braekman JC, Daloze D, Fritz P, Kelecom A, Karlsson R, Losman D (1974) Tetrahedron Lett 747
130. (a) Paquette LA, Ham WH (1986) Tetrahedron Lett 27:2341, (b) Paquette LA, Ham WH (1987) J Am Chem Soc 109:3025
131. Ayanoglu E, Gebreyesus T, Beechan CM, Djerassi C (1979) Tetrahedron J Am Chem Soc 35:1035
132. (a) Mehta G, Murty AN (1984) J Chem Soc Chem Commun 1058, (b) Mehta G, Murty AN (1987) J Am Chem Soc 52:2875
133. (a) Kinney WA, Coghlan MJ, Paquette LA (1985) J Am Chem Soc 107:7352, (b) Kinney WA, Coghlan MJ, Paquette LA (1984) J Am Chem Soc 106:6868
134. Fenical W, Schulte GR, Finer J, Clardy J (1978) J Org Chem 43:4628
135. Gadwood RC, Lett RM, Wissinger JE (1984) J Am Chem Soc 106:3869
136. Schmidt FJ, Hollenbeak KH, Vanderah DJ (1978) Tetrahedron 34:2719
137. Fenical W, Schulte GR, Finer J, Clardy J (1978) J Org Chem 43:3630
138. Gadwood RC (1985) J Chem Soc Chem Commun 123
139. Hayasaka K, Ohtsuka T, Shirahama H, Matsumoto T (1985) Tetrahedron Lett 26:873
140. Paquette LA, Ham WH, Dime DS (1985) Tetrahedron Lett 26:4983
141. Feldman KS, Wu M-J, Rotella DPJ (1989) Am Chem Soc 111, 6457–6458
142. (a) Kasin M, Sheikh YM, Durham LJ, Djerassi C, Tursch B, Daloze D, Braekman JC, Losman D, Karlsson R (1974) Tetrahedron Lett 2239, (b) Sheikh YM, Singy G, Kaisin M, Eggert H, Djerassi C, Tursch B, Daloze D, Braekman JC (1976) Tetrahedron 32:1171, (c) Sheikh YM, Djerassi C, Braekman JC, Daloze D, Kaisin M, Tursch B, Karlsson R (1977) Tetrahedron 33:2115
143. Crisp GT, Scott WJ, Stille JK (1984) J Am Chem Soc 106:7500
144. Oppolzer W, Battig K (1982) Tetrahedron Lett 23:4669
145. Huguet J, Karpf M, Dreiding AS (1982) Helv Chim Acta 65:2413
146. (a) Piers E, Karunaratne V (1984) Can J Chem 62:629, (b) Piers E, Karunaratne V (1989) Tetrahedron 45:1089
147. (a) Stevens KE, Paqette LA (1981) Tetrahedron Lett 22:4393, (b) Paquette LA, Stevens KE (1984) Can J Chem 62:2415
148. Curran DP, Chen M-H (1985) Tetrahedron Lett 26:4991
149. Mehta G, Reddy DS, Murty AN (1983) J Chem Soc Chem Commun 824
150. Mehta G, Murthy AN, Reddy DS, Reddy AV (1986) J Am Chem Soc 108:3443
151. (a) Little RD, Carroll GL (1981) Tetrahedron Lett 22:4389, (b) Little RD, Carroll GL, Petersen JL (1983) J Am Chem Soc 105:928
152. Iyoda M, Kushida T, Kitami S, Oda M (1987) J Chem Soc Chem Commun 1607
153. Liu HJ, Kulkarni MG (1985) Tetrahedron Lett 26:4847
154. Grubbs RH, Stille JR (1986) J Am Chem Soc 108:855
155. Fujita T, Ohtsuka T, Shirahama H, Matsumoto T (1982) Tetrahedron Lett 23:4091
156. (a) Birch AM, Pattenden G (1982) Tetrahedron Lett 23:991, (b) Birch AM, Pattenden G (1983) J Chem Soc Perkin Trans 1 1913
157. Shibasaki MS, Mase T, Ikegami S (1986) J Am Chem Soc 108:2090
158. (a) Ladlow M, Pattenden G, Teague SJ (1986) Tetrahedron Lett 27:3279, (b) Pattenden G, Teague SJ (1988) J Chem Soc Perkin Trans 1 1077
159. Fenical W, Groweiss A, Clardy J, Cun-heng H, Kanghou L, Zhongnian Y, Zhongde W (1985) Tetrahedron Lett 26:2379
160. Zalkow LH, Harris RN, Van Derveer D, Bertrand JA (1977) J Chem Soc Chem Commun 456
161. Paquette LA, Leone-Bay A (1983) J Am Chem Soc 105:7352
162. Iwata C, Takemoto Y, Doi M, Imanishi T (1988) J Org Chem 53:1623
163. Beechan CM, Djeerassi C, Finer JS, Clardy J (1977) Tetrahedron Lett 2395

164. Braekman JC, Daloze D, Dupont A, Tursch B, Declercq JP, Germain G, Van Meerssche M (1981) Tetrahedron 37:179
165. Oppolzer W, Strauss HF, Simmons DP (1982) Tetrahedron Lett 23:4673
166. Collins PA, Wege D (1979) Aust J Chem 32:1819
167. (a) Piers E, Jung GL (1985) Can J Chem 63:996, (b) Piers E, Jung GL (1987) Can J Chem 65:1668
168. (a) Fallis AG, Kingston JF, Antczak K (1985) Can J Chem 63:993, (b) Fallis AG, Kingston JF, Antczak K, Hanson AW (1987) Can J Chem 65:114
169. Oppolzer W, Begley T, Ashcroft A (1984) Tetrahedron Lett 25:825
170. (a) Gonzalez AG, Darais J, Martin JD, Perez C (1974) Tetrahedron Lett 1249, (b) Gonzalez AG, Darias J, Martin JD, Perez C, Sims JJ, Lin GHY, Wing RM (1975) Tetrahedron 31:2449, (c) Gonzalez AG, Martin JD, Perez C, Ramirez MA, Ravelo F (1980) Tetrahedron Lett 21:187
171. Gonzalez AG, Martin JD, Perez C, Ramirez MA, Ravelo F (1980) Tetrahedron Lett 21:187
172. Gonzalez AG, Martin JD, Melian MA (1976) Tetrahedron Lett 2279
173. Imre S, Islimyeli S, Ostune A, Thomson RH (1981) Phytochemistry 20:833
174. Gonzalez AG, Martin JD, Perez C, Ramirez MA, Ravelo F (1981) Tetrahedron Lett 22:5071
175. Hagadone MR, Scheuer PJ, Holm A (1984) J Am Chem Soc 106:2447
176. Bornemann V, Patterson GLM, Moore RE (1988) J Am Chem Soc 110:2339 and references cited therein.
177. Piers E, Yeung BWA, Rettig SJ (1987) Tetrahedron Lett 28:5521
178. Hart DJ, Lai CS (1989) Synthesis Lett 49
179. Caine D, Deutsch H (1978) J Am Chem Soc 100:8030
180. Hsieh S-L, Chiu C-T, Chang NC (1989) J Org Chem 54:3820
181. Piers E, Winter M (1982) Liebigs Ann Chem 973
182. Schiehser GA, White JD (1980) J Org Chem 45:1864
183. Yamamoto H, Sham HL (1979) J Am Chem Soc 101:1609
184. Corey EJ, Behforouz M, Ishiguro M (1979) J Am Chem Soc 101:1608
185. Corey EJ, Ishiguro M (1979) Tetrahedron Lett 2745
186. Frater G, Wenger J (1984) Helv Chim Acta 67:1702
187. De Rosa S, Minale L, Riccio R, Sodano G (1976) J Chem Soc Perkin Trans 1 1408
188. Djura P, Stierle DB, Sullivan B, Faulkner DJ, Arnold E, Clardy J (1980) J Org Chem 45:1435
189. Sarma AS, Chattopadhyay P (1982) J Org Chem 47:1727
190. (a) Fenical W, McConnell O (1975) Experientia 31:1004, (b) Fenical W, Sims JJ, Squatrito D, Wing RM, Radlick P (1973) J Org Chem 38:2383, (c) Cimino G, DeLuca P, DeStefano S, Minale L (1975) Tetrahedron 31:271
191. Welch SC, Rao SCP (1977) Tetrahedron Lett 505
192. (a) Ravi BN (1976) PhD Thesis, The University of Hawaii at Manoa, Honolulu Hawaii, (b) Scheuer PJ, Ravi BN (November 1976) presented in part at the 12th Regional Western Meeting of the American Chemical Society, Phoenix, Arizona
193. Trammell GI (1978) Tetrahedron Lett 1525
194. Roll DM, Scheuer PJ, Matsumoto GK, Clardy J (1983) J Am Chem Soc 105:6177
195. Harada N, Sugioka T, Ando Y, Uda H, Kuriki T (1988) J Am Chem Soc 110:8483
196. Schulte G, Scheuer PJ, McConnell OJ (1980) J Org Chem 45:552
197. Taschner M, Shahripour A (1985) J Am Chem Soc 107:5570
198. Paquette LA, Schaefer AG, Springer JP (1987) Tetrahedron 43:5567
199. Maurer B, Hauser A, Thommen W, Schulte-Elte KH, Ohloff G (1980) Helv Chim Acta 63:293
200. Gonzalez AG, Darias J, Martin JD (1977) Tetrahedron Lett 3375
201. Gonzalez AG, Darias J, Martin JD, Melian MA (1978) Tetrahedron Lett 481
202. Majetich G, Ringold C (1987) Heterocycles 25:271
203. Vig OP, Trehan IR, Malik N, Kumar R (1977) Ind J Chem 15B:900
204. Oppolzer W, Briner PH, Snowden RL (1980) Helv Chim Acta 63:967
205. Pettit GR, Herald CL, Allen MS, Von Dreele RP, Vanell LD, Kao JP, Blake W (1982) J Am Chem Soc 104:3923
206. Pettit GR, Herald CL, Judd GF, Bolliger G, Thayer PS (1975) J Pharm Sci 64:2023
207. Hoye TR, Kurth MJ (1979) J Am Chem Soc 101:5065
208. Hoye TR, Caruso AJ, Dellaria Jr JF, Kurth MJ (1982) J Am Chem Soc 104:6704
209. White JD, Nishiguchi T, Skeean RW (1982) J Am Chem Soc 104:3923
210. Gosselin P, Rouessac F (1983) Tetrahedron Lett 24:5515

211. Shieh H-M, Prestwich G (1982) Tetrahedron Lett 23:4643
212. Tanaka A, Otsuka S, Yamashita K (1984) Agric Biol Chem 48:2535
213. Braekman JC, Daloze D, Bertau R, Macedo de Abreu P (1982) Bull Soc Chim Belg 91:791
214. Jefford CW, Jaggi D, Bernardinelli G, Boukouvalas J (1987) Tetrahedron Lett 28:4041
215. For leading references see: (a) Piers E, Friesen RW (1986) J Org Chem 51:3405, (b) Paquette LA, Lin H-S, Belmont DT, Springer JP (1986) J Org Chem 51:4807
216. Ochi M, Asao K, Kotsuki H, Miura I, Shibata K (1986) Bull Chem Soc Japan 59:661
217. Crews P, Klein TE, Hogue ER, Myers BL (1982) J Org Chem 47:811
218. Piers E, Friesen RW (1986) J Org Chem 51:3405
219. Piers E, Friesen RW (1988) J Chem Soc Chem Commun 125
220. Pattenden G, Robertson FM (1986) Tetrahedron Lett 27:399
221. Mehta G, Krishnamurthy N (1987) Tetrahedron Lett 28:5945
222. Piers E, Llinas-Brunet M (1989) J Org Chem 54:1483
223. Wratten SJ, Faulkner DJ, Hirotsu K, Clardy J (1978) Tetrahedron Lett 4345
224. Corey EJ, Magriotis PA (1987) J Am Chem Soc 109:287
225. Baker JT, Wells RJ, Oberhansli WE, Hawes GB (1976) J Am Chem Soc 98:4010
226. (a) Gerwick WH, Fenical W, Sultanbawa MUS (1981) J Org Chem 46:2233, (b) Gerwick WH, Fenical W (1983) J Org Chem 48:3325
227. (a) Gerwick WH, Fenical W, Van Engen D, Clardy J (1980) J Am Chem Soc 102:7991, (b) Jacobs RS, White S, Wilson L (1981) Fed Proc Fed Am Soc Exp Biol 40:26
228. Salomon RG, Sachinvala ND, Raychaudhuri SR, Miller DB (1984) J Am Chem Soc 106:2211
229. For an aesthetically pleasing use of this well-known phenomenon see: Evans DA, Chapman KT, Carreira EM (1988) J Am Chem Soc 110:3560 and reference cited to earlier work.
230. Tanaka M, Tomioka K, Koga K (1985) Tetrahedron Lett 26:3035
231. Tanaka M, Tomioka K, Koga K (1985) Tetrahedron Lett 26:6109
232. Salomon RG, Basu B, Roy S, Sharma RB (1989) Tetrahedron Lett 30:4621
233. (a) Look SA, Fenical WF, Matsumoto G, Clardy J (1986) J Org Chem 51:5140, (b) Look SA, Fenical W (1987) Tetrahedron 43:3363
234. Look SA, Fenical WF, Jacobs RS, Clardy J (1986) Proc Natl Acad Sci USA 83:6238
235. Broka CA, Chan S, Peterson B (1988) J Org Chem 53:1586
236. Corey EJ, Carpino P (1989) J Am Chem Soc 111:5472
237. Murai A, Abiko A, Masamune T (1984) Tetrahedron Lett 25:4955
238. Yamaguchi Y, Uyehara T, Kato T (1985) Tetrahedron Lett 26:343
239. (a) Nishizawa M, Takenaka H, Hayashi Y (1985) J Am Chem Soc 107:522, (b) Nishizawa M, Takenaka H, Hirotsu K, Higuchi T, Hayashi Y (1984) J Am Chem Soc 106:4290, (c) Nishizawa M, Takenaka H, Hayashi Y (1986) J Org Chem 51:806
240. (a) Nakano T, Hernandez MI (1982) Tetrahedron Lett 23:1423, (b) Nakano T,Hernandez MI, Gomez M, Medina JD (1989) J Chem Res (S) 54, (c) Imamura PM, Sierra MG, Ruveda EA (1981) J Chem Soc Chem Commun 734, (d) Mischne MP, Sierra MG, Ruveda EA (1984) J Org Chem 49:2035, (e) Nakano T, Hernandez MI, Martin A (1984) J Chem Res (S) 262, (f) Nakano T, Hernandez MI (1983) J Chem Soc Perkin Trans 1 135
241. Asao K, Iio H, Tokoroyama T (1989) Tetrahedron Lett 30:6401
242. Asao K, Iio H, Tokoroyama T (1989) Chem Lett 1813
243. Ichikawa Y (1988) Tetrahedron Lett 29:4957
244. Iio H, Asao K, Tokoroyama TJ (1985) Chem Soc Chem Commun 774
245. Gonzalez AG, Martin JD, Rodriguez ML (1973) Tetrahedron Lett 3657
246. Sun HH, Ferrara NM, McConnell OJ, Fenical W (1980) Tetrahedron Lett 21:3123
247. Mori K, Uno T (1989) Tetrahedron 45:1945
248. Sun HH, Waraszkiewicz SM, Erickson KL, Finer J, Clardy J (1977) J Am Chem Soc 99:3516
249. Marshall JA, Wuts PGM (1978) J Am Chem Soc 100:1627
250. Greene AE (1980) J Am Chem Soc 102:5337
251. Kashman Y, Groweiss A (1980) J Org Chem 45:3814
252. Kodama M, Okumura K, Kobayashi Y, Tsunoda T, Ito S (1984) Tetrahedron Lett 25:5781
253. (a) Fattorusso E, Magno S, Novellino E (1983) Experientia 39:1275, (b) Paul VJ, Fenical W (1984) Tetrahedron 39:2913
254. Nakatsu T, Ravi BN, Faulkner DJ (1981) J Org Chem 46:2453
255. Whitesell JK, Fisher M, Jardine P (1983) J Org Chem 48:1557
256. Trost BM, Matsuda K (1988) J Am Chem Soc 110:5233
257. Isoe S, Ge Y, Yamamoto K, Katsumura S (1988) Tetrahedron Lett 29:4591

258. Kashman Y, Hirsch S, Koehn F, Cross S (1987) Tetrahedron Lett 28:5461
259. Snider BB, Ke Y (1989) Tetrahedron Lett 30:2465
260. Kobayashi M (1988) Chem Pharm Bull 36:488
261. Nagaoka H, Iwashima M, Abe H, Yamada Y (1989) Tetrahedron Lett 30:5911
262. Walker RP, Faulkner DJ (1981) J Org Chem 46:1098
263. Nishizawa M, Yamada H, Hayashi Y (1986) Tetrahedron Lett 27:187
264. Piers E, Marais P (1989) J Chem Soc Chem Commun 1222
265. Murphy PT, March G, Wells RJ (1981) Tetrahedron Lett 22:1555
266. Kido F, Abe T, Yoshikoshi A (1986) J Chem Soc Chem Commun 590
267. Huckestein M, Kreiser W, Ruschenbaum V (1987) Helv Chim Acta 70:445
268. Fukuzawa A, Takaya Y, Matsue H, Masamune T (1985) Chem Lett 1263
269. Fukuzawa A, Sato H, Masamune T (1987) Tetrahedron Lett 28:4303
270. (a) Ishitsuka M, Kusumi T, Kakisawa H (1982) Tetrahedron Lett 23:3179, (b) Kirkup MP, Moore RE (1983) Phytochemistry 22:2527
271. Nagaoka H, Kobayashi K, Matsui T, Yamada Y (1987) Tetrahedron Lett 28:2027
272. Nagaoka H, Kobayashi K, Yamada Y (1988) Tetrahedron Lett 29:5945
273. Nagaoka H, Kobayashi K, Okamura T, Yamada Y (1987) Tetrahedron Lett 28:6641
274. Enoki N, Furusaki A, Suehiro K, Ishida R, Shirahama H, Matsumoto T (1983) Tetrahedron Lett 24:4341
275. Kato N, Takeshita H, Tanaka S, Kataoka HJ (1989) Chem Soc Perkin Trans 1 1833
276. (a) De Silva ED, Scheuer PJ (1980) Tetrahedron Lett 21:1611, (b) De Silva ED, Scheuer PJ (1981) Tetrahedron Lett 22:3147
277. Leading references on the biological activity of the manoalides can be found in :Katsumura S, Fujiwara S, Isoe S (1988) Tetrahedron Lett 29:1173
278. Katsumura S, Fujiwara S, Isoe S (1985) Tetrahedron Lett 26:5827
279. Katsumura S, Fujiwara S, Isoe S (1988) Tetrahedron Lett 29:1173
280. De Silva ED, Scheuer PJ (1980) Tetrahedron Lett 21:1611
281. Katsumura S, Fujiwara S, Isoe S (1987) Tetrahedron Lett 28:1191
282. Garst ME, Tallman EA, Bonfiglio JN (1986) Tetrahedron Lett 27:4533
283. (a) Hofheinz W, Schoenholzer P (1977) Helv Chim Acta 60:1367, (b) Gregson RP, Ouvrier D (1982) J Nat Prod 45:1367
284. Hofheinz W, Schoenholzer P (1977) Helv Chim Acta 60:1367
285. Takeda K, Sato M, Yoshii E (1986) Tetrahedron Lett 27:3903
286. Nakagawa M, Ishihama M, Hamamoto Y, Endo M (October 1986). Abstracts of Papers of the 28th Symposium on The Chemistry of Natural Products, Sendai, Japan p 200 [(1987) Chem Abstr 106:96126b]
287. Hagiwara H, Uda H (1988) J Chem Soc Chem Commun 815
288. Sullivan B, Faulkner DJ (1982) Tetrahedron Lett 23:907
289. Piers E, Wai JSM (1987) J Chem Soc Chem Commun 1342
290. Yunker MB, Scheuer PJ (1978) J Am Chem Soc 100:307
291. Sum FW, Weiler L (1979) J Am Chem Soc 101:4401
292. Kigoshi H, Ojika M, Shizuri Y, Niwa H, Yamada K (1982) Tetrahedron Lett 23:5413
293. De Napoli L, Fattorusso E, Magno S, Mayol L (1982) Phytochemistry 21:782
294. Kigoshi H, Ojika M, Shizuri Y, Niwa H, Yamada K (1986) Tetrahedron 42:3789
295. Suzuki T, Suzuki M, Furusaki A, Matsumoto T, Kato A, Imanaka Y, Kurosawa E (1985) Tetrahedron Lett 26:1329
296. Hashimoto M, Yanagiya M, Shirahama H (1988) Chem Lett 645
297. Hashimoto M, Harigaya H, Yanagiya M, Shirahama H (1988) Tetrahedron Lett 29:5947
298. Blunt JW, Hartshorn MP, McLennan TJ, Munro MHG, Robinson WT, Yorke SC (1978) Tetrahedron Lett 69
299. Sakemi S, Higa T, Jefford CW, Bernardinelli G (1986) Tetrahedron Lett 27:4287
300. Suzuki T, Suzuki M, Furusaki A, Matsumoto T, Kato A, Imanaka Y, Kurosawa E (1985) Tetrahedron Lett 26:1329
301. Sakemi S, Higa T, Jefford CW, Bernardinelli G (1986) Tetrahedron Lett 27:4287
302. (a) Nashimoto M, Kan T, Yanagiya M, Shirahama H, Matsumoto T (1987) Tetrahedron Lett 28:5665, (b) Hashimoto M, Kan T, Nozaki K, Yanagiya M, Shirahama H, Matsumoto T (1987) Tetrahedron Lett 29:1143
303. Corey EJ, Ha D-C (1988) Tetrahedron Lett 29:3171
304. Walba DM, Stout GS (1982) Tetrahedron Lett 23:727

305. Carter G:T, Rinehart Jr KL (1978) J Am Chem Soc 100:7441
306. Mori K, Umemura T (1982) Tetrahedron Lett 23:3391
307. (a) Mori K, Umemura T (1981) Tetrahedron Lett 22:4429 (b) Umemura T, Mori K (1987) Agric Biol Chem 51:217
308. Andersen RJ, Le Blanc JJ, Sum FW (1980) J Org Chem 45:1169
309. (a) Constantino MG, Donate PM, Petragnani N (1986) J Org Chem 51:387, (b) Constantino MG, Donate PM, Petragnani N (1982) Tetrahedron Lett 23:1051
310. Albericci M, Braekman JC, Daloze D. Tursch B (1982) Tetrahedron 38:1881
311. Sarma AS, Gayen AK (1985) Tetrahedron 41:4581

Subject Index

MIX
Papier aus verantwortungsvollen Quellen
Paper from responsible sources
FSC® C105338

If you have any concerns about our products,
you can contact us on
ProductSafety@springernature.com

In case Publisher is established outside the EU,
the EU authorized representative is:
Springer Nature Customer Service Center GmbH
Europaplatz 3, 69115 Heidelberg, Germany

Printed by Libri Plureos GmbH
in Hamburg, Germany